The Laboratory Revolution and the Creation of the Modern University, 1830–1940

The Laboratory Revolution and the Creation of the Modern University, 1830-1940

Edited by
Klaas van Berkel and
Ernst Homburg

Routledge
Taylor & Francis Group
LONDON AND NEW YORK

First published in 2023 by Amsterdam University Press Ltd.

Published 2025 by Routledge
4 Park Square, Milton Park, Abingdon, Oxon OX14 4RN
605 Third Avenue, New York, NY 10158

Routledge is an imprint of the Taylor & Francis Group, an informa business

ISBN: 9789463720434 (hbk)
ISBN: 9781041188339 (pbk)
ISBN: 9781003706724 (ebk)
NUR 686

Cover illustration: Archives, Department of Chemistry, University of Zürich (top image) and AIP Emilio Segrè Visual Archives (bottom image)

Cover design: Coördesign, Leiden

DOI: 10.5117/9789463720434

Table of Contents

Part III Laboratory Values

List of Figures and Charts

Figures

Charts

Preface

In this volume we have brought together more than a dozen new studies about the rapid, even revolutionary, development of the laboratory in the nineteenth and early twentieth centuries, especially in the context of the expanding universities. The importance of the research laboratory for the development of science in the second half of the nineteenth century is uncontested, as is the revival of the universities as centres of innovative science and scholarship in the same period. The connections between these two revolutionary developments are seldom studied in detail though. This is partly due to a current lack of interest in laboratory studies as well as in institutional history, but a main reason is also that histories of universities are commonly written by authors with a background in the humanities. This collection of essays tries to bridge the gap and bring representatives from the sciences and the humanities together in a concerted effort to integrate laboratory studies and the history of universities. We have cast a wide net ranging from detailed studies of particular universities and laboratories to general accounts of the laboratory ethos emerging in the nineteenth century, as well as of the rise of the laboratory as a publishing house. Of course, the treatment of this theme is not exhaustive, and we therefore hope that this volume will stimulate others to continue the study of the co-creation of the modern research laboratory and the modern research university.

Klaas van Berkel and Ernst Homburg
November 2022

Part I

The Laboratory Revolution: Origins and Impact

1 The Joint Emergence of the Teaching-Research Laboratory and the Modern University: An Introduction

Klaas van Berkel and Ernst Homburg

Abstract

The tremendous impact of the Laboratory Revolution on the universities as centres of science and learning has thus far received too little attention. Yet the rise of the modern teaching and research laboratory within the university dramatically changed the outlook, the social structure, and the very idea of the university. This introduction offers a brief survey of the long road to the Laboratory Revolution, a review of recent historiography, and an outline of each of the contributions to this volume.

Keywords: Laboratory Revolution, university science, historiography, Justus Liebig, scientific ethos

Introduction

In the public imagination a scientist today is someone in a laboratory. A man or a woman in a white coat, with safety glasses and a tube in his or her hand. Over the course of the nineteenth century, the laboratory became the ultimate place where new knowledge is created. By the end of that century, the former workplace of chemists, situated on the fringes of the learned world, had turned into a central and indispensable element in the infrastructure of science and science education. A true 'laboratory revolution' had taken place, changing both the sciences and the universities.

Over the past decades, the rise of the laboratory, its growth in numbers, its architectural presentation, and its internal organization have been studied by many historians. Another subject of detailed research has been

Berkel, Klaas van, and Ernst Homburg (eds), *The Laboratory Revolution and the Creation of the Modern University, 1830-1940*. Taylor & Francis Group, 2023
DOI: 10.5117/9789463720434_CH01

the tremendous changes laboratory research brought about in the kind of knowledge we strive for.[1] One important but understudied aspect of the Laboratory Revolution, however, is the impact it had on the university as a centre of science and learning. Of course, not all laboratories were university laboratories. Private laboratories remained important for a long time, and today industrial and testing laboratories employ more researchers than university laboratories. But the rise of university laboratories deserves special treatment because it is there that new generations of researchers are trained and educated. In the nineteenth century simple lecture halls gave way to purpose-built laboratories, which would dominate the cityscape. Even academic disciplines that ostensibly needed no laboratory space to develop, such as astronomy and linguistics, each acquired their own laboratories. Other branches of the humanities, like history, employed the idea of a laboratory metaphorically by saying that their libraries, archives, and seminars were their workplaces, their laboratories. Finally, the nature of the academic community changed tremendously as a result of the rise of the laboratory, with each laboratory becoming a small or not-so-small self-contained community of professors, technical assistants, students, and administrative personnel. The rise of the laboratory was a major factor in the creation of the modern research university between 1850 and the first half of the twentieth century.

The Long Road to the Laboratory Revolution

The laboratory has a long history. Originally it was the workshop of an (al)chemist or an apothecary, where medicines were prepared or other chemical substances were made. In the seventeenth century several universities in Germany, Italy, the Netherlands, and some other Western European countries established chemical laboratories within their faculties of medicine that complemented anatomical theatres and botanical gardens as teaching facilities for students of medicine. In several instances the lectures on chemistry took place in the private laboratory of the professors. During the seventeenth and eighteenth centuries, the term 'laboratory' was used almost exclusively for a place where chemical operations such as distillation were performed.[2] Until the early nineteenth century, laboratories within

1 Robert E. Kohler, 'Lab History: Reflections', *Isis*, 99 (2008), 761-68.

2 The only exception we are aware of is Leiden University, where the *Theatrum physicum* was also called *Laboratorium physicum* in several eighteenth-century sources. See: Cornelis de

the universities were in many respects similar to artisanal laboratories. Both were dominated by fireplaces, furnaces, distillation apparatus, and related chemical equipment.[3]

Here we will focus on the situation in the universities. In many of the private and public teaching laboratories of the seventeenth and eighteenth centuries, chemistry professors allowed some of their students to get experience through the practice of chemistry. This was the case, for instance, in Marburg, Utrecht, Leiden, and Glasgow. These lessons in practical chemistry were then not part of a regular curriculum, but a favour granted—often against payment—by the professor.[4]

During the last three decades of the eighteenth century, a few laboratories started to offer practical training in chemistry to far larger groups of the students. Starting in 1779, several private pharmaceutical institutes that offered practical courses were erected in Germany by pharmacists and chemists such as Johann Christian Wiegleb, Johann Bartholomäus Trommsdorff, Sigismund Friedrich Hermbstädt, and, later, Justus Liebig. During the early decades of the nineteenth century, several of these institutes became integrated into the local university.[5]

Pater, 'Experimental physics', in Th.H. Lunsingh Scheurleer and G.H.M. Posthumus Meyjes, eds., *Leiden University in the Seventeenth Century: An Exchange of Learning* (Leiden: Universitaire Pers Leiden/ E.J. Brill, 1975), pp. 308–27, esp. pp. 315, 318-19, 321-22.

3 Frederic L. Holmes, *Eighteenth-Century Chemistry as an Investigative Enterprise* (Berkeley, CA: Office for History of Science and Technology, 1989); Ernst Homburg, 'The Rise of Analytical Chemistry and its Consequences for the Development of the German Chemical Profession (1780–1860)', *Ambix*, 46 (1999), 1–32; Ursula Klein, 'Die technowissenschaftlichen Laboratorien der Frühen Neuzeit', *NTM*, 16 (2008), 5–38; Ursula Klein, 'The Laboratory Challenge: Some Revisions of the Standard View of Early Modern Experimentation', *Isis*, 99, 769–82; Ursula Klein, 'Chemical and Pharmaceutical Laboratories before the Professionalization of Chemistry', in Marta C. Lourenço and Ana Carneiro, eds., *Spaces and Collections in the History of Science. The Laboratorio Chimico Overture* (Lisbon: Museum of Science of the University of Lisbon, 2009), pp. 3–12.

4 Owen Hannaway, 'Johann Conrad Barchusen (1666–1723) – Contemporary and Rival of Boerhaave', *Ambix*, 14 (1967), 96–111; Frederic Lawrence Holmes, 'Laboratory, Chemical', in J.L. Heilbron et al., eds., *The Oxford Companion to the History of Modern Science* (Oxford: Oxford University Press, 2003), pp. 441–2; Robert G.W. Anderson, 'The Creation of the Chemistry Teaching Laboratory', in Lourenço and Carneiro, eds., *Spaces and Collections in the History of Science*, pp. 13–23, esp. p. 14; Marieke M.A. Hendriksen and Ruben E. Verwaal, 'Boerhaave's Furnace. Exploring Early Modern Chemistry through Working Models', *Berichte zur Wissenschaftsgeschichte*, 43 (2020), 385–411 (pp. 392–93).

5 Dieter Pohl, 'Zur Geschichte der pharmazeutischen Privatinstitute in Deutschland von 1779 bis 1873' (PhD diss., Marburg, 1972); A. Wankmüller, 'Pharmazeutische Privatinstitute und Universitäten zu Beginn des 19. Jahrhunderts', *Deutsche Apotheker-Zeitung*, 113 (1973), 636-39, 673-76; H.R. Abe, 'Zur Geschichte der ersten pharmazeutischen Lehranstalten Deutschlands', *Medicamentum* (Berlin), 17 (1976), 93-95.

At about the same time some universities and mining academies set up chemistry courses for future mining officials and supervisors of mines. From 1767 to 1768 a large new laboratory was erected at the University of Uppsala for the professor of chemistry Torbern Bergman, in which excellent facilities for experimental research and for the teaching of chemistry were combined. According to Marco Beretta, 'Bergman's laboratory was attended by hundreds of students, many from foreign countries, and it soon became a model followed outside Sweden'. It inspired chemistry teachers in Paris and strongly influenced the 1785 renovation of the laboratory of the important mining academy in Schemnitz in the Habsburg Empire. Other large laboratories erected in those years were the *Laboratorio Chimico* of the University of Coimbra (1772) and the chemical laboratory of the University of Göttingen (1784), directed by Johann Friedrich Gmelin. Under Gmelin's successor Friedrich Stromeyer the laboratory was enlarged several times. Stromeyer organized practical chemistry courses for growing numbers of students, especially during the 1820s. In those years practical chemistry could also be studied at several other German universities, and in Scotland and England as well. Only in France would it take several decades for universities to take over the role played by private teaching laboratories.[6]

Parallel to the rise of these teaching laboratories, the practice of chemistry itself changed quite drastically. The investigation of gases (pneumatic chemistry) accelerated the introduction of physical instruments into chemistry. Along with this, new, rather small-scale analytical chemical methods were developed, such as the blowpipe and gravimetric and volumetric techniques. Torbern Bergman and especially Antoine Lavoisier were the leaders of this new approach, which brought chemistry and physics into closer mutual contact.[7] This is illustrated by the growing number of sales catalogues

6 Marco Beretta, 'Laboratories and Technology', in: Matthew Daniel Eddy and Ursula Klein, eds., *A Cultural History of Chemistry in the Eighteenth Century* (London: Bloomsbury Academic, 2022), pp. 71–91, esp. pp. 84–87; Peter Konečný, 'Sites of Chemistry in the Schemnitz Mining Academy and the Eighteenth-Century Habsburg Mining Administration', *Ambix*, 60 (2013), 160–78 (pp. 169–72); Pedro Enrech Casaleiro, 'The Restoration of the *Laboratorio Chimico* at the University of Coimbra', in Lorenço and Carneiro, eds., *Spaces and Collections in the History of Science*, pp. 235–44; Lena Hoppe, *Historische Stätten der Chemie: Das Göttinger Alte Chemische Laboratorium, Göttingen, 17. Oktober 2019* (Frankfurt a.M.: Gesellschaft Deutscher Chemiker, 2019); Homburg, 'The Rise of Analytical Chemistry', pp. 9–18; Alan J. Rocke, 'Academic Chemical Laboratories in Paris, 1823–1894', in Lourenço and Carneiro, eds., *Spaces and Collections in the History of Science*, pp. 25–31.

7 Holmes, 'Laboratory, Chemical'; Peter J.T. Morris, *The Matter Factory. A History of the Chemistry Laboratory* (London: Reaktion Books, 2015), pp. 43–46, 58–59; Beretta, 'Laboratories and Technology', pp. 82–85.

for chemical instruments that appeared during the early decades of the nineteenth century. There was even a new periodical created for disseminating the innovations in laboratory instrumentation and architecture: *Das Laboratorium. Eine Sammlung von Abbildungen und Beschreibungen der besten und neuesten Apparate zum Behuf der practischen und physicalischen Chemie*, which was published from 1825 to 1840.[8] During the early nineteenth century this and other innovations, such as the introduction of gaslight, would also finally revolutionize the architecture of chemical laboratories. Whereas the laboratory between about 1600 and 1820 had remained relatively unchanged, dominated by furnaces, a new type of laboratory emerged after a transition period from 1820 to 1850; this space was dominated by benches and tables, Bunsen burners, glassware, and fume hoods. This so-called 'classical laboratory' (Peter Morris) would dominate chemical university practice until the 1960s.[9]

The lessons taught in practical chemistry at some universities from about 1770 onwards, as mentioned above, focused primarily on training skills in making chemical substances and analysing their composition. One essential ingredient of later academic chemistry was missing: research. As Alan Rocke has argued, Justus Liebig took the crucial step in that direction. He not only engaged selected students in his research programme but also created small teams of students who collectively worked on important research questions. Liebig's creation of a research laboratory within the University of Giessen is often seen as the starting point of the Laboratory Revolution. (See also the section on historiography below.[10]

Turning a teaching laboratory into a teaching-research lab was definitely a revolution in science. It was not just a matter of doing the same kind of research but with more hands than before; it also implied a new conception of what chemical knowledge was about and what was required for breeding up-to-date academic chemists. This new trend in science and academic

8 Brian Gee, 'Amusement Chests and Portable Laboratories: Practical Alternatives to the Regular Laboratory', in Frank A.J.L. James, ed., *The Development of the Laboratory: Essays on the Place of Experiment in Industrial Civilization* (Basingstoke/London: The Macmillan Press, 1989), pp. 37–59; *Das Laboratorium. Eine Sammlung von Abbildungen und Beschreibungen der besten und neuesten Apparate zum Behuf der practischen und physicalischen Chemie*, 44 Hefte (Weimar: Grossherzogl. Sächs. priv. Landes-Industrie-Comptoirs, 1825–40).

9 Holmes, *Eighteenth-Century Chemistry*; Homburg, 'The Rise of Analytical Chemistry'; Holmes, 'Laboratory, Chemical'; Klein, 'The Laboratory Challenge'; Peter J.T. Morris, 'The History of Chemical Laboratories: A Thematic Approach', *ChemTexts: The Textbook Journal of Chemistry*, 7(3) (2021), p. 7.

10 Alan J. Rocke, 'Origins and Spread of the "Giessen model" in University Science', *Ambix*, 50 (2003), 90–115, reprinted as Chapter 2 in this volume; Holmes, 'Laboratory, Chemical', p. 442.

training therefore met with fierce resistance from more traditional chemists. The Viennese professor of chemistry Paul Meissner criticized Liebig for turning chemistry into a search for new substances. Liebig did this, so Meissner argued, only to become a celebrity. He called it 'a French disease' to strive for novelty instead of deep understanding. In his view the aim of science ought to be to develop a 'system', to order known facts in a logical way. Liebig had attacked Meissner for his failure to produce any new chemical 'fact', but the Austrian professor did not care much about discovering new facts. He saw himself as a scholar, not as an experimentalist.[11]

Resistance was also quite common when laboratory research later entered the field of medicine.[12] Hospital physicians were sceptical of the usefulness and the relevance of laboratory studies in medicine. According to them, clinical experience was much more important than expertise in experimental research in a laboratory. But proponents of the laboratory revolution in medicine, like Claude Bernard, claimed that their experimental research on the causes of disease not only delivered important clinical benefits but also resulted in unmediated, and therefore much more reliable, knowledge of 'Nature' than the observation of sick patients in a hospital. The new techniques and instruments in the laboratory eliminated the human, subjective element and made it possible to get an objective idea of the workings of nature. According to Bernard and his followers, nature was most herself in the laboratory, where she spoke clearly in her own voice. Just like in chemistry, the cognitive claims of the advocates of laboratory science went against the established views of science. The rise of the laboratory was therefore not the result of a gradual evolution of techniques and ideas, but of a revolutionary break with existing ideas and conceptions of what science should be.

Despite the resistance from some established scientists and practitioners, the 'model' of the teaching-research laboratory developed by Liebig was introduced in most universities in Europe and North America between about 1840 and 1880, not only in chemistry but also in other disciplines. Alan Rocke and others have demonstrated that shortly after Liebig introduced his new laboratory practices during the 1830s, other professors in chemistry in Germany, Austria, Switzerland, and Great Britain followed in his footsteps:

11 Ernst Homburg, *Van beroep 'Chemiker': De opkomst van de industriële chemicus en het polytechnische onderwijs in Duitsland (1790-1850)* (Delft: Delftse Universitaire Pers, 1993), pp. 314–16, 323–25, 328–34.

12 See especially: Andrew Cunningham and Perry Williams, eds., *The Laboratory Revolution in Medicine* (Cambridge: Cambridge University Press, 1992), esp. pp. 10–11.

Göttingen (1838), Prague (1840), Marburg (1841), Leipzig (1842), Royal College of Chemistry, London (1845), University College, London (1845), Vienna (1845), and Halle (c.1846). By the 1870s almost all university chemistry departments in Europe and the United States had followed these examples.[13]

Compared to this rather rapid dissemination of the 'Giessen model' among university chemistry departments, the diffusion to other disciplines was more gradual. In these disciplines there were 'cabinets', 'theatres', 'botanical gardens', and 'dissection rooms', but no 'laboratories'. During the nineteenth century a kind of two-step process becomes evident: first, the introduction of 'laboratories' or the renaming of cabinets etc. into 'laboratories', and second, the development of these laboratories into full-fledged teaching-research laboratories.

Physics was among the earliest non-chemical disciplines in which the term 'laboratory' was introduced. As Frans van Lunteren notes in his chapter in this book, many of the leading 'experimental physicists' of the first half of the nineteenth century also had a background in chemistry. An example is Gustav Magnus, who started a physics laboratory in Berlin during the 1840s. He was joined by a few 'mathematical physicists', such as Wilhelm Weber in Göttingen and Franz Neumann in Königsberg, who established physics laboratories at about the same time.[14] Most of these early physics laboratories were quite small though. The same applies to the student numbers, resulting from the then very limited career prospects for physicists as compared to chemists. Only after 1860 would larger teaching-research laboratories be created in, for instance, Göttingen, Berlin, and Oxford.[15]

A similar pattern was followed in medicine, particularly in physiology. The first laboratories in this field were also founded in the 1830s and 1840s, for instance by Johannes Müller in Berlin, Jan Purkyně in Breslau, and Jacob

13 Rocke, 'Origins and Spread' (Ch. 2); W.V. Farrar, 'Science and the German University System, 1790–1850', in Maurice Crosland, ed., *The Emergence of Science in Western Europe* (London/ Basingstoke: The Macmillan Press, 1975), pp. 179–92, esp. p. 187; Homburg, *Van beroep 'Chemiker'*, pp. 328–34; Anderson, 'The Creation of the Chemistry Teaching Laboratory', p. 20; Morris, *The Matter Factory*, pp. 109–14; Holmes, 'Laboratory, Chemical', p. 442.

14 Frans van Lunteren, 'The Laboratory Ethos, 1850–1900', see Chapter 3 in this volume; Farrar, 'Science and the German University System', pp. 187–88; Henning Schmidgen, 'The Laboratory', *European History Online* (EGO), published by the Institute of European History (IEG) (2011), p. 5, <http://ieg-ego.eu/en/threads/crossroads/knowledge-spaces/henning-schmidgen-laboratory> (accessed 8 November 2022).

15 Farrar, 'Science and the German University System', pp. 187–89; Graeme Gooday, 'Precision Measurement and the Genesis of Physics Teaching Laboratories in Victorian Britain', *British Journal for the History of Science*, 23 (1990), 25–51; Graeme Gooday, 'Placing or Replacing the Laboratory in the History of Science?', *Isis*, 99 (2008), 783–95; Rocke, 'Origins and Spread' (Ch. 2).

Henle in Heidelberg. It also took until the 1860s and 1870s before large teaching-research laboratories in medicine were erected in Leipzig, Paris, Groningen, Berlin, and Utrecht.[16]

Between 1870 and 1900 most other disciplines followed. Teaching-research laboratories were, for instance, erected by universities in bacteriology,[17] biology,[18] and engineering,[19] to mention some examples documented in the literature.

Historiography

Writing the history of the laboratory was part of the Laboratory Revolution. Whenever a laboratory celebrated its 25th or 50th anniversary, an opportunity presented itself to write the history of the laboratory and, in doing so, re-write the history of the discipline in question. In this way, the rise of the research laboratory was presented as the inevitable result of the progressive development of science, while the once highly contested claims of the advocates of the laboratory took on the character of self-evident and naturally compelling statements. In the 1980s the social history of science

16 Farrar, 'Science and the German University System', pp. 187–89; Timothy Lenoir, 'Laboratories, Medicine and Public Life in Germany, 1830–1849: Ideological Roots of the Institutional Revolution', in Cunningham and Williams, eds., *The Laboratory Revolution in Medicine*, pp. 14–71; Richard L. Kremer, 'Building Institutes for Physiology in Prussia, 1836–1846: Contexts, Interests and Rhetoric', in Cunningham and Williams, eds., *The Laboratory Revolution in Medicine*, pp. 72–109; Schmidgen, 'The Laboratory', p. 6; Sven Dierig, 'Engines for Experiment: Laboratory Revolution and Industrial Labor in the Nineteenth-Century City', *Osiris*, 18 (2003), 116–34; Rocke, 'Origins and Spread' (Ch. 2); Van Lunteren, 'The Laboratory Ethos' (Ch. 3).

17 Farrar, 'Science and the German University System', pp. 188–89; Cunningham and Williams, eds., *The Laboratory Revolution in Medicine*, pp. 3–4.

18 Farrar, 'Science and the German University System', pp. 188–89; Robert E. Kohler, *Landscapes and Labscapes: Exploring the Lab-Field Border in Biology* (Chicago/London: University of Chicago Press, 2002); Helge Kragh, 'From Ørsted to Bohr: The Sciences and the Danish University System, 1800–1920', in Ana Simões, Maria Paula Diogo, and Kostas Gavroglu, eds., *Sciences in the Universities of Europe, Nineteenth and Twentieth Centuries. Academic Landscapes* (Dordrecht: Springer, 2015), pp. 31–47, esp. p. 38; Schmidgen, 'The Laboratory', p. 11; Van Lunteren, 'The Laboratory Ethos' (Ch. 3); Bas Nugteren, 'Of Growing Significance. The Support Staff in the Laboratories and Institutes of Utrecht University during the Interwar Period', Chapter 10 of this volume.

19 Graeme Gooday, 'Teaching Telegraphy and Electrotechnics in the Physics Laboratory: William Ayrton and the Creation of an Academic Space for Electrical Engineering, 1873–84', *History of Technology*, 13 (1991), 73–114; Graeme Gooday, 'Placing or Replacing the Laboratory', p. 792; Wilhelm Borchers, 'Über die Mitarbeit der Hochschulen an der Förderung des Metallhüttenwesens seit Erteilung des Promotionsrechtes', *Chemiker Zeitung*, 36 (1912), p. 465; Gisela Buchheim and Rolf Sonnemann, eds., *Geschichte der Technikwissenschaften* (Basel, Boston, Berlin: Birkhäuser, 1990).

effectively discredited these claims and several studies raised an awareness that laboratories were not to be considered as neutral fora for the study of nature. Bruno Latour and Steve Woolgar's *Laboratory Life: The Construction of Scientific Facts* (1979) and Simon Schaffer and Steven Shapin's *Leviathan and the Air-Pump* (1985) were just two of the studies that opened our eyes to the constructive role of laboratory space in the development of science.[20] The volume edited by Frank A.J.L. James, *The Development of the Laboratory: Essays on the Place of Experiment in Industrial Civilisation* (1989), although restricted in scope, signalled that 'laboratory studies' had developed into a new subdiscipline of the history of science.[21] The advent of this new specialty was also announced in a widely read contribution by Karin Knorr Cetina, 'Laboratory Studies: The Cultural Approach to the Study of Science' (1995).[22]

Thirteen years later Robert E. Kohler, in a Focus section of the journal *Isis* (2008), concluded, though, that laboratory studies still had not been established as a duly recognized specialty in the history of science: 'the lab is a neglected subject.'[23] Notwithstanding quite a few excellent laboratory studies, he looked in vain for a book-length overview of the subject or for helpful entries on the laboratory in the major handbooks and encyclopaedias of the history of science. The main reason for this odd development was, in Kohler's eyes, the decline of institutional history as such. He also attributed the lack of general accounts of the development of the laboratory to the emerging realization that so many different institutions and practices are taken together under the umbrella of 'the laboratory'—the grand university laboratories of the late nineteenth century as well as the country-house labs, the monasteries as well as the eighteenth-century amusement chests and portable laboratories. This discouraged historians from even believing in the possibility of a single unified history of the laboratory. Add to this that the laboratory took on so many different shapes in all the branches of science—medicine, physics, biology, engineering, etc.—and it becomes clear why Kohler had trouble finding general accounts.[24] Kohler could only

20 Bruno Latour and Steve Woolgar, *Laboratory Life. The Social Construction of Scientific Facts* (Beverly Hills, CA: Sage, 1979), republished as *Laboratory Life. The Construction of Scientific Facts* (Princeton, NJ: Princeton University Press, 1986); Simon Schaffer and Steven Shapin, *Leviathan and the Air-Pump: Hobbes, Boyle, and the Experimental Life* (Princeton, NJ: Princeton University, 1985).

21 James, ed., *The Development of the Laboratory.*

22 Karin Knorr Cetina, 'Laboratory Studies: The Cultural Approach to the Study of Science', in Sheila Jasanoff et al., eds., *Handbook of Science and Technology Studies* (Beverly Hills, CA: Sage, 1995), pp. 140–66.

23 Kohler, 'Lab History', p. 761.

24 Morris, *The Matter Factory* is a laudable exception, but, as the title announces, the book only deals with the chemistry laboratory. The book has a very helpful bibliography. Recently the

hope that a renewed attention to the congruence between the values of the laboratory and the dominant values in society would reinvigorate the struggling subspecialty of laboratory studies. Historians should again be curious about 'how the conventions of laboratory life embody those of other important social institutions (court, state, market, media) and how they in turn shape public meanings of knowledge in general'.[25]

Whether by coincidence or not, the basis of Kohler's complaints soon evaporated following a flood of historical studies of laboratories in the following decade. As a result of the 'spatial turn' in the history of science, 'spaces' and 'sites' suddenly became hot topics. In 2009 the collection *Spaces and Collections in the History of Science. The Laboratorio Chimico Overture* was published, edited by Marta C. Lourenço and Ana Carneiro. In November 2010 John Perkins and Antonio García Belmar organized a meeting in London where they initiated a series of international workshops on 'Sites of Chemistry, 1600–2000', leading to special issues of the journal *Ambix* on sites of chemistry in the eighteenth, nineteenth, and twentieth centuries. The wealth of new case studies published in these and other publications was synthesized and brought to a higher level by Catherine Jackson in overviews in handbooks and in Peter Morris's book *The Matter Factory*.[26]

Another way to overcome the restrictions to laboratory studies that Kohler identified might be to put the concept of the Laboratory Revolution, with capitals, at centre stage in the history of the laboratory. The history of science as a whole was stimulated enormously in the post-World War II period when 'the' Scientific Revolution (of the seventeenth century) was

laboratory metaphor has gained popularity among general historians to indicate geographical and social spaces that lend themselves to the study of particular developments. See for instance: Georgiy Kasianov and Philipp Ter, eds., *A Laboratory of Transnational History: Ukraine and Ukrainian Historiography* (Budapest and New York: Central European University Press, 2009) and Klaas van Berkel, 'The Dutch Republic as a Laboratory of the Scientific Revolution', *Low Countries Historical Review*, 25 (2010), 81–105.

25 Kohler, 'Lab History', p. 764.

26 Lourenço and Carneiro, eds., *Spaces and Collections in the History of Science*; John Perkins, ed., 'Sites of Chemistry in the Eighteenth Century', special issue of *Ambix*, 60 (2013), 95–178; Antonio García-Belmar and John Perkins, eds., 'Sites of Chemistry in the Nineteenth Century', special issue of *Ambix*, 61 (2014), 109–86; Antonio García-Belmar and John Perkins, eds., 'Sites of Chemistry in the Twentieth Century', special issue of *Ambix*, 62 (2015), 109–88; Catherine M. Jackson, 'The Laboratory', in Bernard Lightman, ed., *A Companion to the History of Science* (Chichester: Wiley, 2016), pp. 296–309; Catherine M. Jackson, 'Laboratorium', in Marianne Sommer, Staffan Müller-Wille, and Carsten Reinhardt, eds., *Handbuch Wissenschaftsgeschichte* (Stuttgart: J.B. Metzler, 2017), pp. 244–55; Morris, *The Matter Factory*, esp. p. 16. See also: Lissa L. Roberts and Simon Werrett, eds., *Compound Histories: Materials, Governance and Production, 1760–1840* (Leiden/Boston: Brill, 2018).

propagated as the central category in the historical development of science; on a lesser scale, the concept of the Laboratory Revolution might do the same.[27] This volume aims to do just that.

The Laboratory Revolution and the Universities

The university was one of the institutions that helped in creating the modern laboratory and was in turn transformed by the rise of the laboratory. Liebig's engagement of students in his laboratory research was, after all, crucial to the further diffusion of the research laboratory across Europe. At Utrecht University the professor of chemistry Gerrit Jan Mulder also put his students to work in his laboratory. These pioneers inspired others to do the same, and during the nineteenth century one university after another supplied research-minded professors with laboratories and equipment. Whereas Liebig and his contemporaries usually worked in existing buildings that sometimes also housed other departments,[28] the view became widely shared by the middle of the century that only new, purpose-built laboratories could accommodate up-to-date scientific research. In chemistry, which was then still a relatively cheap science (the work was mainly done with relatively inexpensive glass-work), this trend was not a major problem for the authorities, but the costs rose substantially as one science after another followed the example set by chemistry and the numbers of students being educated in the lab grew. Still, the authorities went along with it and paid for the salaries, buildings, and equipment. By the end of the nineteenth century, the new stately laboratories had changed the look of universities in both Europe and the United States.[29]

The rise of the university also had a fundamental impact on the kind of education the students received. Until the nineteenth century lectures in which students listened passively to the professor formed the core of a university education, and dissertations seldom presented new knowledge. During the nineteenth century laboratory work began to replace these lectures as the core element in education. This development started in the science departments, but historians, as one example of scholars who had no need for furnaces or Erlenmeyer flasks, developed a system of 'seminars' (working groups led by the professor) that in a sense imitated the collective

27 See: Floris H. Cohen, *The Scientific Revolution. A Historiographical Inquiry* (Chicago: University of Chicago Press, 1994).

28 A few exceptions (Uppsala, Göttingen) have been mentioned above.

29 For these 'palaces', see: Morris, *The Matter Factory*, pp. 146–69.

work done in the chemical laboratory.[30] At the end of the nineteenth century the new dogma was that the best education is found in the laboratory, where students learn by practising science instead of just listening to the professor.

In the field of the history of universities, this co-creation of the modern research laboratory and the modern research university has seldom attracted much attention.[31] Historians of science interested in the laboratory often take the university for granted as a background for developments in the laboratory,[32] while university historians rarely spend much time discussing what happened inside the laboratory. There is an abundance of literature on the architecture of science, but these studies are usually more concerned with the outer appearance of the buildings than with their constructive details.[33] The laying of foundations and the construction of sewage systems receive scant attention, certainly compared to the attention given to ornaments and murals. Nonetheless, historians of the university seem hardly aware that the laboratory is much more than just one of the facilities needed for teaching and research. In the third volume of the well-known *History of the University in Europe* on the universities in the nineteenth and twentieth centuries, Matti Klinge indeed recognizes the expansion of the university laboratories:

> Being at first separate and often somewhat obscure and dirty annexes to the university itself, the laboratories and institutes of natural sciences became, in the latter half of the century, more respected and, in the new buildings of the turn-of-the-century, achieved an almost sacral status.[34]

30 When the Institute of Historical Research in London was opened in 1921, a pamphlet issued by the institute declared that its aim was to become 'an index to historical knowledge, a focus of historical research, a clearing-house of historical ideas, and a historical laboratory open to students of all universities and all nations'. Quoted in: Debra J. Birch and Joyce M. Horn, eds., *The History Laboratory. The Institute of Historical Research 1921–96*, (London: University of London, Institute of Historical Research, 1996), p.15.

31 A recent and notable exception is: Ku-Ming (Kevin) Chang and Alan Rocke, eds., *A Global History of Research Education: Disciplines, Institutions, and Nations, 1840–1950* (= *History of Universities*, 34/1) (Oxford: Oxford University Press, 2021).

32 But see: Christoph Meinel, 'Artibus Academicis Inserenda: Chemistry's Place in Eighteenth- and Early Nineteenth-Century Universities', *History of Universities*, 7 (1988), 89–115.

33 Sophie Forgan and Graeme Gooday, '"A Fungoid Assemblage of Buildings": Diversity and Adversity in the Development of College Architecture and Scientific Education in Nineteenth-Century South Kensington', *History of Universities*, 13 (1994), 153–92. Peter Galison and Emily Thompson, eds., *The Architecture of Science* (Cambridge, MA: MIT Press, 1999) does indeed pay attention to the more technical details of modern laboratories, but this volume does not deal with university laboratories or with the university context of the laboratory.

34 Walter Rüegg, ed., *A History of the University in Europe. Volume III. Universities in the Nineteenth and Twentieth Centuries (1800–1945)*, 4 vols. (Cambridge: Cambridge University Press, 2004), p. 144.

But what this sacral status entailed and how it changed the university is not discussed in that volume. There is no paragraph or chapter devoted to the laboratory as a driving force behind the creation of the modern research university.[35] This volume is trying to remedy this neglect of how the new laboratories shaped the modern university.

The Book in Brief

This volume brings together what until now have essentially been two separate fields of inquiry: the study of the rise of the laboratory as the ultimate source of reliable knowledge and the history of the university as a teaching-research institution. The stage is set by a key essay by Alan Rocke first published in 2003, which opens the first part of this volume: 'The Laboratory Revolution: Origins and Impact'.[36]

Since the early 1970s authors writing on the Laboratory Revolution—sometimes under a different banner—were in large agreement that the 'Giessen model', or 'German model', had played a crucial role in it.[37] There was less agreement, though, about when this model emerged and what its precise nature was. In terms of the 'founding fathers' of the model: did it start with Wilhelm von Humboldt, or with Friedrich Stromeyer, or with Justus Liebig?[38] Rocke brings this discussion to a higher level by investigating in detail what made the laboratory-based education developed by Liebig in Giessen after 1835 distinct from earlier practices in several ways. He also demonstrates the impact of that specific (Giessen) model not only on the development of chemistry laboratories in Germany and abroad but also on the nature of laboratories in other disciplines, such as physics and physiology. The experimental courses organized by Liebig were more intense and were given to more students than ever before. Moreover, they were financially supported by the university and—importantly—offered the opportunity

35 For a recent improvement of the historiography on this point, see: Chang and Rocke, eds., *A Global History of Research Education*.

36 Originally published as Rocke, 'Origins and Spread'. We thank the author as well as others mentioned in the acknowledgement in Chapter 2.

37 Among the earliest contributions are: R. Steven Turner, 'The Growth of Professorial Research in Prussia, 1818–1848 – Causes and Context', *Historical Studies in the Physical Sciences*, 3 (1971), 137–82; J.B. Morrell, 'The Chemist Breeders: The Research Schools of Liebig and Thomas Thomson', *Ambix*, 19 (1972), 1–46; Farrar, 'Science and the German University System'.

38 See for instance: Geert Vanpaemel, 'The German Model of Laboratory Science and the European Periphery (1860–1914)', in Simões, Diogo, and Gavroglu, eds., *Sciences in the Universities of Europe* (Dordrecht: Springer, 2015), pp. 211–25.

for the more advanced students to work in teams on research questions. Rocke argues that these novel elements not only resulted from general institutional, socio-political, and economic factors (state funding, industrial demand), but also from a few very specific material and theoretical aspects of the emerging discipline of organic chemistry, which came together during the 1830s: the discovery of isomers, the practice of using 'paper tools', and the new *Kaliapparat*, invented by Liebig, used in organic analysis. Taken together, a network of quite heterogeneous factors and circumstances in Giessen created the practices that were eventually associated with all modern research universities, and Liebig's extraordinary charisma and communicative skills greatly helped in disseminating his 'model'.

The impact of the Laboratory Revolution that Liebig initiated implied much more than the national and international dissemination of materially well-equipped, spacious new buildings; it also changed the codes and norms that regulated academic life in general. How the rise of the laboratory led to the introduction of a new scientific ethos in several disciplines is analysed in detail by Frans van Lunteren. He explores the shift in the values and self-image of the scientist by looking at character traits prized most highly in a laboratory setting, such as discipline, self-restraint, precision, perseverance, modesty, and (in the case of the physiologists) emotional control. These epistemic virtues moulded a new scientific identity that took different shapes in different disciplines (the author discusses chemistry, physiology, and physics) but that also showed remarkable resemblances. Contrary to recent literature on the emergence of these virtues, the author shows that the rise of the laboratory specifically was instrumental in forging these new scientific virtues. These new virtues only fully took shape after the introduction of the university teaching and research laboratory in the 1830s and not in the late eighteenth century. On the other hand, he points out that the rise of the modern bureaucratic state, which more or less exemplifies the same virtues, also influenced the development of the new scientific ethos.

In the next two parts of the book, these spatial and normative dimensions of the Laboratory Revolution are explored in greater detail, starting the analysis of *Laboratory Networks* (in part two), which include scientists, suppliers, and administrators. One of the earliest university laboratories outside Germany where the teaching-research model was introduced—even without a direct connection to Liebig—was in Zürich. Peter Ramberg shows that the emergence of academic chemistry in Zürich resembled the path followed in Germany, but it also differed from it because of a complex interplay between the policies of the Swiss Federation and the local needs and finances of the Canton of Zürich. Due to limited financial means at the different political levels and a still limited industrial demand for skilled chemists during the 1840s and 1850s, a situation emerged in which most professors had dual appointments at the Cantonal University and at the new Federal Polytechnic. The laboratories of both schools were closely connected but followed different models: the University was explicitly founded on the German model, with its faculties and academic freedom. The Polytechnic was modelled after the Polytechnic at Karlsruhe, with its divisional structure and more restricted freedom for staff and students. Despite these German influences, the final result differed. Ramberg demonstrates that the physical proximity of the University and Polytechnic in the same city, with dual appointments and a shared chemical laboratory until 1861, resulted in arrangements that were largely absent in the German system. The development of organic chemistry also played a crucial role in Zürich. After 1860 the synthetic dye industry in Switzerland, and elsewhere, required growing numbers of (organic) chemists, and at the same time chemistry professor Johannes Wislicenus and his successors set up a very successful research school in organic chemistry, which attracted many students. Wislicenus also untangled the complex relationship in the field of chemistry between the University, the Polytechnic, and other schools in Zürich. By the late nineteenth century Zürich had become the most productive site for chemistry in Switzerland and one of the most successful in the world, largely due to the close spatial and personal network that included the University and the Polytechnic.

While a close collaboration between different laboratories emerged (albeit in the same discipline) in Zürich, Klaas van Berkel demonstrates in his chapter that laboratories erected during the Laboratory Revolution in Groningen—as well as in other cities—undermined the unity of the local universities. The decades around 1900 were a time of great expansion for the Dutch university system, particularly in experimental science, medicine, and some other fields such as psychology. In the state-funded universities in Utrecht, Leiden, and Groningen, new laboratory buildings were erected to accommodate the new science of the times. The distribution of these laboratories over the city had both intended and unintended consequences. There were often unrecognized effects on ideas and perceptions of science, its practitioners, and these practitioners' place within the university. Discussing the example of Groningen in greater detail, Van Berkel shows that the spatial 'diaspora' of laboratories that started in the late-nineteenth century undermined the contact between professors of different disciplines, and the same was true for the students. It was a disruptive force that contributed to a disintegration of the university as a unified social and intellectual community. Although several measures were taken to counteract the consequences of these spatial changes, basically all of them failed. Only after World War II did the establishment of campuses constitute a more serious attempt to re-unify the university, but even that step was only partially successful.

In the chapters discussed so far, the spatial (Ramberg, Van Berkel) and socio-political (Rocke, Ramberg) networks influencing laboratory science come to the fore, and to a lesser extent material aspects as well, such as the role played by the *Kaliapparat*. Pierre Laszlo puts the mutual relations between material and spatial concerns centre stage in his study of the suppliers of laboratory chemicals and instruments in Paris. Historical studies of laboratories often discuss the 'outputs' of laboratories, such as the number of chemists or other professionals who were 'bred' (Morrell), or the number of discoveries made. But the 'inputs' are often overlooked. Laszlo sheds an unexpected light on the Laboratory Revolution by investigating the role of the commercial laboratory suppliers. He identifies and locates the main Parisian laboratories and discusses in detail the interplay between the Parisian scientists and the ten to twenty local suppliers from 1880 to 1910. Interestingly, most suppliers to laboratories had their shops predominantly in the Latin Quarter, in close proximity to the laboratories. Many suppliers had a background in pharmacy. They occupied the 'trading zone' (Galison) between laboratory chemists, wholesalers, and industrialists, and between chemists and physicians. To some extent, they can be called 'activists', because they were not merely merchants but also proponents of the contemporary ideology of progress through science, reinforced by their participation in the 1889 and 1900 World Fairs. Over time, they not only supplied the Parisian laboratories but increasingly also laboratories and hospitals in the French colonies worldwide. The chapter therefore illustrates the large-scale consequences of the Laboratory Revolution in an unexpected way.

It should be no surprise that spatial factors would play a large role in a geographically 'long' country such as Norway. The traditional university of the country was positioned in a rather eccentric way in the capital Kristiana (Oslo) in the south. When an Institute of Technology was established in 1910, the town of Trondheim was chosen, positioned more or less in the middle of the country to better serve the industry of different regions. Discussing the competition between the University in Kristiania and the Institute in Trondheim with respect to the erection of new, large, and modern laboratories, Annette Lykknes shows that more was at stake than geography. It was also a socio-political and ideological battle. The University of Kristiana had the advantage of being close to the national centre of power. Trondheim, in turn, could profit from its network with industry. Even more important was the debate on the type of chemists that industry needed. Professors from the University argued that a solid background in pure science was also a good preparation for industrial positions. The professors from the Institute, by contrast, claimed that knowledge of industrial chemistry and chemical engineering was what industry needed. The visions of those two opposing parties did of course have consequences for laboratory design. The competition became worse in the hard financial times after World War I. On top of differences in training and qualities of education, the geographical differences in Norway added to this polarization, with questions related to centre and periphery sharpening the dispute. Lykknes argues that in those years, the laboratory even became a symbol of the nation for the chemistry professors in Trondheim. Without a proper chemical laboratory, the industry and the whole country would suffer. The examples show that although Kristiania would get priority in the construction of a new laboratory, the Laboratory Revolution was part of a whole network of interests and ideas on the future of the nation.

Socio-political and ideological issues are also centre stage in Ab Flipse's contribution on the laboratory of the Vrije Universiteit (Free University) in Amsterdam, founded by Dutch Calvinists in 1880. Although the laboratory was generally seen as a symbol of scientific progress, in religious circles it also carried a negative image as a symbol of 'materialism' and disbelief. It therefore took almost four decades before the leaders of the Vrije Universiteit decided that Calvinist physicians were also important for the constituency of their university and that a medical faculty with a laboratory was a crucial next step in developing their university. Founded in 1918, the physiological laboratory started rather smoothly. The scientists connected to the laboratory succeeded in performing a careful balancing act by arguing that it would be possible to create a radically different research strategy on the basis of Calvinist principles. The foundation of this first laboratory was in a sense revolutionary because it initiated the Vrije Universiteit's own 'laboratory revolution'. For the first time, the university came into contact with an academic discipline in the sciences. As a result, more than before, the university adapted to what was already customary elsewhere. The physiological laboratory paved the way for the future science faculty in 1930 and for further growth into a university that comprised all faculties later in the century.

The third part of this volume, *Laboratory Values*, is less concerned with the laboratory itself than with the ethos that has become prevalent in a laboratory setting, especially in the case of academic and semi-academic laboratories. Laboratories changed the face of the university, not only by claiming so many spacious new buildings but also by changing the codes and norms that regulated academic life in general. The new scientific virtues—discussed by Frans van Lunteren in his chapter mentioned above—were propagated mainly by the professors, but these were not the only people who inhabited laboratories. Peter Morris and Bas Nugteren both attend to the academic and technical staff below the rank of full professor. Morris compares German and English laboratories, more precisely the academic laboratories in Heidelberg and London (Imperial College). He shows that the intricate social organization of the laboratories in these countries, different though they were, did not hinder them from becoming productive and dynamic social communities. The technicians and assistants were crucial for the success of a laboratory. Therefore, laboratory studies in which the historian or sociologist only pays attention to the professor paint an unrealistic picture of how science proceeds. The emergence of new epistemic virtues in the late nineteenth century not only depended on the professor's powers of persuasion; virtues like precision, disinterestedness, and the strict following of fixed procedures were also moulded by the practices of a modern laboratory. The new laboratory was a factory-like organization not only because of the use of machines and instruments, but also because of its social structure.

Bas Nugteren focuses on the laboratories at one particular institution: Utrecht University in the first half of the twentieth century. He details the finer social structure in the laboratories while also documenting the gradual but growing appreciation for the contributions of the support staff. Public statements by university authorities, the awarding of royal distinctions for long-term service to the university, and newspaper articles all point to a growing awareness that the support staff was essential to the smooth operation of the laboratory and should be honoured for that. At the same time non-academic staff members also started to manifest themselves strongly in labour unions, who were then recognized by the university administration. However, as Nugteren also shows, the staff was not included in the concept of the *civitas academica*, the imaginary academic community consisting solely of professors, lecturers, and students.

Ida H. Stamhuis adds another perspective to the study of laboratory life by considering the issue of gender. The rise of the laboratory was not without consequences for the gender relations within the university and in science in general. The growing importance of laboratories can, as Margaret W. Rossiter in 2003 claimed, 'be seen as a new level of exclusion, creating new male retreats or preserves to which women gained entry only by special permission'.[39] She spoke specifically about the physical sciences, but her statement is roughly true for laboratories in other disciplines as well. Yet, as Stamhuis demonstrates, the situation is ambivalent. In the case of the emerging field of genetics, the newly created laboratories at the fringes of the university were indeed led by male scientists, but—because relationships were not yet fixed in these emerging fields—they at the same time offered new opportunities for female scientists to make important contributions to science. As so often happens, change started at the periphery rather than the centre of the university.

Quite a few of the laboratories founded during the Laboratory Revolution began publishing their own journal. One of the first laboratory journals was Gerrit Jan Mulder's *Scheikundige onderzoekingen, gedaan in het Laboratorium der Utrechtsche Hoogeschool* ('Chemical investigations undertaken at the laboratory of Utrecht University'), founded in 1842. The rise of this new type of science publication thus more or less coincided with the Laboratory Revolution. In her chapter Dorien Daling discusses a sample of these 'in-house' publications and analyses their functions. She concludes that these journals, which almost ceased to exist after the mid-twentieth century, did not aim to circulate new knowledge as such. These journals often comprised articles that had already been published elsewhere. Instead, the laboratory journals served to legitimize the existence of the laboratories, to further the internal social cohesion of the laboratory, and to offer assistants and PhD students training in writing scientific articles. By setting up exchange subscriptions, these journals also forged closer ties between laboratories in the same discipline all over the world.

In the final chapter, Geert Vanpaemel discusses how the laboratory was (or was not) represented in popular science magazines during the Laboratory Revolution. He focuses on the engravings and photographs in the French magazine *La Nature*, published since 1872. The author concludes that pictorial representations of science in the late nineteenth century retained quite a few elements of an older iconography, especially the images of the alchemical laboratories and workshops of the early modern period. Furthermore, he observes that although there is a growing recognition in *La Nature* of the laboratory as the most important expression of a stable

39 Margaret W. Rossiter, 'A Twisted Tale. Women in the Physical Sciences in the Nineteenth and Twentieth Centuries', in Mary Jo Nye, ed., *The Physical and Mathematical Sciences. The Cambridge History of Science, vol. 5*, 8 vols. (Cambridge: Cambridge University Press, 2003), pp. 54–71, quote on p. 55.

scientific research environment and as an institution of cultural authority, the editors of the journal seemed to be even more interested in technological innovation and industrial machinery. Apparently, the magazine did not perceive the laboratory as the major driving force of science; instead, the modern laboratory was seen as an adaptation of science to the modernization of industry and society.

This last and tentative conclusion harks back to Van Lunteren's remarks in his chapter on the agreement between the new values of laboratory life and the values of the modern bureaucratic state, such as the 'specialization and division of labour, hierarchical layers of authority, selection on the basis of technical skills acquired through training and experience, and, above all, a strong emphasis on impersonal rules and procedures'. This would imply that the new value system introduced by the laboratory, which in turn to a large extent created the modern university, basically reflected the values of a modern industrial and bureaucratic society.

About the Authors

Klaas van Berkel recently retired as Rudolf Agricola Professor of History at the University of Groningen. In his research he focuses on cultural history and the history of science, especially the history of scientific institutions (academies, universities). In 2021 he published, with Guus Termeer, *The University of Groningen in the World. A Concise History* (Amsterdam: Pallas Publications).

Ernst Homburg is professor of history of science and technology emeritus at Maastricht University. His research is mainly on the interaction between science and technology in industry and academia from the nineteenth to the twenty-first century. His most recent book publications are: *Hazardous Chemicals: Agents of Risk and Change, 1800–2000* (New York, Oxford: Berghahn, 2019), edited with Elisabeth Vaupel, and *Een eeuw chemische technologie in Nederland* (Delft: Stichting Hoogewerff-Fonds, 2021), with Ton van Helvoort.

2 Origins and Spread of the 'Giessen Model' in University Science

Alan J. Rocke

Abstract

In seeking to understand the rise of Justus Liebig's model for research and teaching, three interrelated and overlapping factors intrinsic to his specialty of organic chemistry have not been sufficiently brought into the explanatory field: the discovery of isomers, the novel practice of using 'paper tools', and the '*Kaliapparat*' method of organic analysis. The existence of these three interacting factors made it a uniquely positioned context within which to create the practices that eventually were associated with all modern research universities.

Keywords: Liebig, paper tools, research groups, team research, Giessen

Introduction

The chemical teaching-research laboratory that Justus Liebig established in the small German university town of Giessen was long viewed by many as the fount of a powerful innovation that transformed both pedagogy and research practice in universities around the world. This simple mythology, centred in field-internal factors and in an implicit and outmoded 'great man' theory of history, has been fruitfully unpacked, repacked, and attacked during the last generation. What continues to require explanation is how the research/teaching model that is associated with Liebig's name really did get off the ground; it is also necessary to understand more clearly the causes of the remarkable rise of German chemistry (and, more broadly, German science) from its undistinguished condition around 1820 to what many have considered world supremacy by the third quarter of the nineteenth century.

Aside from some scattered early contributions, the re-thinking began in 1960 with contributions from historical sociologists Joseph Ben-David and

Berkel, Klaas van, and Ernst Homburg (eds), *The Laboratory Revolution and the Creation of the Modern University, 1830-1940*. Taylor & Francis Group, 2023
DOI: 10.5117/9789463720434_CH02

Avraham Zloczower, who argued that institutional dynamics and career perceptions in the decentralized German academic marketplace could provide an adequate understanding of the take-off of German university science in the middle decades of the nineteenth century.[1] Then, between 1971 and 1982, R. Steven Turner published a series of path-breaking articles that demonstrated how German universities transformed themselves from conservative teaching-oriented corporative bodies into dynamic research-oriented state institutions by wedding neo-humanist ideology to a powerful new research ethic, and both to a kind of experiential pedagogical philosophy that favoured laboratory practica. Turner's approach was distinct from, but not inconsistent with, that of Ben David and Zloczower; his work broadened the explanatory field by bringing in state interests as well as institutional and cultural factors.[2] In 1976 Peter Borscheid suggested further agency for state interests: he concluded that the spread of the Liebigian pedagogical model in Baden after 1848 owed much to the desire of its ruling elite to promote agricultural chemistry, for the government hoped to calm a restive population by improving crop yields and thus eliminating famine. More recently, Arleen Tuchman has offered a fine parallel analysis for physiology and medicine, which also looks (inter alia) to utilitarian socio-economic interests by government officials of Baden (and, by extension, of other German states) in promoting science education.[3] Other scholars have approached these issues through the role of pharmacists

1 Joseph Ben David, 'Scientific Productivity and Academic Organization in Nineteenth Century Medicine', *American Sociological Review*, 25 (1960), 828–43; Avraham Zloczower, *Career Opportunities and the Growth of Scientific Discovery in Nineteenth-Century Germany, with Special Reference to Physiology*, PhD diss., Hebrew University, 1960 (repr. New York: Arno, 1981). Ben David published several additional contributions on this subject over the next 24 years; his *Scientist's Role in Society: A Comparative Study*, 2nd ed. (Chicago: University of Chicago Press, 1984) recapitulates some of these themes.

2 R. Steven Turner, 'The Growth of Professorial Research in Prussia, 1818–1848 – Causes and Context', *Historical Studies in the Physical Sciences*, 3 (1971), 137–82; R. Steven Turner, 'University Reformers and Professorial Scholarship in Germany, 1760–1806', in Lawrence Stone, ed., *The University in Society*, 2 vols. (Princeton, NJ: Princeton University Press, 1974), vol. 2, pp. 495–531; R. Steven Turner, 'The Prussian Professoriate and the Research Imperative', in H.N. Jahnke and M. Otte, eds., *Epistemological and Social Problems of the Sciences in the Early Nineteenth Century* (Dordrecht: Reidel, 1981), pp. 109–21; and R. Steven Turner, 'Justus Liebig versus Prussian Chemistry: Reflections on Early Institute Building in Germany', *Historical Studies in the Physical Sciences*, 13 (1982), 129–62. See also Charles McClelland, *State, Society, and University in Germany, 1700–1914* (New York: Cambridge University Press, 1980), and Christoph Meinel, '*Artibus Academicis Inserenda*: Chemistry's Place in Eighteenth- and Early Nineteenth-Century Universities', *History of Universities*, 7 (1988), 89–115.

3 Peter Borscheid, *Naturwissenschaft, Staat und Industrie in Baden (1848-1914)* (Stuttgart: Klett, 1976); Arleen Tuchman, *Science, Medicine, and the State in Germany: The Case of Baden, 1815–1871* (Oxford: Oxford University Press, 1993).

and pharmaceutical education, through prosopographical studies of students, or through comparative analyses of research schools.[4]

These valuable examinations of the subject have generally sought *structural* explanations for the rise of German academic science from sociological, cultural, economic, or institutional approaches and have reduced the importance of contingent, cognitive, or field-specific factors, such as individual personalities or their innovations in tools, techniques, or theories. A different complementary approach has recently been proposed in a series of important contributions by Ernst Homburg. Homburg by no means ignores institutional analysis, drawing particular attention to hitherto neglected factors such as the role of German polytechnics and to the demand for the product of both universities and polytechnics that was created by the rising German chemical industry. What is intriguing for our purposes, though, is his argument for locating the initial impetus for the take-off of the field not in Liebig's organic-chemical laboratory in Giessen, as the old mythology asserted, but in Friedrich Stromeyer's earlier inorganic-analytical laboratory in Göttingen. This manoeuvre amounts to a partial return of focus towards personalities and field-specific factors.[5]

In what follows I will argue that Homburg's historiographic strategy has much to recommend it: an eclectic approach that adopts approaches from a variety of sources, including cognitive issues, as well as sensitive incorporation of the social-institutional literature has the best promise for offering the most robust understanding of the subject. However, I want to contest the degree of importance that he places on Stromeyer and on analytical chemistry. Rather, I will return to a carefully qualified version of the older Liebigian mythology, emphasizing the leading roles of personalities, ideas,

4 Bernard Gustin, 'The Emergence of the German Chemical Profession, 1790–1867' (PhD diss., University of Chicago, 1975); Joseph S. Fruton, 'The Liebig Research Group – A Reappraisal', *Proceedings of the American Philosophical Society*, 132 (1988), 1–66; J.B. Morrell, 'The Chemist Breeders: The Research Schools of Liebig and Thomas Thomson', *Ambix*, 19 (1972), 1–46. The most authoritative single source on Liebig is William H. Brock, *Justus von Liebig: The Chemical Gatekeeper* (Cambridge: Cambridge University Press, 1997), who critically summarizes all this literature.

5 Ernst Homburg, *Van beroep 'Chemiker': De opkomst van de industriële chemicus en het polytechnische onderwijs in Duitsland (1790-1850)* (Delft: Delft University Press, 1993); Ernst Homburg, 'The Teaching of Chemistry at the German Polytechnic Schools, 1803–1860', *Mitteilungen der Fachgruppe Geschichte der Chemie der Gesellschaft Deutscher Chemiker*, 13 (1997), 75–93; Ernst Homburg, 'Two Factions, One Profession: The Chemical Profession in German Society 1780–1870', in David Knight and Helge Kragh, eds., *The Making of the Chemist: The Social History of Chemistry in Europe, 1789–1914* (Cambridge: Cambridge University Press, 1998), pp. 39–76; Ernst Homburg, 'The Rise of Analytical Chemistry and Its Consequences for the Development of the German Chemical Profession (1780–1860)', *Ambix*, 46 (1999), 1–32.

and tools. In this regard, I am taking much from Jack Morrell's approach to explaining the success of Liebig's 'chemist breeding'.[6] In addition, I want to highlight a factor stressed by Frederic L. Holmes, William Coleman, Kathryn Olesko, and Joseph Fruton, namely the idea of group research as an adjunct to experiential pedagogy.[7] Among other things, this approach places greater importance back on contingencies of time, place, and discipline.

In particular, I want to examine three interrelated and overlapping factors intrinsic to Liebig's specialty, organic chemistry, that have not sufficiently been brought into the explanatory field. In the empirical realm, the explicit recognition of the concept of isomers in 1830 introduced a disorienting increase in the numbers of organic compounds. In the theoretical realm, the novel practice of using chemical formulae as 'paper tools' has been dated to the same period, 1827–33; this practice provided a simple explanatory and heuristic understanding for isomerism. In the instrumental realm, Liebig's new means of organic analysis provided a fast, easy, and precise method that could cope with the sudden acceleration of the science, and one that was also uniquely favourable to both the pedagogical and research practices that became so closely associated with Liebig's name.

The existence of these three interacting factors, all of which emerged suddenly and essentially simultaneously, made the new field of organic chemistry a uniquely positioned context within which to create the practices that eventually were associated with modern research universities. No other discipline could come close. In this respect, Liebig was simply fortunate to be pursuing the right field at the right time. Finally, for comparative purposes, I want to examine not only the situation in Germany itself, but also France, and, much more briefly, the United States.

Isomerism

At the beginning of the nineteenth century, chemists implicitly assumed that a substance's elemental composition determined its identity. (The idea

6 Morrell, 'Chemist breeders'.

7 Frederic L. Holmes, 'The Complementarity of Teaching and Research in Liebig's Laboratory', *Osiris*, [2] 5 (1989), 121–64; William Coleman, 'Prussian Pedagogy: Purkyne at Breslau, 1823–1839', in William Coleman and Frederic L. Holmes, eds., *The Investigative Enterprise: Experimental Physiology in Nineteenth-Century Medicine* (Berkeley, CA: University of California Press, 1988), pp. 15–64; Kathryn Olesko, 'On Institutes, Investigations, and Scientific Training', in Coleman and Holmes, *The Investigative Enterprise*, pp. 295–332; Kathryn Olesko, *Physics as a Calling: Discipline and Profession in the Königsberg Seminar for Physics* (Ithaca, NY: Cornell University Press, 1990).

that two distinct substances could have the same chemical composition probably must have seemed as unlikely to them as, in this century, the possibility that two unrelated people might have identical fingerprints or even identical genetic material.) However, some of the earliest good elemental analyses began to erode confidence in this assumption. Between 1811 and 1826, chemical twins such as glucose and starch, acetic acid and cellulose, wax and spermaceti, racemic and tartaric acids, and differing species of sugar with identical compositions mystified individual chemists, without however immediately calling for a reassessment of the composition/identity correlation assumption. But the collision between the young chemists Friedrich Wöhler and Justus Liebig over fulminic and cyanic acids (1824–6) brought new attention to the problem, partly because the contretemps was viewed as being implicitly between their famous mentors Jacob Berzelius and Joseph Louis Gay-Lussac. Wöhler's discovery of the interconvertibility of ammonium cyanate and urea and Gay-Lussac's recognition of the identity in composition of racemic and tartaric acids (both in 1828) were the final straws. In 1830 Berzelius named the phenomenon 'isomerism', argued for its generality and importance, and suggested that differing arrangements of atoms within the molecules could provide an explanation.[8]

By 1820 only about 120 organic compounds had been characterized in the literature, and it must have appeared in that year that organic chemistry would remain a small sub-field, real chemical importance continuing to inhere in inorganic materials. This perception changed with the advent of isomerism. Suddenly it was no longer sufficient simply to compile a list of a few dozen substances, all derived from nature and each with a unique composition, in order to summarize all of organic chemistry. Now, when each compositional formula might spawn two, three, five, or ten or more unique substances, including any number of substances not known in nature, the sky was the limit. To be sure, isomerism was not restricted to organic substances, but isomers were known to proliferate particularly in that area. Also to be sure, there were other factors besides isomerism that meant that organic chemistry in particular became an almost infinitely large field; but isomerism was the earliest of such indications. Already by 1829 Wöhler could comment, on the discovery that a purported isomer of

8 J.J. Berzelius, 'Über die Zusammensetzung der Weinsäure und Traubensäure ... nebst allgemeinen Bemerkungen über solche Körper, die gleiche Zusammensetzung, aber ungleiche Eigenschaften haben', *Annalen der Physik*, [2] 19 (1830), 305–35; *Jahresbericht über die Fortschritte der physischen Wissenschaften*, 11 (1832), 44–8; J.R. Partington, *A History of Chemistry*, 4 vols. (London: Macmillan, 1964), vol. 4, pp. 203, 256, 258–60, 272, 751; Maurice P. Crosland, *Gay-Lussac* (Cambridge: Cambridge University Press, 1978), pp. 134–5.

cyanic acid was only a phantom, that 'L[eopold] Gmelin will say, "Thank God, there's one fewer [organic] acid"' that needed to be included in his definitive index of all known compounds.[9]

By the next generation, the volume of organic substances was truly intimidating. Friedrich Beilstein complained, 'Everywhere we see indicated isomerisms and possibly existent compounds, which are becoming so numerous that even all the masses of material now known appears by comparison as but a drop.... Do these billions of compounds really exist? ... I could never imagine that the dear Lord would want to make life so hard for chemists.' Regarding Alexander Butlerov's new structuralist textbook, he commented: 'Our students are already frightened by the mass of material; imagine their faces when they have predicted for them the existence of God knows how many additional compounds.' And Liebig knew where to put the blame when he wrote to A. W. Hofmann, 'Through isomerisms, organic chemistry is gradually becoming enough to make one mad.'[10] Marcellin Berthelot made the point with his usual flair when he calculated that the known *possible* esters of sorbitol, all 1.4×10^{15} of them, would fill 14,000 libraries, each containing a million books, comprising a campus that would require an area the size of Paris. And this space would be required simply to *list the names* of the possible compounds, not even to describe their properties.[11] Of course, many of these predictions were based on crudely ad hoc structural combinatorial exercises, and the number of compounds that were really possible was always far smaller than that. But the number of real compounds was indeed mushrooming, and the point was broadly taken.

Recognition of the inexhaustibility of organic chemistry was not entirely new to mid-century minds. As early as 1819, soon after the first indications of the as-yet unnamed and largely unrecognized phenomenon of isomerism were published, the Bonn chemist Carl Gustav Bischof sat down with pencil and paper and performed some interesting calculations and speculations based on the phenomenon. He assumed that all organic plant materials were composed

9 Wöhler to Liebig, 8 June 1829, in A.W. Hofmann, ed., *Aus Justus Liebig's und Friedrich Wöhler's Briefwechsel*, 2 vols. (Braunschweig: Vieweg, 1888), vol. 1, p. 4.

10 Beilstein to Butlerov, 30 August 1865, in G.V. Bykov and L.M. Bekassova, 'Beiträge zur Geschichte der Chemie der 60-er Jahre des XIX. Jahrhunderts: II. F. Beilsteins Briefe an A. M. Butlerov', *Physis*, 8 (1966), 267–85, on p. 275; Otto Krätz, ed., *Beilsteins-Erlenmeyer: Briefe zur Geschichte der chemischen Dokumentation und des chemischen Zeitschriftenwesens* (Munich: Fritsch, 1972), p. 79; Liebig to Hofmann, 24 January 1868, in E. Heuser and R. Zott, eds., *Justus von Liebig und August Wilhelm Hofmann in ihren Briefen* (Mannheim: Bionomica, 1988), p. 45.

11 Marcellin Berthelot, 'Sur les principes sucrés', in *Leçons de chimie et de physique professées en 1862* (Paris: Hachette, 1863), pp. 248–9.

of various combinations of five proximate components: water, the two oxides of carbon, and the two hydrides of carbon (in the Dalton-Wollaston-Gmelin atomic weights of c.1819, these were HO, CO, CO_2, CH, and CH_2). He then determined that, for instance, 11 distinct isomers (he called them 'Komplexionen') must exist of succinic acid, 41 of tartaric acid, 54 of citric acid, and no fewer than 239 of multiples of the basic carbohydrate formula, CHO.[12] I cite this obscure work not to assert any substantial influence it may have had—for which I know of no evidence for this at all—but simply as an indication that the new phenomenon of isomerism must quickly have opened many eyes and minds to radically expanded possibilities for the science of organic chemistry.

Formulae as Paper Tools

Bischof was able to perform this combinatorial exercise only because he was clever enough to exploit a new theory that had been born and had begun a furious development just within the previous decade: chemical atomism.[13] Among those involved with this development in the second decade of the century (besides Bischof) were Dalton, Davy, Thomson, Wollaston, Prout, Berzelius, Avogadro, Ampère, Gay-Lussac, Dulong, Gilbert, Schweigger, Kastner, Meinecke, L. Gmelin, Mitscherlich, and Döbereiner.[14]

Some of these men did not believe that they were participating in the elaboration of a scientific *theory*, but rather regarded chemical atomism to be co-extensive with the empirical and law-like subject of 'stoichiometry'. I believe that this perception was influenced by the understandable and pervasive urge by scientists to remain above the uncertain world of speculation and conjecture, and I think that it was ultimately misguided. The laws of stoichiometry do not entail the postulation of irreducible smallest gravimetric entities for each element that can be combined in summary fashion in a symbolic formula for every known compound—which is what I call 'chemical atomism'—nor does this postulation require any speculation about the *physical* qualities or form of those chemical smallest parts. Chemical atomism exhibits every legitimate characteristic of 'theory' and

12 Christian Gottfried Nees von Esenbeck, Carl Gustav Bischof, and Heinrich August Rothe, eds., *Die Entwickelung der Pflanzensubstanz physiologisch, chemisch und mathematisch dargestellt, mit combinatorischen Tafeln der möglichen Pflanzenstoffe und den Gesetzen ihrer stöchiometrischen Zusammensetzung* (Erlangen: J.J. Palm und Ernst Enke, 1819), pp. 108–9, 133–63, 178–98.

13 Carl Gustav Bischof, *Lehrbuch der Stöchiometrie* (Erlangen: Palm, 1819).

14 Partington, *History of Chemistry*, passim; Alan J. Rocke, *Chemical Atomism in the Nineteenth Century: From Dalton to Cannizzaro* (Columbus, OH: Ohio State University Press, 1984), chs. 2–5.

is categorically distinct from any collection of empirical generalizations (such as the laws of stoichiometry). The various systems of weights for the chemical elements that were known in the nineteenth century as 'chemical equivalents' or 'equivalent weights' or simply 'equivalents', widely thought to be theory-free, were almost invariably systems of *chemical atomism* by this definition, hence they were *theoretical* and not empirical statements, despite the contrary view of many chemists over the course of the nineteenth century. This fundamental misperception had powerful sequelae.[15]

All scientific theories worthy of the name function both representationally and heuristically, and chemical atomism exhibits these qualities. Dalton and Thomson wrote atomistic formulae for chemical compounds right from birth of the theory (1803–7), and as early as 1813 Berzelius had proposed a system of Latinate abbreviations, a modification of which we still use today. These early uses of atomistic chemical symbols and icons were used predominantly to illustrate and express existing knowledge, especially that of the elemental composition of compounds. For example, writing CO for the lower oxide of carbon, NO for the middle oxide of nitrogen, or $2C_2N_2OAgO_2$ for silver cyanate—which are close to the Berzelian representations—provided a compact, efficient, and easily remembered and reproduced substitute for enumerating the percentages of carbon, nitrogen, oxygen, and silver in these substances. Chemical atomism had predictive or heuristic utility from the start,[16] but early in the theory's history that heurism was not exercised primarily by manipulation of the symbolic system.

This situation changed suddenly in a handful of years surrounding the year 1830, when there occurred a small but important shift that had remained little noted until Ursula Klein began to draw our attention to it in 1998.[17] In 1827, together with a student named Polydore Boullay, Jean-Baptiste Dumas published his first article on the formation and constitution of ether, proposing a constituent of ether which they named hydrogène bicarboné and which Berzelius called etherin. This publication initiated an intensive theoretical

15 Rocke, *Atomism*, esp. chs. 1 and 11; Britta Görs, *Chemischer Atomismus: Anwendung, Veränderung, Alternativen im deutschsprachigen Raum in der zweiten Hälfte des 19. Jahrhunderts* (Berlin: ERS, 1999).

16 Alan J. Rocke, 'Chemical Atomism and the Evolution of Chemical Theory in the Nineteenth Century', in Ursula Klein, ed., *Tools and Modes of Representation in the Laboratory Sciences*, (Boston, MA: Kluwer, 2001), pp. 1–11 (pp. 1–2).

17 Ursula Klein, 'Paving a Way through the Jungle of Organic Chemistry', in M. Heidelberger and F. Steinle, eds., *Experimental Essays – Versuch zum Experiment* (Baden-Baden: Nomos, 1998), pp. 251–71; Ursula Klein, 'The Creative Power of Paper Tools in Early Nineteenth-Century Chemistry', in *Tools and Modes of Representation*, pp. 13–34; Ursula Klein, *Experiments, Models, Paper Tools: Cultures of Organic Chemistry in the Nineteenth Century* (Stanford, CA: Stanford University Press, 2003).

dialogue with his German rival, Liebig, and with the most famous of all European chemists, Berzelius, culminating in the famous benzoyl radical paper (Liebig and Wöhler, 1832), and the theory of the ethyl radical (1833).

These events have been much studied. However, what Klein has now correctly discerned is that this foundational series of papers represents the first time (neglecting Bischof's similar but excessively speculative essay) that manipulations of chemical formulae on paper were used in a productive generative fashion to construct and to justify the modelling of both chemical reactions and the constitutions of chemical compounds. This was a new epistemic technique that went far beyond mere shorthand representation. In short, formulae were now being used as 'paper tools' in the fullest sense of the word 'tool'. As Klein puts it:

> The notion of paper tools implies the assertion that scientists often apply representations or signs systems in general for the same epistemic purpose and in a similar way to laboratory instruments in the strict sense: to produce new representations of invisible objects or processes. Paper tools are material devices in the broadest sense ... Unlike laboratory instruments, they do not interact physically with the object under investigation ... Even so, paper tools are visible marks which can be manipulated on paper to create representations of a scientific object.[18]

From the time of the Dumas/Liebig/Berzelius episode of 1827–33, atomistic chemical formulae were routinely used in the generative heuristic fashion that Klein describes. Klein has also correctly pointed out that it was precisely *organic* chemistry for which this epistemic technique was most valuable, for organic reactions are in general far more dynamic than inorganic ones and tend to generate cascades of products rather than simple precipitates. The heuristic manipulation of formulae gave organic chemists a handle on the complexities with which they were forced to deal; it also provided them an extraordinarily productive theoretical tool, a means to create endless ideas for investigation and endless new substances to try to create.

The *Kaliapparat*

Along with this new *theoretical* tool, and, coincidentally, arriving precisely at the same time, was a new *material* tool for determining organic composition:

18 Klein, 'Creative Power', p. 28.

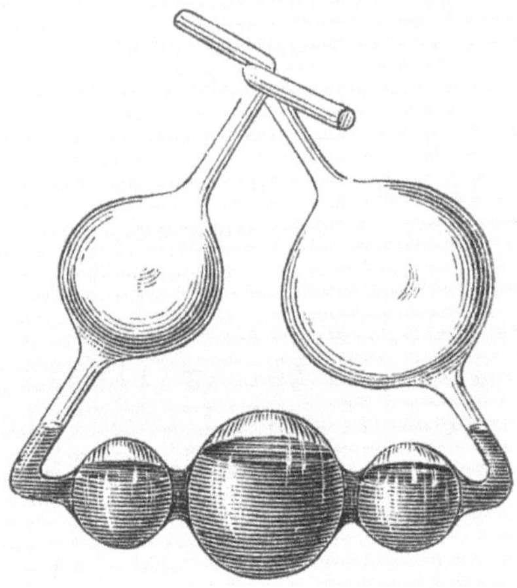

Figure 2.1 Liebig's Kaliapparat.

Liebig's five-bulbed potash apparatus, the *Fünf-Kugel-Apparat* or *Kaliapparat*. Subsequent rhetoric of Liebig and of students such as A.W. von Hofmann have elevated the invention of this device to a status equivalent almost to the birth of organic chemistry as a science. As a matter of fact, at the time of Liebig's invention organic elemental analysis had already been enjoying a standard practice and standard procedure, developed principally by Gay-Lussac and by Berzelius, for nearly two decades.[19] Nonetheless, there were sufficient legitimate advantages to the Liebig method that it did indeed mark a new era in organic analysis.

The most obvious of these advantages was the fact that Liebig could now measure both water vapour and carbon dioxide *gravimetrically* and in a *single* operation. (These two substances are the sole combustion products in organic compounds containing only carbon, hydrogen, and oxygen; measuring their weights provides an immediate basis for determining percentage elemental composition, hence the formula, of the sample.) Gay-Lussac and Berzelius had relied on direct collection of *gaseous* carbon dioxide. Compared to Liebig's innovation, this earlier method had three disadvantages: it required significantly more skill and care, it was intrinsically less precise, and it imposed an upper limit on sample size. Liebig's method was designed to operate with large samples, and it was sufficiently simple and precise that analyses could for the first time be placed in the hands of relatively untutored students. A second feature is less obvious but no less

19 Holmes, 'Liebig's Laboratory'; Alan J. Rocke, 'Organic Analysis in Comparative Perspective: Liebig, Dumas, and Berzelius, 1811–1837', in F.L. Holmes and T. Levere, eds., *Instruments and Experimentation in the History of Chemistry* (Cambridge, MA: MIT Press, 2000), pp. 273–310.

important. Observing the evolution of the combustion gases through the liquid lye in the glass bulbs provided an abundance of indicators that gave superb feedback on the progress and quality of the run. Flawed combustions were evident, the reasons for failures were usually apparent, and they could generally just as easily be corrected.[20]

Almost nothing is known about the invention process, but correspondence reveals that Liebig completed the device during the autumn of 1830. He immediately informed his friends Berzelius and Wöhler, and his article on the subject appeared in January 1831.[21] By the end of 1831, an apparatus was taken to Paris by Liebig's French student Charles Oppermann, and the operation was demonstrated to Dumas, Pelouze, and Gay-Lussac. These French chemists immediately adopted the new method (Gay-Lussac's own son Jules was by then another of Liebig's students in Giessen). By happenstance, this was just the time (November 1831 and January 1832) that Wöhler had two separate opportunities to spend some time with his friend in Giessen. Wöhler was highly impressed by the improvement in organic analysis and wrote Berzelius about his experience.[22] In April and again in October 1832, Liebig had opportunities personally to demonstrate the operation to the Berlin chemists Mitscherlich and Magnus, and they carried the knowledge home with them. During the late summer of that year, Wöhler spent six more weeks in Giessen, where he and Liebig operated daily with the method; the 21-year-old Bunsen, who visited Giessen at this time, was probably inducted then into the new methodology.[23] Berzelius did not have the benefit of first-hand instruction but had no difficulty in reproducing the operation from written instructions, and he also recognized its superiority.[24] In short, within two years of its invention, the *Kaliapparat* had spread through nearly all the active centres of organic-chemical research and had nearly completely displaced the earlier method.

In his fine analysis of the early development of Liebig's laboratory, the late Frederic L. Holmes expressed a somewhat qualified view of the success of the *Kaliapparat*. He affirms that the device was adopted rapidly across Europe and that it 'quickly changed research practices in the field at large'.

20 This is all predicated on the purity of the sample—a crucial assumption, as Liebig emphasized!

21 Justus Liebig, 'Ueber einen neuen Apparat zur Analyse organischer Körper, und über die Zusammensetzung einiger organischen Substanzen', *Annalen der Physik*, [2] 21 (1831), 1–47; Justus Carrière, ed., *Berzelius und Liebig: Ihre Briefe von 1831-1845* (Munich: Lehmann, 1898), pp. 3–4; O. Wallach, ed., *Briefwechsel zwischen J. Berzelius und F. Wöhler*, 2 vols. (Leipzig: Engelmann, 1901), vol. 1, p. 327.

22 Wallach, *Briefwechsel*, pp. 381–7, 399–400, 448.

23 Wallach, *Briefwechsel*, p. 431; Carrière, *Briefe*, pp. 41, 43.

24 Carrière, *Briefe*, pp. 43, 46, 49–51, 60, 66, 68; Wallach, *Briefwechsel*, p. 609.

Nonetheless, he believes that the disadvantages of the previous method have been overemphasized by previous scholars who were perhaps too influenced by the Liebig mythology; he also believes that the new method 'did not reduce the production of chemical knowledge in his laboratory to a routine process at which any young person of modest ability could succeed'.[25] My inclination is towards a more favourable assessment, supporting Liebig's broad claims for the simplicity, reliability, precision, and ease of use of his apparatus.[26]

Part of the evidence for this opinion is indirect in nature. Many chemists in the 1830s and 1840s successfully learned to perform the new method by simply reading about it, and controversies over conflicting analyses, common before 1830, largely ceased after that time. There is also more direct evidence. Actual laboratory experience with historically accurate reproductions of the *Kaliapparat* method supports Liebig's claim that even beginners could quickly learn to produce reliable and amazingly precise analyses.[27] This experience also suggests that novices did not even need personal instruction from a practised chemist to do so, providing they had sufficiently detailed written instructions, as they did after 1837. Once allied with a good method to analyse the nitrogen content of organic substances, which Dumas provided in 1833, organic analysis became truly routine.

Equipped with a newly improved analytical method and empowered by a productive new theoretical approach to the exploration of organic reactions and constitutions, organic chemists after 1830 suddenly discovered themselves in possession of a 'kit' that would enable them to master the dismaying proliferation of new organic compounds. The first institutional laboratory that achieved a significant approach to such mastery was Liebig's in Giessen.

The Rise of Liebig's Laboratory

We have known for many years now that Liebig's Giessen laboratory, contrary to mythologists and hagiographers, was far from the first in Germany to

25 Holmes, 'Liebig's Laboratory', pp. 132–6, 142.

26 Rocke, 'Organic Analysis'.

27 Many such 1832-style combustions have been carried out in Melvyn Usselman's laboratory at the University of Western Ontario. His valuable work, carried out in conjunction with undergraduate chemistry honours students, has provided insights concerning the above-mentioned feedback mechanisms consequent on observing the combustion stream as it bubbles through the potash bulbs. I am grateful to Professor Usselman for allowing me to look over the shoulder of Christina Reinhardt, the first of these students, as she performed Liebig combustions. A summary of all this work will soon be published.

offer exercises as part of a course of chemical study. A partial list of Liebig's predecessors would include Stromeyer at Göttingen (1810), C.G. Gmelin at Tübingen (1819), J. Döbereiner at Jena (c.1820), J.N. Fuchs at Landshut (1820), N. Fischer at Breslau (1821), and Bischof at Bonn (1825).[28] Moreover, we also now know that Liebig's first plan was not to create a university research school, but rather to run a laboratory-based pharmaceutical institute, in the pattern set, for instance, by Johann Trommsdorff in Erfurt (1795) and Carl Göbel in Jena (1820).[29]

In conjunction with two colleagues, Liebig began his institute in summer semester 1826; an all-day laboratory practicum began the following semester, the intensive character of which Liebig claimed at the time to be novel.[30] Giessen officials initially declined to incorporate this institute into the regular university curriculum. The data and indirect indications suggest that his enrolments (i.e. number of pupils in his private institute plus the few university chemistry and pharmacy students) remained around ten at any one time in the late 1820s and early 1830s; this was a not unusual number of *Praktikanten* in a German university at this time. Nonetheless, the success of Liebig's institute (his partners did not remain part of the

28 Julius Schiff, 'Das erste chemische Institut der Universität Breslau', *Archiv für die Geschichte der Naturwissenschaften und der Technik*, 9 (1920), 29–38; Ferdinand Henrich, 'Zur Geschichte des chemischen Unterrichts in Deutschland,' *Chemiker-Zeitung*, 47 (1923), 585–7; Georg Lockemann, 'Der chemische Unterricht an den deutschen Universitäten im ersten Viertel des neunzehnten Jahrhunderts', in Julius Ruska, ed., *Studien zur Geschichte der Chemie*, (Berlin: Springer, 1927), pp. 148–58; Hugo Döbling, *Die Chemie in Jena zur Goethezeit* (Jena: Fischer, 1928), pp. 116–17, 210; Georg Lockeman and Ralph E. Oesper, 'Friedrich Stromeyer and the History of Chemical Laboratory Instruction', *Journal of Chemical Education*, 30 (1953), 202–4; Wilhelm Prandtl, 'Johann Nepomuk Fuchs', *Journal of Chemical Education*, 28 (1951), 136–42; Arthur Lüttringhaus and Christamaria Baumfelder, 'Die Chemie an der Universität Freiburg i. Br. Von den Anfängen bis 1920', *Beiträge zur Freiburger Wissenschafts- und Universitätsgeschichte*, 18 (1957), 23–76; Dieter Pohl, 'Zur Geschichte der pharmazeutischen Privatinstitute in Deutschland von 1779 bis 1873' (PhD diss., Marburg, 1972), pp. 178–80; and Turner, 'Liebig versus Prussian Chemistry', p. 152. Ferdinand Wurzer at Marburg, sometimes included in such lists, did not really have an open practicum: Christoph Meinel, *Die Chemie an der Universität Marburg seit Beginn des 19. Jahrhunderts* (Marburg: Elwert, 1978), p. 20. The evidence for a practicum for Leopold Gmelin at Heidelberg is ambiguous: U. Thomas, *Die Pharmazie im Spannungseld der Neuorientierung: Philipp Lorenz Geiger (1785-1836), Leben, Werk und Wirken – Eine Biographie* (Stuttgart: Deutscher Apotheker-Verlag, 1985), pp. 157–8.

29 The standard monographic treatments of this subject are Pohl, 'Zur Geschichte' and Rudolf Schmitz, *Die deutschen pharmazeutisch-chemischen Hochschulinstitute* (Stuttgart: Boehringer, 1969).

30 'Ankündigung eines demnächst zu eröffnenden chemisch-pharmaceutischen Instituts in Giessen', n.d. (c. January or February 1826), Sondersammlungen des Deutschen Museums, no. 3050.

enterprise for more than a couple of years), combined with his effective advocacy, resulted in two welcome decisions by the university in 1834–5: they modestly expanded the space, and they now integrated the pharmacy institute into the university. Enrolment suddenly increased to around twenty at a time. In the statistics of the University of Giessen, this inflection point appears even more dramatic than it was, for Liebig's private students, hitherto unrecorded in any surviving document, were now counted as university students.[31]

As Holmes has demonstrated in detail, another kind of change took place after 1835: more systematically than he had ever done before, Liebig began to take his most experienced *Praktikanten* and give them research assignments. In 1835 and 1836 these were very small numbers, one to three, and the projects were 'gap-filling' exercises. In 1837–8 the number increased to around five to seven. In summer 1838 there were 33 total *Praktikanten*, and Liebig now began to develop research projects for which organized manpower of a larger magnitude could assist: a coordinated and well-articulated attack across a broad research front.[32] By 1843, in a further enlarged space and in a new branch laboratory for beginners, there were no fewer than 68 practicum students, and by this time he had a well-established senior research group, including many foreigners. Liebig cleverly drew attention to his dramatic success by writing two arresting polemical articles on 'the state of chemistry in Austria' (1838), and 'the state of chemistry in Prussia' (1840).[33] The Liebig laboratory was attracting worldwide attention and was a regarded as a distinctly new phenomenon.[34]

Friedrich Schödler, writing in 1875, clearly pointed to the difference between Liebig's innovation and previous practices: 'While the beginners in Liebig's laboratory sought out the elements (in the full sense of the word) and learned to spell the ABCs of chemistry, in his private laboratory the master himself pursued scientific problems in conjunction with chosen advanced students.' Some scientific advances could be settled by single master workers, Schödler suggested, but many other important questions

31 Brock, *Liebig*, pp. 56–65.

32 Holmes, 'Liebig's Laboratory', pp. 146–62.

33 For which, see Turner, 'Liebig versus Prussian Chemistry' and Brock, *Liebig*, pp. 65–70.

34 In 1844, Otto Erdmann—a nearly exact contemporary of Liebig, but not associated with him geographically or collegially—stated that the Liebig research school had blossomed extraordinarily, drawing 'the most general attention' of the scholarly world; the conduct of the Giessen laboratory was novel in important respects and was being emulated everywhere (Erdmann, 'Das chemische Laboratorium der Universität Leipzig', *Journal für praktische Chemie*, 31 (1844), 65–75 (p. 66). I will return to this point below.

Figure 2.2 Interior view of Liebig's Analytical Laboratory in Giessen, 1842.

could not. 'When Liebig asked himself, What role does fat play as a nutritional substance?—or, Does the material in the ash residue following combustion of a plant have significance for its life?—such tasks as these could never be decided or solved by a lucky stroke, but rather required a comprehensive series of tedious investigations that looked at the matter from all sides, in order to arrive at a satisfactory answer.'[35]

Liebig was a principal player in the recognition of the phenomenon of isomerism and in the transition from the use of formulae as mere '*Denkökonomie*' to their use as paper tools to guide investigation and to develop chemical theory. And he was *more* than 'a' principal player in the invention and spread of the *Kaliapparat*. In sum, he was the single best-positioned scientist in the world to follow up this productive nexus. Holmes and Morrell have already put some of these factors together. Holmes wrote, 'The growing productivity of Liebig's laboratory was clearly not due to any single factor such as his famed method for organic analysis. It was the combined outcome of the simplifications that this method provided, of the rapid expansion in the numbers of known organic compounds to which one could apply investigative strategies that had succeeded in analogous situations, and of the way in which Liebig learned to organize the research efforts of a growing

35 Friedrich Schödler, 'Das chemische Laboratoriums unserer Zeit', *Westermanns illustrierte deutsche Monatshefte*, 38 (1875), 21–47 (pp. 30–31).

number of advanced students.'[36] And in his classic paper a generation ago, Morrell compiled a list of various advantages which Liebig enjoyed, some of which I have followed and have striven to elaborate, and others which I have chosen not to highlight, despite their obvious importance.[37]

Explosive Expansion of the Field

A crucial empirical discovery about the nature of organic compounds (isomerism), a new theoretical technique of great heuristic power (formulae as paper tools), and a new apparatus to identify the entities upon which one was operating (the *Kaliapparat*)—all three of these landmark events occurred at roughly the same time, around 1830. Did the combination result in an acceleration of the rate of discovery of new organic substances?

The numbers of known compounds at various points in the nineteenth century provided by Otto Krätz suggested an inflection point in the curve at about 1860, when the doubling time of about twenty years shortened to around nine—an acceleration which in 1993 I suggested may have been spurred by the acquisition of structure theory.[38] But a much fuller recent 'scientometric' analysis by Joachim Schummer of the number of known chemical substances as a function of time casts a different and revealing light on the matter. Inter alia, we learn from this valuable paper, first, that in 1820 there were not many more than 100 known organic compounds but over 1000 inorganic ones; second, that from that time on, organic compounds rapidly increased in number with a constant doubling time of about nine years, while the number of known inorganic substances, starting from around 1830, exhibited a much slower constant doubling time of about 20–25 years; and third, that as a consequence, since the 1860s the curve for the number of all known compounds has been decisively determined by that for organic compounds, which comprise today 99% of all chemical substances. Schummer's paper suggests that the apparent inflection point that can be derived from the Krätz figures merely represents the point of

36 Holmes, 'Liebig's Laboratory', p. 159.

37 Morrell, 'Chemist Breeders'. Among Morrell's factors is the ready availability of a publishing outlet, the *Annalen der Pharmacie*, which Liebig acquired shortly after my red-letter date of 1830.

38 Otto Krätz, 'Der Chemiker in den Gründerjahren', in Eberhard Schmauderer, ed., *Der Chemiker im Wandel der Zeiten* (Weinheim: Verlag Chemie, 1973), pp. 259–84, at pp. 269–70. I used Krätz's raw numbers to calculate doubling times, from which I discerned the inflection point and speculated on its cause: Alan J. Rocke, *The Quiet Revolution. Hermann Kolbe and the Science of Organic Chemistry* (Berkeley, CA: University of California Press, 1993), p. 2.

transition when organic compounds take the lead.[39] The important result for us is that organic compounds started a process of rapid geometric increase as early as the 1820s rather than the 1860s. This fact is consistent with the arguments made in this paper.

In 1875 Schödler wrote, regarding the progress of pure science:

> In the last fifty years, chemistry has enjoyed a very special advantage: it has crossed, as it were, into hitherto untouched California gold fields. One only needed to dig in order to uncover riches. To be sure, the prospector required for this purpose a mental divining rod to direct him to the right spot, but since such a tool was available, in no other field has the wealth so increased during this period as in chemistry.... What once were only dozens [of academic chemists], are now just as many hundreds. The obvious question must be asked: is it not inevitable that chemistry will now advance with giant steps, and by this massive attack continually reveal novel and important knowledge?[40]

Wöhler and Bunsen Follow Liebig's Lead

After Wöhler was hired at Göttingen (spring 1836), he used the laboratory left him by Stromeyer and taught a regularly scheduled chemistry *Praktikum* (two hours a day, four days a week); in addition, he allowed advanced *Praktikanten* to work all day every day in the lab. Unfortunately, enrolment

39 Joachim Schummer, 'Scientometric Studies on Chemistry, I: The Exponential Growth of Chemical Substances, 1800–1995' and 'II, Aims and Methods of Producing New Chemical Substances', *Scientometrics*, 39 (1997), 107–23, 125–40. Schummer actually discerns two stages of growth in the number of organic compounds, 1800–1825 and 1825–1870. I believe that his data cannot support discrimination of such detail – first, because he has a limited number of data points from which to work (just nine across the nineteenth century); second, because the data derive from six different compilers, which suggests that there may be significant variability of data-collection method in the different compilations; third, because there are intrinsic and unavoidable uncertainties in the raw data across several axes (when a given compound was actually discovered as opposed to when it appeared in the compilation, discovery of genuine compounds missed, phantom compounds erroneously included, what is a 'compound', what is an 'organic' and what is an 'inorganic' compound, changes of nomenclature, and so on); and fourth, because at a point when the total numbers are so few (just a few dozen organic compounds known at the beginning of the century), the statistics in per cent can be greatly influenced by a rather small and possibly random increment in absolute numbers (a potential perturbing factor which no longer applies by about 1820). This minor quibble, if valid, in no way diminishes the importance of this fine paper.

40 Schödler, 'Laboratorium', pp. 30, 45.

statistics for chemistry courses survive only from 1842. However, by using a combination of the Göttingen matriculation register, passages from Wöhler's extensive correspondence, and evidence from papers published by Wöhler and his students, it is possible to obtain at least an approximate idea of what was happening during the first six years of his Göttingen tenure. Whatever the earliest numbers of *Praktikanten* were, Wöhler remarked on a noticeable increase in summer semester 1838.[41] Further increase resulted in a total of 28 by spring 1840 (of whom two were engaged in real research) and 40 by the end of 1841.[42] The number of 'advanced *Praktikanten*'—those who were permitted to work all day, and with whom Wöhler could collaborate—was very small at first, apparently zero to three at any one time.[43] But in summer semester 1841 we see a sudden increase in this category to eight advanced students,[44] and the following semester there were no fewer than fourteen.[45] The first Wöhler chemistry PhD was granted to Friedrich Voelckel in 1841. In the following year a new laboratory was built, adjacent to the old, in Hospitalstrasse.

All of this suggests that Wöhler's trajectory as regards chemical practical education and the gradual building of group research followed the same general path as Liebig's, with a lag of something like two or three years. Wöhler may well not have been consciously imitating the Liebig model, for students 'voted with their feet', and the demand was clearly building in many places. Moreover, in Göttingen Stromeyer's model was

41 Wöhler to Berzelius, 27 May and 22 November 1838, in Wallach, *Briefwechsel*, vol. 2, pp. 30, 70–2 ('again' and 'exceptionally large number' of students in the laboratory).

42 Wöhler to Berzelius, 22 May 1840 and 13 January 1842, Wallach, *Briefwechsel*, vol. 2, pp. 175 and 276–7.

43 For example, Wöhler commented on 10 February 1840 that he had no students that winter semester to whom he could entrust experiments (Wallach, *Briefwechsel*, vol 2, p. 164), but he occasionally mentioned student research to Berzelius all through his early years at Göttingen.

44 The eight were Friedrich Weppen, Georg Schnedermann, Friedrich Voelckel, Hermann Kolbe, Otto Griepenkerl, August Vogel, August Beringer, and Wilhelm Knop.

45 Wöhler to Berzelius, 3 November 1841, in Wallach, *Briefwechsel*, vol. 2, p. 266. The material in this paragraph summarizes the discussion in Rocke, *Quiet Revolution*, pp. 18–20. Since then, Günther Beer has published 'Dissertationen unter Friedrich Wöhler an der Universität Göttingen', in Johannes Büttner and Wilhelm Lewicki, eds., *Stoffwechsel im tierischen Organismus: Historische Studien zu Liebigs Thier-Chemie* (Seesen: HisChymia, 2001), pp. 399–412. A wider discussion of what is known about Wöhler's practicum in Göttingen is provided in Gustav-Adolf Ganss, *Geschichte der pharmazeutischen Chemie an der Universität Göttingen* (Göttingen, 1937), pp. 46–64. The best English-language biography of Wöhler is Robin Keen, 'The Life and Work of Friedrich Wöhler' (PhD diss., University College London, 1976), repr. as: Robin Keen, *The Life and Work of Friedrich Wöhler (1800–1882)*, ed. Johannes Büttner (Nordhausen: Verlag Traugott Bautz, 2005).

even closer to hand than Liebig's. However, the timing of the Liebig and Wöhler trajectories—especially including the use of selected students in research programmes, which was a possible but not a necessary result of higher enrolments, and which was genuinely novel in European science—as well as some explicit statements on Wöhler's part suggest that he had a clear opinion regarding who the leader of this movement was. A few years after these events, Wöhler wrote to Liebig complaining of his own workload at Göttingen: 'You are the one who is really to blame, by raising chemistry to its great reputation through your achievements and writings, that we must slave as we do, since now the whole world wants to do chemistry. But the damage you have inflicted must be borne.'[46]

Robert Bunsen, eight years younger than Liebig and eleven years younger than Wöhler, was a fellow traveller in this movement. A student of Stromeyer in Göttingen, he spent nearly two years on a *Wanderjahr* in France, Germany, and Austria, including a month in Giessen in August 1832,[47] just when Liebig and Wöhler were collaborating on the benzoyl radical paper. Then, after three years as *Privatdozent* in Göttingen, he moved to the *Technische Hochschule* in Kassel as replacement for Wöhler when Wöhler answered the call as Stromeyer's successor. Bunsen was finally hired *Extraordinarius* at Marburg in 1839; two years later he became *Ordinarius*.

There was apparently a small and primitive research laboratory in Kassel used by both Wöhler and Bunsen during their respective periods there (combined dates 1831–9), but there is no evidence that either had a significant *Praktikum*, significant enrolments, or anything approaching a research group.[48] Similarly, there is evidence of early laboratory teaching in Marburg—Bunsen's predecessor Ferdinand Wurzer had been accepting individual *Praktikum* students since 1811 and acquired a semi-satisfactory university laboratory in 1825—but only on a very small scale and never integrated into the university's official functions. In the spring of 1840 Bunsen instituted a university-sanctioned *Praktikum*, his (and Marburg's) first enterprise of this character. Lockemann, who has carried out one of the most detailed and authoritative studies of Bunsen, stated that Bunsen began his Marburg *Praktikum* 'following Liebig's

46 Wöhler to Liebig, 10 May 1851, in Hofmann, *Briefwechsel*, vol. 1, p. 364.

47 Georg Lockemann, *Robert Wilhelm Bunsen* (Stuttgart: Wissenschaftliche Verlagsgesellschaft, 1949), p. 34; Lockemann commented that not only Wöhler but also Jules Gay-Lussac was present in Liebig's laboratory while Bunsen was there. Lockemann used family letters in his research on Bunsen's life and is reliable on such details.

48 Heinrich Debus described the Kassel laboratory as it appeared in the late 1830s: Debus, *Erinnerungen an Robert Wilhelm Bunsen* (Kassel: Fischer, 1901), pp. 6–8.

example'.[49] There is no question, however, but that Bunsen regarded Stromeyer as the finest of teachers; 'bei dem hat man auch etwas gelernt', he told Edmund von Lippmann in 1878.[50] From this period on, Bunsen was one of the leading lights of German chemistry and one of the country's most popular teachers.

Wöhler and Bunsen were the most eminent members of the German chemical community to adopt the new model, but they were not alone. Otto Linné Erdmann, an experienced chemist at the University of Leipzig, was given a fine new laboratory, began a newer-style practicum there in 1843, and described it in some detail the following year. He first noted that the earliest German academic chemical practica were introduced in the 1810s at Berlin (?) and Göttingen. However, Liebig's research school at Giessen had recently provided a new model that was being rapidly adopted 'überall'. The novelties, Erdmann commented, included the idea of *all-day* practica and the incorporation of a pedagogy that mixed research with instruction. This made the students 'Zeugen und Mitarbeitern an den Forschungen des Lehrers', encouraging them to do their own research. His new laboratory institute, he wrote, would follow this pattern.[51]

Another example is Hermann Kolbe, who enthusiastically adopted the Liebig pattern when he was called to Marburg in 1851, after Bunsen was hired at Breslau. The following year, Bunsen was called to Heidelberg to replace Wöhler's teacher there, Leopold Gmelin. This began an elaborate chain of successions that was fundamentally to transform the German academic chemical community. In this new post-1848 world, the newer model quickly proliferated throughout all the German states.[52]

49 In Bunsen's first semester in Marburg he had ten *Praktikanten*, but the next three semesters only six to eight at a time. From 1842 until he left Marburg in 1851, he had semester laboratory enrolments of anywhere from 10 to 30 (the actual capacity of the laboratory was about 20). The standard source on Wurzer's and Bunsen's periods in Marburg is Meinel, *Die Chemie an der Universität Marburg*, pp. 12–51 and, for enrolment statistics, p. 471; see also Debus, *Bunsen*, pp. 9–28, and Lockemann, *Bunsen*, pp. 75–9. The quoted phrase is in Lockemann, p. 75.

50 Related in Henrich, 'Geschichte', p. 586.

51 Erdmann, 'Das chemische Laboratorium ... Leipzig', pp. 65–6. Schmitz, *Hochschulinstitute*, p. 233 states not only that this laboratory was 'nach Giessener Muster erbaut und eingerichtet', but also that it was conducted 'nach dem Vorbild des Liebig-Instituts in Giessen'.

52 Regarding the period before 1848, Jeffrey Johnson's impressive summary of renovations and new constructions of nineteenth-century German university chemical laboratory institutes includes only Giessen and Göttingen, whereas no fewer than 34 projects were carried out from 1851 to 1895 (Johnson, 'Academic Chemistry in Imperial Germany', *Isis*, 76 (1985), 500–24, esp. the table on p. 502).

Pedagogy

The type of laboratory-based education that developed in Giessen after 1835 was distinct from earlier patterns in several ways: first, in its intensity (the larger number of laboratory hours per week expected); second, in the broader targeted student population (*all* students expected to take the practicum, and in principal, all students in all of the sciences, not just elite professionalizing chemists invited into a laboratory as a kind of patronage exercised by the director); third, in the university's acceptance of practica in regular course lists and in their financial support of the enterprises; and fourth, in the creation of a small subset of advanced *Praktikanten* who formed the basis for group research and research groups. These developments were made possible by higher student demand, doubtless created in part by exogenous market forces. But the market simply provided raw pedagogical material—in other words, it is only a necessary and not a sufficient factor to explain the *particular* kind of education offered these students.

University reform, neo-humanism, and the influence of a sensationalist-experientialist pedagogical philosophy stressing 'active learning' were all factors in the rise of the new-style *Praktikum* pedagogical philosophy in early nineteenth-century Germany.[53] Liebig was far from alone in promoting it. In the field of physiology and as early as the 1830s, for example, Johannes Müller in Berlin and Jan Purkyně in Breslau had moved beyond experiments as simply lecture demonstrations and were putting selected students behind microscopes and dissection apparatus. In describing these developments, however, Coleman cautions that this was 'but one and a still quite tentative step' towards the later nineteenth-century academic laboratory institute, for both men were in effect exercising patronage for a select few students, not for the rank and file; it was a 'small affair' and never realized its potential.[54] A better case in the field of physiology can be made for the efforts of Jacob Henle at the University of Heidelberg after 1843.[55]

In the field of physics, Wilhelm Weber's practicum at Göttingen (begun in 1833), Franz Neumann's mathematical-physical seminar at Königsberg (1834), and Gustav Magnus's practicum at Berlin (1843) have often been mentioned as the earliest efforts along these lines. However, these examples cannot really compare to what was happening in chemistry in Giessen,

53 See works cited in notes 2 and 7 above. Johann Heinrich Pestalozzi was a prominent Enlightenment source for this experiential pedagogical philosophy.

54 Coleman, 'Purkyne', pp. 38–40.

55 Tuchman, 'Science, Medicine, and the State'.

Göttingen, Marburg, and Leipzig in the 1830s and early 1840s. The physics seminars all remained quite small and none exhibited the full constellation of characteristics outlined above (i.e. intensive study, broad clientele, university sanction, and group research activity). Mature examples of this model in physics did not really emerge until after the Revolution of 1848.[56]

The older model consisted in solitary professorial research, experimental demonstrations for the lectures, and elite laboratory patronage for professionalizing chemists; the newer pattern was closer to mass laboratory education combined with real group research. Chemistry as a field had two additional intrinsic advantages that aided it in moving to this model. One advantage was the comparative inexpensiveness of the apparatus, which greatly facilitated putting it in the hands of a broad student clientele. Ernst Homburg, building on earlier work by Brian Gee, has recently developed this point as regards inorganic analytical methodology. Homburg points out that as early as about 1780, French and Swedish chemists had already developed microchemical apparatus and techniques (the blowpipe, analytical reagents used in connection with test tubes and slides, and an emphasis on precise gravimetry) that transformed laboratories from large workshops into precision workplaces.[57] This transformation resulted in a corresponding reduction in the cost of increasing the number of workers in an academic lab from a small handful to a small throng.

I do not want to suggest that running an academic chemical laboratory was ever *cheap*; the cost of consumable chemicals was always a real issue, not to mention breakage of glassware. However, most expensive items (such as balances or platinum crucibles) were one-time acquisitions that could be shared by many *Praktikanten* working simultaneously. Moreover, as regards organic chemistry in particular, the crucial analytical device which I have emphasized here, the *Kaliapparat*, was actually manufactured by individual chemists. Although it required patience, practice, and a rather high level of glassblowing skill, many chemists learned to make the device routinely, and the raw material was nothing but glass tubing. Compare this with the instrumentation in a physicist's laboratory or with microscopes in physiology. By comparison to other fields, academic organic chemistry labs could be 'scaled up' rather readily.

56 Olesko, 'Commentary' and *Physics as a Calling*; David Cahan, 'The Intellectual Revolution in German Physics, 1865–1914', *Historical Studies in the Physical Sciences*, 15 (1985), 1–65 (pp. 6–12).

57 Homburg, 'Rise of Analytical Chemistry;' Brian Gee, 'Amusement Chests and Portable Laboratories: Practical Alternatives to the Regular Laboratory', in Frank A.J.L. James, ed., *The Development of the Laboratory* (New York: American Institute of Physics, 1989), pp. 37–59.

The second advantage of chemistry, and particularly organic chemistry, over other fields of science in the early nineteenth century was an effective appeal to practice. The science of physics c.1840 apparently had only a distant relation to everyday life and national economy, and the fact that students in *Vormärz* physics courses were mostly prospective Gymnasium teachers, especially considering the modest level of physics teaching in Gymnasia, meant that both the clientele and the rhetorical purchase of faculty on university administrations was limited. Although the rhetoric was much more expansive in the case of physiology, the direct benefits to clinical medicine of academic laboratory practica could also not readily be demonstrated.[58]

Chemistry was different, really unique among the sciences. In general, chemical professorships moved from medical faculties to philosophical faculties, mostly over the first half of the nineteenth century. The very real resistance to this movement was based on the preconception that chemistry was *too* applied and consisted mostly of compounding of pharmaceuticals, boiling of soap, and preparation of heavy chemicals such as potash, soda, salt, and mineral acids.[59] By contrast, Liebig and others who were powerfully influenced by the Romantic and neo-humanist movements were at great pains to stress that chemistry was a true *Wissenschaft*, independent but complementary to other sciences such as physics, mathematics, and even philology and history. The purpose of all-day *Praktika*, they argued, was to educate, not to train. Those who learned about both empirical phenomena *and* theory by active learning in the laboratory had learned how to *think*, not simply how to mix drugs. Such chemists would also be far better able to *apply* their learning, by comparison to those who had trained in a craft, apprentice-style, merely by rote.

But by the same token, our chemists were nothing if not sensitive to the place of chemistry in practical life and in the national economy. Liebig and Wöhler were both raised in technically oriented families (Liebig's father was a materials merchant, Wöhler's an agronomist), and Bunsen was exquisitely sensitive to applications (without, on a point of principal, ever taking out a personal patent). By 1840 Liebig had grown tired of theoretical disputes in basic research, and ever after that date pursued applications—in agriculture, in physiology and medicine, and in food technology. In the 1860s what troubled him most about modern structural organic chemistry was its apparent inapplicability to physiology and medicine.[60] Liebig constantly stressed

58 Tuchman, *Science, Medicine, and the State*, p. 85.

59 Meinel, '*Artibus Academicis Inserenda:* Chemistry's Place'.

60 Liebig to Erlenmeyer, 27 March 1861, in Emil Heuser, ed., *Justus von Liebig und Emil Erlenmeyer in ihren Briefen von 1861-1872* (Mannheim: Bionomica, 1988), p. 11.

the economic and human importance of chemistry, and by the 1840s his words were heeded. In the 1850s, the 'Liebig model' became extraordinarily influential in elite political circles across the German states.[61]

An important qualification needs to be added. As much as chemists were in an ideal position to move in the direction of the newer *Praktikum*/group research model, nothing compelled them to do so. In the very homeland of neo-humanist university reform, Berlin, they declined. Part of the problem was a failure of the university administration before 1848 to sanction such activity. Mitscherlich's laboratory was not in the university but in the *Akademie der Wissenschaften*, where he was a member. Heinrich Rose's practicum, begun as early as 1822, was never supported by the university, and he had to use his own lodgings for this purpose. Magnus used his residence, as well. In 1841, Mitscherlich and Rose had ten laboratory students between them; in 1843 this number declined to three. Although they deplored the miserly support from their administration, Mitscherlich, Rose, and Magnus all regarded the lecture-demonstration model as perfectly sufficient for university science education.[62] This variation of response even in the contemporary academic chemical community highlights the important agency of those who did choose to embrace the new model—Liebig above all, but also Wöhler, Bunsen, and a few others.

The Stromeyer Question

We need now to address an important issue raised by Ernst Homburg, namely the putatively crucial role played by Friedrich Stromeyer, who began a regular university-sanctioned practicum at Göttingen as early as 1810.[63] Homburg is absolutely right about the importance of this personality for German chemistry in the first quarter of the nineteenth century, and about his unjust latter-day neglect. He is also undoubtedly correct that this neglect must be related to Berzelius's hostility, and to Liebig's exaggerated and self-promoting rhetoric. But it is less certain that we are justified in replacing

61 Borscheid, *Naturwissenschaften, Staat und Industrie*; Tuchman, *Science, Medicine, and the State*.

62 Turner, 'Liebig versus Prussian Chemistry', pp. 133–8, 144–7, 152–4; Max Lenz, *Geschichte der königliche Friedrich-Wilhelms-Universität zu Berlin*, 3 vols. (Halle: Buchhandlung Waisenhauses, 1910–18), vol. 2, pp. 227–30, 509–10.

63 Homburg, 'Rise of Analytical Chemistry'; Homburg, 'Two Factions, one Profession'. I use the date when Stromeyer began his 'exercitationes chemiae practicas in Laboratorio academico', rather than 1805, when he began a 'privatissimis lectionibus Chemiae practicae cursum'.

Liebig's name with Stromeyer's when we speak of the transformation of German laboratory science and pedagogy.

Homburg's impressive research has revealed that at least 22 and possibly as many as 25 of Stromeyer's students became professors (*Ordinarien* or *Extraordinarien*) at European universities, *technische Hochschulen*, or mining academies. However, if we restrict the category to professors of chemistry at German universities, we have a shorter list: L. Gmelin (Heidelberg), Mitscherlich (Berlin), Wackenroder (Jena), Blücher (Rostock), Kühn (Leipzig), Bunsen (Marburg and Heidelberg), and Himly (Göttingen and Kiel). Of these, Gmelin, Mitscherlich, and Bunsen predominate in importance. But each of these three outstanding chemists had constitutive formative stimuli from multiple sources. Gmelin was educated by his father and his cousin, in addition to Stromeyer, and spent nearly a year learning from Vauquelin, Gay-Lussac, and other Parisian chemists. Mitscherlich also spent time in Paris as well as in Berlin and was decisively influenced by his period in Stockholm with Berzelius. Bunsen, too, spent nine months in Paris; in addition, he was strongly influenced by contacts with Liebig, Wöhler, and Berzelius. In short, of the three personalities who were, by far, Stromeyer's most illustrious pupils, it is not possible to say that it was Stromeyer's imprint that was most decisive.[64]

And what about Stromeyer's early *Praktikum*? It is assuredly true that Stromeyer is the single best candidate for the 'Liebig model' before Liebig, since his course was extremely well populated by students, it was properly sanctioned and supported by the university, and it included more than just professionalizing chemists. However, it differed in several respects from what later developed in Giessen and elsewhere: the subject was quantitative inorganic analysis, the course was conducted only one or two days a week, and there was no intent to combine instruction and research. It also seems to have had relatively little influence in the great transition of the late 1830s, described above. Stromeyer did believe, probably rightly, that he was the first to introduce a regular university-sanctioned chemistry practicum in Germany[65]—his model may well have been the École polytechnique in its earliest years—but he never made any wider pedagogical or philosophical

64 Indeed, one of the many merits of Homburg's essays is appropriately to redirect attention regarding sources of the rise of German chemistry, to French and Swedish chemists during the period around 1780–1825. This very point could be made regarding Stromeyer himself, who was educated partly in France.

65 See Henrich, 'Geschichte', p. 586 (citing a paraphrased statement by Stromeyer, published by Friedrich Saalfeld, *Geschichte der Universität Göttingen in dem Zeitraume von 1788 bis 1820* (Hannover: Helwingschen Hofbuchhandlung, 1820), p. 330).

claims, as far as we know. Of the four criteria for the newer model—intensity of instruction, breadth of clientele, university support, and development of group research—only the second and third were unequivocally present. And the most important and distinctive element of all, the development of group research practices, was, as far as is known, completely absent.[66]

The fact that group research was absent in Stromeyer's pedagogy is not surprising. Stromeyer was training student analysts who worked on samples that were *known* 'unknowns'. His educational enterprise was not intended to explore scientific novelties or puzzles, and the students were not normally exposed to truly unidentified materials. Liebig's case was different. As he found that student organic analyses with his new device could be virtually as good as his own, it was a natural and almost inevitable step for him to begin to make use of those hands in advancing a broad research front. Connected with this, the explosion of new compounds on which to operate provided almost a mandate to create research groups that included students. Only *groups* could make progress in such a large and rapidly expanding field.

Homburg has also substantially increased our appreciation for the surprising number of German universities that provided some sort of practicum education in chemistry during the 1820s; however, contemporary descriptions of the nature of these efforts are often insufficient to judge their proper placement in this story.[67] In his examination of a slightly later period around 1835–48—the period in which the research-group model was first seen—he concludes that 'though Liebig's influence did play a role, the growth of student numbers and the formation of an informal group of advanced students took place quite independently of the Giessen example.'[68] It is unclear how the parts of this sentence on either side of the comma can

66 Lockemann and Oesper, 'Friedrich Stromeyer'; Lockemann, 'Chemischer Unterricht'; Ganss, *Geschichte*, pp. 33–45. Lockemann, who has studied Stromeyer more carefully than anyone except probably Homburg, concluded that Liebig had been wrong to claim that practical chemical laboratory instruction began only with his efforts in Giessen. However, he still regarded Liebig as the 'true founder' of laboratory instruction in Germany, because of the totality of his accomplishments and because of the influence he exercised ('Unterrichte', p. 157). Similarly, Henrich, who argued keenly for Stromeyer's importance, was also careful to state that Liebig expanded and developed the existing model, in particular towards the education of future research chemists ('Geschichte', p. 587).

67 For example, H.F. Kilian mentions several examples of chemical exercises at German universities being conducted in 1826 but does not provide the details that would enable us to distinguish the older pattern from the newer: Kilian, *Die Universitaeten Deutschlands in medicinisch-naturwissenschaftlicher Hinsicht* (Heidelberg and Leipzig: Karl Groos, 1828), passim.

68 Homburg, 'Rise of Analytical Chemistry', pp. 14–15; Homburg, 'Two Factions, One Profession', p. 68.

be simultaneously true, and the evidence with which I am familiar is not, in my judgment, sufficient to establish the second.[69] What is beyond question is that Homburg has, in several original and provocative essays, raised and helped to answer many interesting and productive questions. However, in my view at least, the 'Stromeyer question' still requires further investigation.

Exportation to Other Countries

Many have written on the exportation of the Giessen model to other countries. Perhaps the best-known instance of this is Hofmann's establishment of the Royal College of Chemistry in London in 1845; the RCC became the nucleus of what eventually developed into Imperial College of Science, Technology, and Medicine.[70] Because this story is well known, I will pass over it in near silence and move to France and the United States.

As early as around 1830, leading French chemists were looking with envy upon their German rivals; by the 1860s many perceived a real crisis in the state of French science.[71] In the spring of 1838, a few weeks before he opened his soon-to-be-famous private laboratory in the Rue Cuvier, Dumas wrote Liebig to tell him that the reason that he had taken this step (and why he had taken a new professorship at the Faculty of Medicine) was simply to match Liebig's suddenly accelerating research output; only by utilizing collaborative student research, as Liebig was just then beginning to do, could he hope to keep pace.[72] But the Rue Cuvier laboratory lasted only ten years and never was able to match Liebig's output. In 1837, 1840, 1846, and 1847 Dumas wrote pleading official reports to the ministry of public instruction,

69 As far as the *number* of students is concerned, Homburg's argument rests on the hitherto under-appreciated role of early nineteenth-century trade and polytechnic schools and mining academies; he has also usefully drawn attention to the surprisingly vigorous labour market for chemists in medicine, pharmacy, public health, mining, and manufacturing (see his 'Two Factions, One Profession', and his *Van beroep 'Chemiker'*, both cited in note 5). As important as they are, neither of these points relates to the rise of the research-group model.

70 See Gerrylynn K. Roberts, 'The Establishment of the Royal College of Chemistry: An Investigation of the Social Context of Early-Victorian Chemistry', *Historical Studies in the Physical Sciences*, 7 (1976), 437–85.

71 The discussion of this and the next paragraph summarizes material in Alan J. Rocke, *Nationalizing Science: Adolphe Wurtz and the Battle for French Chemistry* (Cambridge, MA: MIT Press, 2001), pp. 104–16, 127–9, 143–52, 269–77, and 388–97.

72 Dumas to Liebig, n.d., postmark 21 April 1838, and n.d., but annotated as May 1838: 'Alors seulement, je serai [after starting up his new laboratory] en mesure de reprendre des expériences en concurrence avec les votres. Je ne puis pas aller votre pas dans ce moment.' Munich, Bayerische Staatsbibliothek, Liebigiana IIB; complete quotation in Rocke, *Nationalizing Science*, p. 113.

deploring the lack of official laboratory facilities in Paris and often urging a model strikingly reminiscent of the Giessen facility and organization. It is not necessary, he wrote, 'to wait for a question to be resolved by the individual work of one of the [the faculty's] professors extended over several years, when it can do so in a few weeks under his direction by the collective effort of a dozen beginners in science'.[73]

Nor was Dumas alone in seeking to transplant the new German model into Paris. Pelouze, Gerhardt, Wurtz, and other leading French chemists were powerfully influenced by Liebig and the Giessen model. These three chemists all knew the Giessen laboratory first-hand; Gerhardt and Wurtz had been students there, and Pelouze had carried out collaborative work with Liebig in Giessen. Pelouze started up a private laboratory in 1838 with a capacity of about ten workers, simultaneously with Dumas's Rue Cuvier opening; between 1845 and 1857 he ran a much larger enterprise, with a capacity of about 30. When Dumas's laboratory closed in the revolutionary conditions of 1848, both Gerhardt and Wurtz opened private laboratory schools in which they could do research (and, at least potentially, collaborative work). Both of these enterprises had failed by 1853, perhaps killed off by Pelouze's effective competition. Pelouze, Gerhardt, and Wurtz all explicitly invoked Liebig's name as their model. However, none of these enterprises was really comparable to the Giessen institute; they were all private laboratory training schools rather than higher educational/research institutions. The only Parisian research laboratory that was really comparable to Liebig's was the one that Wurtz started privately and unofficially in the Faculty of Medicine in 1854. Only after 23 years of Wurtz's pleading with government functionaries did this lab finally win official sanction.

The French story, in short, is that despite their best efforts, they were not able to create a 'Giessen on the Seine', at least not before the Third Republic. Margaret Rossiter, whose modified phrase this is, studied a parallel failure to establish 'Giessen on the Charles' by an American named Eben Norton Horsford. Horsford studied in Giessen in 1845–6 and was fortunate to be called to a professorship at Harvard in the latter year. The university's president, Edward Everett, had many years earlier become the first American to earn a PhD from a German university, had met Liebig personally in 1844, and was enthusiastic about German university education. He hired Horsford largely because of the Liebig connection; simultaneously he persuaded the

73 J.B. Dumas, 'Rapports addressés à M. le Ministre de l'instruction publique par M. Dumas, Doyen de la Faculté des Sciences', *Moniteur universel*, 28 October 1846, p. 2448–50; full quotation in Rocke, *Nationalizing Science*, p. 128.

Boston industrialist Abbott Lawrence to endow a new laboratory for Horsford at the sum of $50,000. Horsford modelled his institute closely and explicitly on Liebig's. But the endeavour never really succeeded, and Horsford resigned his position in 1863. (Horsford's assistant then was Charles W. Eliot who, upon Horsford's resignation, studied abroad for two years, including a semester with Hermann Kolbe in Leipzig; in 1869 Eliot became president of Harvard.)

Rossiter's exemplary study traces the cause of the failure of Horsford's endeavour to the cultural and economic differences between Germany and the United States, and her analysis is certainly correct. However, one factor, which she characterizes early on as 'crucial' but does not include in her later summary, is the fact that Lawrence's gift was divided, so that in the end only $25,000 actually went to support Horsford's laboratory. The essential difficulty was always lack of sufficient funds, and I think that there can be little doubt that the endeavour *would* have succeeded, despite the differences between the two countries, had the original gift remained undivided.[74] The case of Hofmann and the Royal College of Chemistry is almost the inverse of this, for the College limped along on far too little funding, and nearly capsized, before it was definitively rescued by the proceeds of the Great Exhibition of 1851.

Historians are curiously resistant to introducing contingent factors into their explanatory fields (probably because this would seem to carry them further from social science and towards 'mere' storytelling). I also am susceptible to this weakness. In my recently published analysis of the trajectory of French chemistry in the nineteenth century, I repeatedly invoked contingency during the body of the book, but it dropped completely out of my concluding analysis, which was confined exclusively to institutional, cultural, and structural factors. In particular, in describing the French failure to keep pace with German laboratory education and research, I had 'forgotten' that I had stressed the existence of powerful and efficacious movements in the 1840s and again in the 1860s, which sought to renovate French laboratory education and research more towards the German model. The first of these was unfortunately derailed by revolution, the second by the outbreak of war. It took a perceptive reviewer to note the discrepancy.[75]

In any case, the first truly successful transplant of German-style graduate education to the United States came with the establishment in 1876 of the Johns Hopkins University, an explicitly science-based research university

74 Margaret Rossiter, *The Emergence of Agricultural Science: Justus Liebig and the Americans, 1840–1880* (New Haven, CT: Yale University Press, 1975), pp. 68–88, on p. 75.

75 Seymour Mauskopf, review of *Nationalizing Science, Chemical Heritage*, 19(3) (2001), 36–8.

directly modelled on the German pattern.[76] One of the first hires at Johns Hopkins was Ira Remsen, fresh from a PhD from the Liebig institute in Munich. The first president was Daniel Gilman, who ever after looked to the German founding mythology. The Sheffield Scientific Laboratory at Yale had acquired its first gifts as early as 1847. 'But for twenty years previous to 1847,' Gilman intoned at the Sheffield semi-centennial, 'a force had been at work in a little country town of Germany destined to affect the education of Christendom, and at the same time to enlarge the boundaries of human knowledge, first in chemistry and the allied branches, then in every other one of the natural sciences. The place was Giessen; the inventor, Liebig; the method, a laboratory for instruction and research.'[77]

Concluding Observations

Gilman was exercising his eloquence at a celebratory event and can be excused for rhetorical excess. But these words simply summarized a founding myth that was accepted uncritically around the world. Recent historical work has done much to provide context and needed reassessments of this mythology. However, reassessment blends insensibly into deconstruction, and deconstruction into destruction.

It does seem that some have gone too far. One example is Harry Paul, who 30 years ago sought to counter the conventional wisdom of the day that France had lost scientific supremacy to Germany during the nineteenth century. Paul thought that earlier analysts had betrayed a 'fetishism' towards German science and towards the 'icon' before which historians had had to 'genuflect', Liebig. His opinion was that there was no sustainable comparative case for the decline of nineteenth-century French science relative to that of Germany, and therefore Liebig could not have caused a phantom event.[78] More radically, Pat Munday has sought to argue that 'the realities of Liebig's life (and character) were more nasty and brutish then his immediate heirs (and perhaps we still) would have liked to believe.'[79]

76 Owen Hannaway, 'The German Model of Chemical Education in America: Ira Remsen at Johns Hopkins', *Ambix*, 23 (1976), 145–64.

77 Daniel Gilman, *University Problems in the United States* (New York: Century, 1898), p. 120.

78 Harry W. Paul, 'The Issue of Decline in Nineteenth-Century French Science', *French Historical Studies*, 7 (1972), 416–40 (pp. 418, 450).

79 Pat Munday, 'Social Climbing through Chemistry: Justus Liebig's Rise from the niederer Mittlestand to the Bildungsbürgertum', *Ambix*, 37 (1990), 1–19.

To be sure, as William Brock has put it, 'Liebig could be irascible, pig-headed, quarrelsome, opinionated, and sometimes devious.'[80] He was also one of the most important scientists of the nineteenth century, an incredibly hard worker, an extraordinarily creative thinker, and so skilled as to be a virtual magician in laboratory operations (and Brock adds, rightly, that with all his character flaws 'he never became quirky, obstreperous, [or] an embarrassment'). He was also capable of inspiring almost universal love and admiration by his students. And it is surely relevant to note that two of the most saintly personalities among his contemporary colleagues, Friedrich Wöhler and Robert Bunsen, found qualities in Liebig that inspired their highest regard.

But one quality has been too little stressed: along with everything else, Liebig was supremely fortunate. He was fortunate to find himself in Germany, which had the right combination of movements just at this time; he was fortunate to find himself at the University of Giessen, whose administration did indeed endorse his activities;[81] he was fortunate to find himself close to the starting point of the branch of science to which he devoted his efforts, organic chemistry. Above all, he was fortunate to have chosen chemistry at all, and organic chemistry in particular, for (as I have argued here) organic chemistry around 1830 was uniquely positioned to provide a home for the style of research and education which Liebig so skilfully helped to develop and which has spread so universally throughout the world.

Acknowledgements

I would like to thank William H. Brock and Melvyn Usselmann for helpful suggestions, and Laura Ritchie Morgan for valuable bibliographic assistance.

From the editors: Originally published as Alan J. Rocke, 'Origins and Spread of the "Giessen model" in University Science', *Ambix*, 50 (2003), 90–115. © 2003 Society for the History of Alchemy and Chemistry, reprinted by permission of Taylor & Francis, https://www.tandfonline.com/10.1179/amb.2003.50.1.90, on behalf of Society for the History of Alchemy and Chemistry (SHAC). We thank the author, SHAC, the editor of *Ambix*, and the publisher for their permission to republish the article in this edited volume.

80 Brock, *Liebig*, p. viii.

81 A point made effectively in William H. Brock, 'Breeding Chemists in Giessen', *Ambix*, 50 (2003), 25–70.

About the Author

Alan J. Rocke is Distinguished University Professor Emeritus at Case Western Reserve University, Cleveland, OH, USA. His area of research is the history of science, especially the history of nineteenth-century chemistry in Europe. Recently he published: 'Ideas in Chemistry: The Pure and the Impure', *Isis*, 109 (2018), 577–86, and edited, together with Gisela Boeck, *Lothar Meyer: Modern Theories and Pathways to Periodicity* (Cham: Birkhäuser, 2022).

3 The Laboratory Ethos, 1850–1900

Frans van Lunteren

> We shall be able to employ in scientific education, not only the trained attention of the student ... but the keenness of his eye, the quickness of his ear, the delicacy of his touch, and the adroitness of his fingers.[1]
>
> J.C. Maxwell, 1871

Abstract

This essay discusses the new pedagogical regimes and the associated values that accompanied the nineteenth-century rise of the university laboratory. Although these pedagogical practices and the related epistemic virtues differed for different scientific fields, there was a general trend towards a stronger emphasis on discipline, accuracy, and persistence. It is argued that this process helped to shape the emerging scientific disciplines and to create a new scientific persona that embodied these virtues. In the late nineteenth century, diligent, meticulous, disciplined work rather than genius characterized the ideal scientist. These claims are illustrated for the new laboratory sciences of chemistry, physiology, and physics.

Keywords: laboratory, pedagogical regimes, epistemic virtues, scientific persona, discipline formation

Introduction

The nineteenth-century laboratory revolution transformed the university into a new kind of institution that trained the hand as well as the mind. Because laboratories have become such an integral part of modern society,

1 J.C. Maxwell, 'Introductory Lecture on Experimental Physics', in W.D. Niven, *The Scientific Papers of James Clerk Maxwell*, 2 vols. (Cambridge: Clay, 1890; repr. New York: Dover, 1965), vol. 2, p. 242.

Berkel, Klaas van, and Ernst Homburg (eds), *The Laboratory Revolution and the Creation of the Modern University, 1830-1940*. Taylor & Francis Group, 2023
DOI: 10.5117/9789463720434_CH03

it is all too easy to overlook the radical nature of this transformation. Indeed, the introduction of large-scale laboratory training in the university curriculum implied a major transgression of the time-honoured division between intellectual and manual work that had characterized teaching and learning processes since antiquity. The pioneers of such pedagogic novelties had to fight an uphill battle. When in 1825 the newly appointed professor of chemistry in Giessen, Justus Liebig, discussed his plans for a pharmaceutical teaching laboratory, the senate made clear that such an institute could not be incorporated within the university. The university's task was to train 'civil servants, *not* apothecaries, soap makers, beer-brewers, dyers and vinegar-distillers'.[2] And although the success of Liebig's laboratory eventually resulted in the absorption of the institute into the university, academic resistance to intensive and systematic hands-on training remained strong.

In 1840 Liebig launched a vigorous attack on what he saw as the deplorable state of chemistry teaching at Prussian universities, none of which had followed the Giessen example. At fault, he believed, was the Prussian tendency to regard chemistry as an art or craft rather than as an autonomous branch of science. And while he did not want to downplay the utilitarian importance of chemistry (far from it), he emphasized that scientific chemistry, as taught and practised in the university laboratory, was not a mere set of artisanal recipes, rules and classificatory principles, as it involved laws and causal principles. Moreover, this particular branch of science could only be learned in a chemical laboratory, where large numbers of students could work from morning until evening for an extended period of time.[3] In their response the Prussian universities stressed the task of universities to provide students with a liberal education of a predominantly theoretical nature. Liebig's practical, time-consuming, and specialized training was ill-fitted for such aims. No doubt, they were also concerned about the costs of the required facilities. It would take more than two decades before the Prussian state would agree to fund large chemical teaching laboratories.[4]

Even as late as 1871, James Clerk Maxwell, the first director of the newly planned physics laboratory in Cambridge, feared that the new institute would

2 William H. Brock, *Justus von Liebig. The Chemical Gatekeeper* (Cambridge: Cambridge University Press, 2002), pp. 43, 47.

3 J. Liebig, *Ueber das Studium der Naturwissenschaften und über den Zustand der Chemie in Preussen* (Braunschweig: Vieweg und Sohn, 1840); R.S. Turner, 'Justus Liebig versus Prussian Chemistry: Reflections on Early Institute Building in Germany', *Historical Studies in the Physical Sciences*, 13 (1982), 129–62 (pp. 131–2).

4 Turner, 'Justus Liebig versus Prussian Chemistry', pp. 133–8.

be seen as disruptive to the university.[5] At the opening of his inaugural lecture, he therefore set out to consider 'by what means we, the University of Cambridge, may, as a living body, appropriate and vitalise this new organ'.[6] By presenting laboratory training as part of a liberal education—hence the organic metaphor—and as complementary to other forms of education, he hoped to dispel such fears. Maxwell's concerns were far form unique. One year earlier, the German chemist Rudolph Fittig, in his inaugural lecture at the university of Tübingen, had referred to a 'circle of scholars' who still questioned the teaching of chemistry at a university. In their view the proper place for chemical training was the polytechnic or a separate vocational school. Like Maxwell and Liebig, Fittig strongly defended the educational values of laboratory training.[7]

By the end of the century, university laboratories—and, concomitantly, laboratory instruction—had become as natural as the botanical gardens and anatomical collections of olden days. What had originally been a chemical workplace—hot, smoky, secretive, solitary, artisanal—had now become a spacious and well-lit training centre for large numbers of budding chemists, physicians, physicists, botanists, and even psychologists. As hands-on teaching institutions, which had to accommodate large numbers of students, laboratories demanded new pedagogical regimes and new educational goals. Simultaneously these university laboratories became the dominant research institutions, with advanced students making up the greater part of the workforce. The collaborative nature of much of this research and the specific skills required for using new and complex instruments fostered a new self-image among the scientists. As teaching and research became more closely related, the new scientific ethos in many ways reflected the pedagogical values attributed to laboratory training. This essay aims to explore this mid-century shift in values and self-image by looking at those character traits or 'epistemic virtues' that were valued most highly in the laboratory setting and at the ways that laboratories moulded new kinds of scientific identities.[8]

5 Isobel Falconer, 'Cambridge and Building the Cavendish Laboratory', in Raymond Flood, Mark McCartney, and Andrew Whitaker, eds., *James Clerk Maxwell: Perspectives on his Life and Work* (Oxford: Oxford University Press, 2014), pp. 67–98.

6 Maxwell, 'Introductory Lecture', p. 241.

7 R. Fittig, *Das Wesen und die Ziele der chemischen Forschung und des chemischen Studiums* (Leipzig: Von Quandt & Händel, 1870), pp. 3, 6–8, 16.

8 For epistemic virtues, see: Jeroen van Dongen and Herman Paul, eds., *Epistemic Virtues in the Sciences and the Humanities*. Boston Studies in the History and Philosophy of Science, 321 (Cham: Springer, 2017).

Of course, laboratories changed over time, they varied in size and scope, and a botanical laboratory differs in many respects from a physical or chemical laboratory. One thus needs to be sensitive to time-bound and disciplinary distinctions and peculiarities. In a sense, every single laboratory may be characterized as having had its own specific 'moral economy'.[9] Yet for all the diversity, both before and after the rise of the laboratory, some leitmotifs can be singled out in late nineteenth-century moral discourse, which differed considerably from those that were in vogue in early nineteenth-century research. But even then, it may be hard to establish a causal relationship between the demands of the laboratory setting and the moral rules observed or aspired to in the laboratory. Did new epistemic values simply reflect new laboratory practices? Did they bring about such practices? Or should we look for a common cause elsewhere? Here, I will not worry too much about such chicken-and-egg problems but mainly look for examples of apparent resonance between the exigencies of laboratory life and the virtues recommended in those settings.

A related question concerns the relationship between virtue discourse and everyday practices—in other words, between rhetoric and reality. Again, I will not be too concerned about the closeness of their match. To the extent that the professed virtues seem to lack a solid basis in real practices and subjects, they can be seen as constitutive of a 'scientific persona', an ideal model of scientific selfhood. Moreover, as several historians have emphasized, a scientific persona tends to mould the self.[10] Through processes of training or self-fashioning, scientists tend to cluster around a set of shared virtues as well as skills and methods. In this way, they form moral communities, often coinciding with specific disciplines or research fields. Such shared virtues may help to consolidate or strengthen specific practices and teaching regimes and vice versa.

In what follows, I will start with some general considerations, taking into account related contemporaneous developments such as the gradual emergence of scientific disciplines, the growing connections between science and industry, and the scientists' campaigns for public and political recognition. Subsequently I will look at more specific laboratory practices,

9 For the use of the concept 'moral economy' in a laboratory context, see: Robert E. Kohler, *Lords of the Fly: Drosophila Genetics and the Experimental Life* (Chicago: University of Chicago Press, 1994); Lorraine Daston, 'The Moral Economy of Science', *Osiris*, 10 (1995), 2–24.

10 For the concept of scientific or scholarly persona and its relationship to epistemic or intellectual virtues, see: Lorraine Daston and H. Otto Sibum, 'Introduction: Scientific Personae and Their Histories', *Science in Context*, 16 (2003), 1–8; Herman Paul, 'What Is a Scholarly Persona? Ten Theses on Virtues, Skills and Desires', *History and Theory*, 53 (2014), 348–71.

concerns, and values in several different disciplines, in particular chemistry, physiology, and physics. My findings are largely based on the abundant literature on nineteenth-century laboratories in these different disciplines, on the one hand, and inaugural lectures and laboratory manuals, on the other. Moreover, I have limited myself to the more prominent laboratories and their spokesmen, so I cannot claim general validity for my findings. But they do, I believe, paint a clear and consistent picture.

Laboratory, Discipline, and Restraint

What characterizes a late-nineteenth-century laboratory, setting aside some disciplinary peculiarities? What made them such privileged sites of knowledge? Above all, they were places of control. They allow one to control circumstances and ways of interfering with the system that one wants to understand by isolating some specific features and fencing off disturbing influences. One can recreate the system at will and repeat experiments under exactly the same circumstances. One can even produce circumstances that cannot be found outside the walls of the laboratory, such as extremely high pressures or low temperatures. One can apply all kinds of complex and sophisticated machinery and profit from such facilities as water, gas, and electricity. And one can install a set of rules or protocols. Together, this enables one to do things that would be extremely hard to realize outside the confines of the laboratory. It allows for increased simplicity and purity, repeatability, greater precision, safety, and the identification of causal relationships. As we will see, different disciplines emphasized different aspects of the features listed above.

If control characterized the late-nineteenth century laboratory, then discipline often marked those who worked in the laboratory. To many researchers, laboratory work came to imply strict procedures, careful preparations, standardized operations, skilful handling of delicate instruments, meticulous notetaking, and scrupulous data-analysis. Accordingly, students needed to be subjected to the new laboratory regime; they needed to be disciplined. Ideally, they learned to work in a routine and mechanized manner. In laboratory training, method was often seen to be more important than results, carefulness and patience more important than ingenuity or creativity.[11] Laboratory work came to mean patience, perseverance, and

11 For the emphasis on method rather than results in science teaching from 1850 onwards, see: E.W. Jenkins, 'The "Nature of Science" in the School Curriculum: The Great Survivor', *Journal*

LEITFADEN

DER

PRAKTISCHEN PHYSIK

ZUNÄCHST FÜR DAS PHYSIKALISCHE PRAKTICUM

IN GÖTTINGEN

VON

F. KOHLRAUSCH

A. O. PROFESSOR IN GÖTTINGEN.

LEIPZIG,

VERLAG VON B. G. TEUBNER.

1870.

Title page of F. Kohlrausch, *Leitfaden der praktischen Physik* (Leipzig: B. G. Teubner, 1870).

meticulousness—in other words, self-imposed constraint. Laboratory training, then, was not restricted to developing skills in the handling of sophisticated kinds of machinery or delicate specimens. It just as much aimed to create a certain habitus through the internalization of virtues such as diligence, patience, carefulness, and precision. From the 1860s onward new laboratory manuals showed students the ropes, informing them of proper laboratory procedures and practices.

of Curriculum Studies, 45 (2013), 132–51, as well as several expressions of that sentiment in this paper. Paul Forman regards the 'primacy of procedure', or a 'commitment to "methodism"', as a hallmark of modernity, see: Paul Forman, 'The Primacy of Science in Modernity, of Technology in Postmodernity, and of Ideology in the History of Technology', *History and Technology*, 23 (2007) 1–152 (p. 3).

Self-imposed restraint included observational judgment. As Lorraine Daston and Peter Galison have amply shown, around mid-century several researchers became more distrustful of traditional ways of interpreting and rendering visual information. The human proclivity to mould the phenomena on one's own speculative notions posed a constant danger for the observer. This growing 'fear of the imagination' ushered in a craving for 'mechanical objectivity'.[12] As a concomitant, mechanical means of reproduction of visual information—e.g. photography, self-registering instruments—were sometimes preferred to handmade illustrations that would unavoidably contain subjective elements. Behind this predilection Daston and Galison project a new self, regarded to be active, wilful, ambitious, imaginative, and passionate and therefore in need of being curbed, or even eliminated from the process of registering and representing phenomena. It took willpower to do so. Sacrifice and self-denial demanded strength as much as humility.[13]

Remarkably, in detailing the rise of mechanical objectivity, Daston and Galison hardly refer to the sites in which these new virtues were cultivated and disseminated most of all: the laboratories. Admittedly, the mid-century shift in mentality was, on the one hand, far from universal and, on the other, part of a much wider phenomenon that could also be found elsewhere.[14] Still, it is tempting to see the rise of the laboratory and the rise of mechanical objectivity as more than just simultaneous events. Daston and Galison's prime examples of the new scientific persona, like Hermann von Helmholtz, Claude Bernard, and Thomas Huxley, who pioneered epistemological labels like 'objective' and 'subjective' in their modern sense, were all physiologists.[15] When they were promoting objectivity, physiology was on its way to surpassing chemistry as the laboratory science par excellence. Both Helmholtz and Bernard were clear-cut laboratory researchers, whereas Huxley introduced laboratory work at the Royal School of Mines and also published a laboratory manual.[16] In the following paragraphs we will see multiple examples of mechanical objectivity in a laboratory context, although mostly in physiology

12 Lorraine Daston and Peter Galison, *Objectivity* (New York: Zone Books, 2007), pp. 34–5, 115–90.

13 Daston and Galison, *Objectivity*, pp. 216–33.

14 For Leopold Ranke and Emil Durkheim, see: Herman Paul, 'The Scientific Self: Reclaiming Its Place in the History of Research Ethics', *Science and Engineering Ethics*, 24 (2017), 1379–92 (p. 1383).

15 Daston and Galison, *Objectivity*, pp. 205–16.

16 Graeme Gooday, '"Nature" in the Laboratory: Domestication and Discipline with the Microscope in Victorian Life Science', *British Journal for the History of Science*, 24 (1991), 307–41 (pp. 334–40).

and closely related fields. This is not to suggest that the rise of the laboratory directly brought about the emergence of mechanical objectivity but merely that the laboratory proved fertile ground for mechanical objectivity, one that offered multiple opportunities for giving shape to related practices and norms.

The gradual emergence of the modern system of scientific disciplines around mid-century and especially the pedagogical innovations that supported this process are a simultaneous development that may shed some light on the new self-image of the scientist and the accompanying epistemic virtues. To clarify this point, we need to contrast the dominant new scientific persona with previous images of the ideal researcher. As several historians have emphasized, the true hallmark of genius in the early nineteenth century was the rare ability to fathom nature's hidden truths or essences, usually manifested through remarkable discoveries.[17] Humphry Davy, the embodiment of Romantic science, carefully cultivated his image as a genius, both on the stage and in his poetry. Davy, in Simon Schaffer's apt phrasing, 'made glorious discovery both the ideal prize and the fundamental mechanism of scientific change'.[18] Moreover, Davy frequently deployed his own body as an instrument, registering and even publishing the mental sensations induced by the breathing of several gases, thereby maximizing the contrast with the late nineteenth-century scientist, anxious to avoid everything that smacked of subjectivity.[19]

In this Romantic framework, discoveries were rarely seen as the outcome of methodical work or the laborious accumulation of data. Instead, they were connected with a lightning flash of inspiration. As David Brewster stressed in his *Life of Newton* (1835): 'The impatience of genius spurns the restraints of mechanical rules, and never will submit to the plodding drudgery of inductive discipline.'[20] Compare this picture of Newton with that of Samuel Smiles, more than thirty years later:

> When asked by what means he had worked out his extraordinary discoveries, he modestly replied, 'By always thinking unto them' ... It

17 Daston and Galison call the corresponding epistemic virtue 'truth-to-nature'. Daston and Galison, *Objectivity*, pp. 55–113.

18 Simon Schaffer, 'Scientific Discoveries and the End of Natural Philosophy', *Social Studies of Science*, 16 (1986), p. 409; Jan Golinski, 'Humphrey Davy: The Experimental Self', *Eighteenth-Century Studies*, 45 (2011), pp. 17–20. Where Schaffer and Golinski connect this picture to the emergence of disciplines, I see it as preceding this process. For a similar view of Romantic science, see: Richard Holmes, *The Age of Wonder* (London: HarperCollins, 2009).

19 Golinski, 'Humphrey Davy', pp. 17–20.

20 ° David Brewster, quoted in: Schaffer, 'Scientific Discoveries', p. 410.

> was in Newton's case, as in every other, only by diligent application and perseverance that his great reputation was achieved.[21]

Indeed, by this time, the ideal scientist was much closer to a diligent, disciplined, meticulous plodder than to a creative and inspired genius. Many prominent scientists readily agreed with Smiles. When turning 70, Helmholtz emphasized that his innovative ideas had not, as some believed, been the result of sudden flashes of brilliance but were developed 'slowly from small beginnings through months and years of tedious and often groping work from unprepossessing seeds'.[22] And one only has to glance through late-nineteenth century obituaries, usually singling out the more laudable traits of the deceased, to be struck by the ubiquity of a rather disenchanted view of what characterizes the true scientist.[23] One could be forgiven for mistaking the object of praise for a civil servant.

So how does this shift play into the contemporaneous emergence of disciplines? To put it quite simply, the latter process hinged on the introduction of disciplinary training programs, which hardly existed in the early nineteenth century. Now, neither genius nor the making of discoveries is something that can be trained, unlike the careful preparation of materials, skills in handling delicate instruments, the proper use of notebooks, methods of data analysis, the writing of research reports, and, more generally, discipline. Therefore, as an ideal, the Romantic image of genius was far less suitable for the disciplinary research schools connected to the new laboratories. A genius should not be disciplined but rather allowed an open field. The new model of a patient, persevering, meticulous and objective persona fitted the new pedagogical regimes much better. This does not mean that scientists like Huxley, Bernard, Maxwell, or Helmholtz and Emil du Bois-Reymond were less ambitious or more modest than their Romantic predecessors, for they clearly were not. In a sense, their modesty was as much a pose as Davy's genius, although in both cases it was also a fully internalized creed. Scientific personae tend to mould the self.[24] The resulting ethos neatly matched the demands of the newly emerging scientific disciplines which called for well-trained cadres rather than a handful of brilliant luminaries.

21 S. Smiles, quoted in: Daston and Galison, *Objectivity*, p. 229.

22 Daston and Galison, *Objectivity*, pp. 229–30.

23 I encountered a similar picture in the recommendations for membership of the Royal Academy of Sciences in the Netherlands. F.H. van Lunteren, "Wetenschap voor het Vaderland. J.D. van der Waals en de Afdeling Natuurkunde", in Klaas van Berkel, ed., *De Akademie en de Tweede Gouden Eeuw* (Amsterdam: Edita, 2004), pp. 43–106 (pp. 50–1).

24 See: Paul, 'What Is a scholarly Scholarly Persona?', pp. 355–7.

As I have argued elsewhere, the emergence of scientific disciplines—replacing older categories such as natural philosophy, natural history, and (mixed) mathematics—more or less coincided with the emergence of university laboratories in the third quarter of the nineteenth century and not, as is often claimed, with the decades around 1800.[25] Following Michel Foucault, many others have noted the close relationship between the disciplining of novices and the emergence of disciplines but have tended locate these processes in the late eighteenth and early nineteenth centuries.[26] However, it is only in the second half of the nineteenth century that we find specific career patterns for budding scientists, mostly in education but also in industry, resulting in a rapid rise in the numbers of science students.[27] Accordingly, university programs became more specialized, more standardized, and more like a vocational training. Indeed, early nineteenth-century pedagogical innovations, such as the seminar, were primarily aimed at professional careers and hence focused on teaching skills.[28] Their contribution to the emergence of disciplines was an unintentional side effect. But it was above all in the new teaching laboratories that students were truly disciplined.

Another obvious connection between the laboratory and the virtues of discipline and mechanical objectivity can be found in the factory, the birthplace of mechanization and standardization. The analogy between the factory and the laboratory was frequently noted at the time. Thus, Huxley declared the modern university to be 'a factory of new knowledge. Its professors must be at the top of the wave of progress. Research and criticism must be the breadth of their nostrils; laboratory work the main business of the scientific student.'[29] The connection between the industrialization of Europe and the rise of the laboratory has not been missed by historians,

25 Frans van Lunteren, 'Het ontstaan van het systeem van bètadisciplines: de natuurkunde', *Studium*, 6 (2013), 91–112. Cf. Schaffer, 'Scientific Discoveries', pp. 406–13; Jan Golinski, *Science as Public Culture: Chemistry and Enlightenment in Britain,1760–1820* (Cambridge: Cambridge University Press, 1992), pp. 13–24. Admittedly, this process already started in the late eighteenth century with the dissolution of the Republic of Letters, but it took more than a century before the full disciplinary framework was installed.

26 Golinski, 'Humphrey Davy', pp. 15–16; Schaffer, 'Scientific Discoveries', pp. 406–7; cf. Michel Foucault, *Discipline and Punish: The Birth of the Prison* (Harmondsworth: Penguin Books, 1977).

27 For chemistry, see: Ernst Homburg, 'The Rise of Analytical Chemistry and its Consequences for the Development of the German Chemical Profession (1780–1860)', *Ambix*, 46 (1999), pp. 1–32 (pp. 18–19).

28 Kathryn M. Olesko, *Physics as a Calling. Discipline and Practice in the Königsberg Seminar for Physics* (Ithaca: Cornell University Press, 1991), pp. 1, 21–25.

29 Th.H. Huxley to E. Ray Lankester, 11 April 1892, in *Life and letters of Thomas Henry Huxley*, 2 vols. (New York: D. Appleton & Co, 1900; repr. New York: AMS Press, 1979), vol. 2, pp. 328–9.

but the extent to which the laboratory came to be modelled on the factory has received less attention.[30] Germany's leading physiologist, Emil du Bois-Reymond, took the analogy further than most of his contemporaries. As he informed the Prussian ministry, a physiological institute should be run 'like a factory'.[31] His model was the electro-technical firm of his friend Werner Siemens, but also the physiological institute of Carl Ludwig in Leipzig.

Often, the link between industry and the (university) laboratory was even more direct. Following Liebig, many professors of organic chemistry used their laboratory facilities to find useful applications, and they often maintained close relations with the rising chemical industry. Something similar can be said of Louis Pasteur. Both Maxwell and William Thomson utilized their laboratories to serve the interests of the telegraph industry.[32] The most prominent German physics institute, the Berlin *Physikalisch-Technische Reichsanstalt*, was even founded to form a link between academic science and industry.[33] We find a similar purpose in the foundation of the German Chemical Society.[34] In the Netherlands Heike Kamerlingh Onnes made overtures to the refrigeration industry.[35] In general, the needs of industry—purity, control, precision, and standardization—neatly matched the cherished values of the laboratory. This was more expedient as ever more students pursued a career in the chemical and electro-technical industries.

The emergence of scientific disciplines, the rise of a science-based industry, and scientist's craving for social recognition and authority are not mutually independent phenomena. All three mark a change in the professional identity of university graduates. After all, the emergence of disciplines was closely connected to new facilities that enabled the training of special skills, characterizing the modern scientist. Both these facilities

30 For chemistry see: Homburg, 'The Rise of Analytical Chemistry', pp. 20–5. For physiology, see: Arleen M. Tuchman, 'From the Lecture to the Laboratory: The Institutionalization of Scientific Medicine at the University of Heidelberg', in William Coleman and Frederick L. Holmes, *The Investigative Enterprise. Experimental Physiology in Nineteenth-Century Medicine* (Berkeley, CA: University of California Press, 1988), pp. 65–99 (pp. 65–6, 85).

31 E. du Bois-Reymond, quoted in: Sven Dierig, 'Engines for Experiment: Laboratory Revolution and Industrial Labor in the Nineteenth-Century City', *Osiris*, 18 (2003), 116–34 (p. 121).

32 Simon Schaffer, 'Late Victorian Metrology and Its Instrumentation: A Manufactory of Ohms', in Robert Bud and Susan Cozzens, eds., *Invisible Connections: Instruments, Institutions, and Science* (Bellingham, WA: Society of Photo-Optical Engineers, 1992), pp. 23–56.

33 David Cahan, *An Institute for an Empire: The Physikalisch-Technische Reichsanstalt, 1871–1918* (Cambridge: Cambridge University Press, 1989).

34 Christoph Meinel, 'August Wilhelm Hofmann – "Reigning Chemist-in-Chief"', *Angewandte Chemie*, 31(10) (1992), 1265–82 (pp. 1274–5).

35 Dirk van Delft, *Freezing Physics. Heike Kamerlingh Onnes and the Quest for Cold* (Amsterdam: Edita KNAW, 2007), pp. 455–67.

and the professional careers, based upon new and specialized training methods, required social and political recognition. Only in the second half of the nineteenth century did a scientific training at a university open up the possibility of such a career, either in education or in industry. As a result, there was a need to stress both the educational and material benefits of science to society and a need for professional training and a certain degree of organization. It is hardly surprising that this shift was accompanied by the emergence of a new scientific ethos, more aligned with scientist's new role in society.

Chemistry: Utility, Discipline and Safety

In the nineteenth-century laboratory revolution, chemistry, the laboratory science par excellence, led the way. In early nineteenth century Germany, many chemical chairs had moved from the medical faculty to the philosophical faculty. Therefore, chemical professors could target others aside from medical students, who would often be unable and unwilling to take an extensive course in the chemical laboratory. As R.S. Turner has shown, through a large part of the nineteenth century, the student clientele of the chemical laboratories consisted primarily of pharmacists, who also made up the majority in Liebig's laboratory.[36] The gradual emancipation from medicine enabled the chemists to develop into an autonomous discipline and to innovate their teaching. It was the chemists who started intensive laboratory training for beginning students, who initiated group research for advanced students, who scaled up the number of *Praktikanten*, and who started to 'drill' their students, mainly through repetitive exercises in chemical analysis. By mid-century physiologists and physicists were following their lead.

Liebig was far from the first German chemist to train his students in analytical methods.[37] It was, above all, the intensive character of his laboratory training that Liebig claimed to be novel.[38] Even so, he was by far the most visible campaigner for laboratory training at the universities. Moreover, the perceived success of the Giessen laboratory in drawing students from all

36 Turner, 'Justus Liebig versus Prussian Chemistry', p. 143.

37 Ernst Homburg, 'The Rise of Analytical Chemistry and its Consequences for the Development of the German Chemical Profession (1780–1860)', *Ambix*, 46 (1999), 1–32 (pp. 9–18).

38 Alan J. Rocke, 'Origins and Spread of the "Giessen Model" in University Science', *Ambix*, 50 (2003), 90–115 (p. 100). See also Chapter 2 in this volume.

quarters of the world certainly helped in stimulating a gradual following, if not in Prussia than at least in other German states. In their demand for training laboratories the chemists faced the following predicament: In order to entice the German states to fund such costly facilities, it seemed pertinent to emphasize the utilitarian prospects of chemistry. However, too much stress on the industrial ties of academic chemistry could alienate the universities, which cherished their task of providing a liberal education to the state's intellectual elites.

In his paper on the state of chemistry in Prussia, Liebig directly confronted this dilemma by emphasizing the truly scientific nature of chemistry, a field grounded in a search for laws and causal relationships. From such knowledge useful applications would follow almost automatically. Scientific chemistry formed the basis for countless industrial applications and hence for the prosperity of the state, but also for other sciences such as physiology and medicine, which were doomed to remain superficial without a chemical foundation.[39] Such chemical knowledge could only be learned in the laboratory. And the primary task of the *Praktikum* was to educate young people, by teaching them how to think chemically, not to teach them some particular craft. It involved 'research into the causes of chemical transformations ... [and] use of the derived causal laws to make the trans-empirical apparent and understandable to the eye of intellect'.[40] As such laboratory work could and should be seen as part of a liberal education.[41] These arguments would later be echoed by younger chemists.[42]

By the time Liebig wrote his Prussia paper, he had extended his laboratory so as to have separate laboratory spaces for *Praktikanten*, beginning students who were trained in the basics of chemical analysis, and more experienced students to whom he gave research assignments and who contributed significantly to the research output of the laboratory.[43] Meanwhile, Liebig's example was followed by his pupils August Wilhelm Hofmann in London (at the London College of Chemistry) and Carl Remigius Fresenius in Wiesbaden

39 J. Liebig, *Ueber das Studium der Naturwissenschaften und über den Zustand der Chemie in Preussen* (Braunschweig: Vieweg und Sohn 1840), pp. 15–17, 23–5.

40 J. Liebig, quoted in: R.S. Turner, 'Justus Liebig versus Prussian Chemistry', p. 132.

41 Liebig, *Ueber das Studium der Naturwissenschaften*, pp. 21–6. For Liebig's campaign, see also: Ernst Homburg, *Van beroep 'Chemiker'. De opkomst van de industriële chemicus en het polytechnische onderwijs in Duitsland* (*1790-1850*) (Delft: Delftse Universitaire Pers, 1993), pp. 319–28.

42 E.g. Hermann Kolbe, *Das chemische Laboratorium der Universität Marburg* (Braunschweig: Vieweg und Sohn, 1865), pp. 17–28; Rudolph Fittig, *Das Wesen und die Ziele der chemischen Forschung und des chemischen Studiums. Rede* (Leipzig, 1870).

43 Rocke, 'Origins and Spread', p. 101.

(at the agricultural institute), but also by others such as Friedrich Wöhler in Göttingen, Robert Bunsen in Marburg, Hugo Erdmann in Leipzig, and Hermann Kolbe, who succeeded Bunsen in Marburg in 1851 when the latter was called to Breslau (and subsequently Heidelberg). By this time laboratory manuals were widely available. One of the first, Fresenius's textbook on quantitative analysis, spelled out the required virtues for proper laboratory work:

> Knowledge and ability must be combined with ... a sense of honesty and a severe conscience. Every analyst occasionally has doubts about the accuracy of his results ... In these cases it requires a strong conscience to repeat the analysis and not to make a rough estimate of the loss or apply a correction. Anyone not having sufficient will-power to do this is unsuited to analysis no matter how great his technical ability or knowledge.[44]

Ernst Homburg has discussed the new training methods in analytical chemistry in terms of Monte Calvert's distinction between 'school culture' and 'shop culture', the latter referring to the kind of 'on-the-job' training that would come to distinguish the practical chemical crafts from chemical occupations that required academic training. According to Homburg, 'the laboratory techniques of analytical chemistry demanded manual and cognitive skills which could not be learned "on the job", as part of conventional production of chemicals and drugs.'[45] In his view, these educational reforms resulted in new industrial positions for academically trained scientists. Although he locates the breakthrough of the new training methods in Germany in the 1820s, he also points out that it would take another three decades before one could speak of large-scale industrial career patterns for university-trained chemists.[46] By that time, synthetic organic chemistry was gradually producing a new job market in the dye industry.

The real boom in the building of chemical laboratories in Germany coincided with this latter period, the 1860s and 1870s. This time, it was Prussia that took the lead. In this period Germany was going through an unparalleled process of modernization and industrialization in which Prussia was leading the way. By now, the Prussian authorities fully recognized the economic potential of chemical training and research. This was also

44 Fresenius (1872), quoted in: Ferenc Szabadváry, *History of Analytical Chemistry* (New York: Pergamon Press, 1966), p. 176.

45 Homburg, 'The Rise of Analytical Chemistry', pp. 6–7, quotation on p. 7.

46 Homburg, 'The Rise of Analytical Chemistry', p. 19.

the period in which the chemical industry had started its rapid ascent, culminating in the domination of the world market in synthetic dyes. The new political climate opened the way to a second generation of university laboratories, which differed in several ways from their predecessors. Apart from a spectacular change in scale, we see an increasing differentiation both in types of research and in levels of laboratory training. Because of these increasing demands, chemical laboratories could no longer be arranged within existing buildings.[47]

Among the first 'modern' chemical laboratories were those built in the 1860s for—and designed by—Liebig's pupil Hofmann, one in Bonn and the other in Berlin. The Bonn chemical institute contained three teaching laboratories for undergraduates, advanced students, and young chemists, respectively. Each of the laboratories could accommodate up to twenty students. They were equipped with fume cupboards and rows of benches, each of which was supplied with washbasins and gas and water taps. As Hofmann explained, the specific design served several purposes:

> Not only was it possible to fit up each laboratory in a manner suitable to the wants of each particular class, but the situation of the rooms themselves could be so adapted to the remaining parts of the building as to offer the greatest facilities to each division. And higher still must be rated the advantages as regards readier supervision and increased means of maintaining discipline in all parts of the institution afforded by an arrangement of this kind.[48]

Even larger than both the Bonn and Berlin institutes was Kolbe's new institute in Leipzig, which opened its doors in 1868. The several teaching laboratories could accommodate 130 students at a time, as much as Bonn and Berlin combined.[49]

With the growth of the institutes, training regimes also became stricter. By this time, most medical students would take at least a one-term *Praktikum*.[50] As the number of undergraduate students was rapidly increasing, the practice of 'drilling' undergraduates in chemical analysis became more and more

47 Catherine M. Jackson, 'Chemistry as the Defining Science: Discipline and Training in Nineteenth-Century Chemical Laboratories', *Endeavour*, 35 (2011), 55–62 (pp. 56–7).

48 Hofmann, quoted and translated in: Jackson, 'Chemistry as the Defining Science', p. 57.

49 Hermann Kolbe, *Das chemische Laboratorium der Universität Leipzig* (Braunschweig: Vieweg & Sohn, 1872), p. xix; on Kolbe, see: Alan J. Rocke, *The Quiet Revolution: Hermann Kolbe and the Science of Organic Chemistry* (Berkeley, CA: University of California Press, 1993).

50 Turner, 'Justus Liebig versus Prussian Chemistry', p. 142.

customary.[51] This habit was probably born of necessity given the decreasing teacher-student ratio, but it also served the higher aim of having students rapidly acquire several useful skills. Peter Morris has suggestively linked such new regimes to the example of 'the grammar drills of Latin teaching'.[52] To instil such discipline among the students, constant supervision was a main requirement. With increasing numbers of students, teaching assistants became indispensable. In some cases, the director's office was provided with a 'spy window' that overlooked the main teaching laboratory. This novelty was first introduced by Kolbe when he renovated the Marburg institute in 1863. It recalls the hatch in Liebig's laboratory and, more germane to the notion of discipline, Jeremy Bentham's 'panopticon', used by Foucault as a metaphor for the modern disciplinary society.[53]

As Catherine Jackson has reminded us, the urge to instil discipline in the students was not simply a matter of pedagogical requirements. An additional factor, largely missing in other laboratory sciences such as physiology and physics, was safety. The risk of fire or exploding glass demanded care and vigilance. Obnoxious or even poisonous fumes, such as the ubiquitous hydrogen sulphide, posed a constant threat to students' health. Ventilation was always a major concern in the design of chemical laboratories.[54] Liebig had pioneered fume cupboards in his Giessen laboratory, a practice followed elsewhere, but there were never sufficient cupboards for all students. Moreover, they were inconvenient as they forced the researchers to work at arm's length. As Kolbe pointed out:

> There are some people who are extremely difficult to convince that they should use these fume extractors without fail, despite repeated advice and even where the fume extractor is only a few steps away.[55]

Above all, the new chemical palaces testified to the growing prestige of chemists. And although the latter still emphasized the scientific nature of the field and the essential difference between pure chemistry and applied chemistry, they increasingly presented themselves as the lifeblood of the German economy. They no longer felt the need to stress the causal and theoretical aspects of chemistry to underline its scientific credentials.

51 Morris, *The Matter Factory*, pp. 106–7.
52 Morris, *The Matter Factory*, pp. 107–8.
53 Morris, *The Matter Factory*, pp. 167–8.
54 Jackson, 'Chemistry as the Defining Science', pp. 57–60; Morris, *The Matter Factory*, pp. 97–107.
55 Kolbe, quoted and translated in: Jackson, 'Chemistry as the Defining Science', p. 57.

Instead, both Kolbe and Hofmann came to adopt a factual, empiricist stance. They showed little enthusiasm for the more speculative and hypothetical ideas of such colleagues as August Kekulé and J.H. van 't Hoff.[56] Hofmann himself maintained close ties with the chemical industry. Hofmann was also instrumental in the foundation of the German Chemical Society in 1867, an organization that he would rule for 25 years. He did not hide his intentions:

> The new Society is actually designed to provide an opportunity for the mutual exchange of ideas between representatives of the speculative and applied branches of chemistry in order to seal anew the alliance between science and industry.[57]

By ceaselessly playing the industrial card, while insisting on the scientific nature of their discipline, German chemists had finally managed to realize their aims: the directorship of large research institutes (at least at some of the major German universities), acceptance of laboratory training as an indispensable part of a scientific chemical education, and, more generally, political 'acceptance of the professional and autonomous status of chemistry as a discipline'.[58]

Physiology and physics would both benefit from this precedent. They followed in footsteps of chemistry and their new institutes soon rivalled the chemical institutes. They, too, insisted on the importance of laboratories for training young scientists. However, compared to chemistry, they had two disadvantages. They could hardly point to useful applications, and they lacked the comparatively cheap apparatus, which the chemists could supply to large numbers of students. These circumstances can partly account for the fact that their rhetorical strategies differed in subtle ways from those of the chemists. The physiologists, above all, aimed to show their utility to medical practice, on the one hand, and the superiority of the laboratory as a site of medical knowledge compared to the clinic, on the other. Their particular emphasis on mechanical objectivity at least partly aimed to distance their research from the more 'subjective' clinical research. The physicists, like the chemists, tended to rub shoulders with industry. But they differed from the chemists in putting much more emphasis on educational values. This

56 See, for instance, Kolbe on chemical training, *Das chemische Laboratorium*, pp. xxxiv–clvi; Meinel, 'August Wilhelm Hofmann', pp. 1267–8. For Kekulé, see: Alan J. Rocke, *Image and Reality: Kekulé, Kopp, and the Scientific Imagination* (Chicago: University of Chicago Press, 2010).

57 Hofmann, quoted in: Meinel, 'August Wilhelm Hofmann', p. 1274.

58 Turner, 'Justus Liebig versus Prussian Chemistry', p. 129.

probably reflected the fact that, throughout the nineteenth century, the majority of physicists were facing a teaching career.

The Physiological Laboratory: Empiricism, Objectivity, and the Mechanization of Experiment

If the nineteenth-century laboratory revolution started with the chemical laboratory, it was the new discipline of physiology that came to compete with chemistry as the prime laboratory science. In 1877 Emil du Bois-Reymond undauntedly referred to physiology as 'the queen of the sciences'.[59] His own work had contributed greatly to the prestige of experimental physiology. Together with his companions in arms, Carl Ludwig, Ernst Brücke, and Hermann Helmholtz, he had started a reform movement that aimed to free physiology from all vital principles. Their goal was to reduce all living phenomena to physics and chemistry and, accordingly, to give physical and chemical research a dominant place in physiology.[60] All three would eventually lead prominent German institutes, where many medical students would receive laboratory instruction, and all three would create powerful research schools. On top of that, both Helmholtz and Du Bois-Reymond would become major public figures, whose portraits were offered for sale in the shop windows of Berlin.[61]

Even more renowned than these German luminaries was their French colleague Claude Bernard. As an assistant to François Magendie, professor of medicine at the College de France, and a scholar generally seen as one of the pioneers of experimental physiology, Bernard rapidly earned a reputation for himself as a master of medical experimentation. This reputation was so strong that the government created a chair of general physiology for him in 1854 at the Sorbonne. The following year he also succeeded Magendie at the College de France. Since physiology was still a marginal discipline at the time, the Sorbonne did not provide him with a laboratory. This omission was only rectified in 1868 when the chair was transferred from the Sorbonne to the

59 E. du Bois-Reymond, *Der Physiologische Unterricht sonst und jetzt* (Berlin: August Hirschwald, 1878), p. 21.
60 For the ideological background of the reform movement, see Timothy Lenoir, 'Laboratories, Medicine and Public Life in Germany, 1830–1849', in Andrew Cunningham and Perry Williams, eds., *The Laboratory Revolution in Medicine* (Cambridge: Cambridge University Press, 1992), pp. 14–71.
61 Gabriel Finkelstein, 'M. du Bois-Reymond Goes to Paris', *British Journal for the History of Science*, 36 (2003), 261–300 (p. 263).

Muséum nationale d'Histoire naturelle. By then he had published his highly influential *Introduction à l'étude de la médecine expérimentale* (Introduction to the study of experimental medicine, 1865), one of the first scientific works to stress the notion of 'objectivity' in its modern sense.[62] In this work he emphasized the importance of the experimental method, borrowed from physics and chemistry, as the only way forward in medicine. In order to become truly scientific, medicine needed to be based on physiology. Therefore, clinical observation, until recently the main source of medical knowledge, no longer sufficed.[63]

Yet experimenting on living bodies, being much more complex than inorganic substances, was 'incomparably harder'.[64] Therefore, as in all scientific research, 'minutiae of method are of the highest importance'. Indeed, 'one must be brought up in laboratories and live in them to appreciate the full importance of all the details of procedure in investigation.'[65] As a matter of course, experiments also involved observations, and experimenters therefore must be good observers. They

> must be photographers of phenomena; their observations must accurately represent phenomena. We must observe without any preconceived idea; the observer's mind must be passive, that is, must hold its peace; it listens to nature and writes at nature's dictation.[66]

Admittedly, in setting up an experiment, an experimenter's mind must be active. In this phase the experimenter 'reflects, tries out, gropes, compares, contrives'. But as soon as the result of the experiment appears, the experimenter himself 'must now disappear', that is, 'change himself into an observer'. It is only when the results have been noted precisely 'that his mind will come back'. Preconceived ideas, so essential to the design of the experiment, must never colour the observations. In Bernard's view the difficulty of this disappearance act, or rather of this constant switching, was 'one of the great stumble blocks of the experimental method'.[67] Self-doubt, modesty, and even humility characterize the true scientific spirit.[68]

62 Daston and Galison, *Objectivity*, pp. 205–14.
63 Claude Bernard, *An Introduction to the Study of Experimental Medicine* (New York: Dover, 1957), pp. 1–3, 168.
64 Bernard, *Introduction*, p. 2.
65 Bernard, *Introduction*, pp. 14–15.
66 Bernard, *Introduction*, p. 22.
67 Bernard, *Introduction*, pp. 22–3.
68 Bernard, *Introduction*, pp. 36, 43.

One way to realize Bernard's ideal of the passive or even absent observer was to use photography. Another was to introduce self-registering instruments. One of the pioneers in this latter approach was the German physiologist Carl Ludwig. Starting his career in Marburg, he later moved to Zürich, where he published his *Lehrbuch der Physiologie des Menschen* (1852). This first modern textbook on physiology was dedicated to his friends Brücke, Helmholtz, and Du Bois-Reymond.[69] Continuing his career in Vienna, he finally arrived in 1865 at the University of Leipzig. Here he built up his large physiological institute that opened its doors in 1869. This was one of the first institutes that was fully devoted to experimental physiology, and, with its modern facilities, it served as a model for later physiological laboratories.[70]

As a young branch of science, experimental physiology lacked established methods and instruments. Like his friends Du Bois-Reymond and Helmholtz, Ludwig proved particularly resourceful in designing his own instruments. Still in Marburg, he investigated the relationship between respiration rate and blood pressure in living animals. As the fluctuations of the level of mercury in the manometer made a careful reading rather difficult, he developed what came to be known as the kymograph. As Robert Brain and Norton Wise have pointed out, the instrument and the resulting graphical method were at least partly rooted in James Watt's indicator diagrams, which aimed to measure the work performed by his steam engines.[71] These and similar graphical methods were rapidly taken up by other physiological laboratories, including laboratories for plant physiology and, later, psychological laboratories. As Soraya de Chadarevian has suggested, the graphical method 'conveyed disciplinary identity to the practitioners involved in the physical studies of physiological phenomena'.[72] It also seems plausible that, as in the case of photography, the availability—and widespread use—of ways to exclude the human observer contributed to the new ethos noted by Daston and Galison.

69 Carl Ludwig, *Lehrbuch der Physiologie des Menschen*, 2 vols. (Heidelberg: C.F. Winter, 1852–56), vol. 2.

70 Timothy Lenoir, 'Science for the Clinic: Science Policy and the Formation of Carl Ludwig's Institute in Leipzig', in Coleman and Holmes, eds., *The Investigative Enterprise*, pp. 139–78.

71 Robert Brain and Norton Wise, 'Muscles and Engines: Indicator Diagrams in Helmholtz's Physiology', in Lorenz Krüger, ed., *Universalgenie Helmholtz: Ruckblick nach 100 Jahren* (Berlin: Akademie Verlag, 1994), pp. 124–45; repr. in: Mario Biagioli, ed., *The Science Studies Reader* (New York: Routledge, 1999), pp. 51–66.

72 Soraya de Chadarevian, 'Graphical Method and Discipline: Self-Recording Instruments in Nineteenth-Century Physiology', *Studies in History and Philosophy of Science*, Part A, 24(2) (1993), 267–91, quotation on p. 284.

It is important to realize that the introduction of self-recording instruments did not supplant the dexterity of the experimenter. On the contrary, the new and delicate instruments, developed for physiological experiments, were seen to require considerable training for users to acquire the skills to produce fruitful results. In her study of a controversy between Charles Darwin and Julius Sachs, Soraya de Chadarevian showed how the rise of the botanical laboratory, with its sophisticated instrumentation and standardized procedures, reduced a gentleman scientist like Darwin, who experimented in his country house garden, to a mere amateur in the modern sense of the word. In Sachs's view, Darwin simply lacked the laboratory skills and methods that characterized the modern scientist.[73]

The mechanization of physiological research did not end with the graphical methods of registering measurements. The experiment itself could also be mechanized. Again, it was Ludwig who paved the way. As Du Bois-Reymond remarked at the opening of his own new laboratory: Ludwig's new institution proved momentous in that

> Ludwig, with creative boldness, was the first to make the progressive technology of our age useful for science, by providing all workrooms, as in a factory, not only with water, but also with mechanical power from a gas engine.[74]

Small gas engines gradually became a requisite for all German physiological laboratories despite their high costs. In all these cases, machines outstripped human assistants. They also increased the resemblance of scientific laboratories to factories.

It is tempting to connect the feigned modesty and strict empiricism of the experimental physiologists to the mechanistic view of life. Empiricism and proper methodology served to cleanse medicine of speculative elements, such as vital principles, providing a sound scientific basis for a true understanding of the living organism. Indeed, both Helmholtz and Du Bois-Reymond contrasted the sound empirical nature of their physiological research with the speculative character of traditional medicine. In 1869 Helmholtz looked back on his student days:

73 Soraya de Chadarevian, 'Laboratory Science versus Country-House Experiments. The Controversy between Julius Sachs and Charles Darwin', *British Journal for the History of Science*, 29 (1996), 17–41 (pp. 34–6)

74 Du Bois-Reymond (1877), quoted and translated in: Dierig, 'Engines for Experiment', p. 122.

> It was not difficult to recognize that the old predominant theorizing methods of practicing medicine were altogether untenable; with these theories, however, the facts on which they had actually been founded had become so inextricably entangled that they also were mostly thrown overboard. How a science should be built up anew had already been seen in the case of the other sciences; but the new task assumed colossal proportions.[75]

In the spirit of the new of the new inductive scientific ethos, Helmholtz also stressed the collective nature of science:

> Each one must be convinced that it is only in connection with others that he can further the great work, and that therefore he is bound, not only to investigate, but to do his utmost to make the results of his investigation completely and easily accessible.[76]

In his staunch empiricism, Du Bois-Reymond resembled his friend Helmholtz. He even projected his own methodological doubts on their former teacher Johannes Müller. Although he admitted that the latter had been a vitalist, he suggested that Müller had gradually moved towards an agnostic stance, captured in his frequent expression 'ignoramus'.[77] Much later, in his 1872 address to the Berlin academy, he would put the expression to good use while discussing the limits of science. Even when science would arrive at a full mechanical understanding of nature, some things would forever remain hidden to our cognitive faculties, such as the nature of matter and that of consciousness: 'ignorabimus!'[78]

Yet Helmholtz's epistemological views were much more subtle and profound than those of Du Bois-Reymond. Helmholtz had inherited the view from his mentor Müller that the character of a representation of an object is determined by the physiological structure of the sensory organs rather than purely by the properties of the object itself. Over time, Helmholtz had developed a sophisticated theory of perception, partly in connection with his study of the physiology of the sensory organs. In Helmholtz's view the mind does not construct a coherent picture of the sensations that form its

75 Helmholtz (1869), quoted and translated in: Edward Jurkowitz, 'Helmholtz and the Liberal Unification of Science', *Historical Studies in the Physical Sciences*, 32 (2002), 291–317 (p. 297).

76 Helmholtz (1862), quoted in: Jurkowitz, 'Helmholtz and the Liberal Unification of Science', p. 291.

77 Jurkowitz, 'Helmholtz and the Liberal Unification of Science', p. 298.

78 Emil du Bois Reymond, *Über die Grenzen des Naturerkennens* (Leipzig: Von Veit, 1872), p. 34.

experiences through some innate faculty, as Immanuel Kant would have it, but rather through a series of unconscious mental adjustments. Thus, perceptual space is the product of a learning process, a mental generalization of our orientation with respect to objects in space. The same kind of inductive process should lead the scientist to empirical generalizations, which, though not absolutely valid, would still be highly useful.[79]

In his opening lecture of 1877, Du Bois-Reymond emphasized the importance of the new institute for training physicians. It was, perhaps, not so much the physiological knowledge itself which they needed but rather the mental training which physiology could provide. Like Bernard, Du Bois-Reymond stressed the need to distrust preconceived ideas and the related importance of the laboratory experience for students. As he pointed out, the most widespread fault among scientific novices was their inclination to memorize changeable theories rather than unshakeable facts and to merely look for confirmation of these theories. He knew no better cure for this danger than to build from the start, as much as possible, on purely sensory evidence. Students should learn to see for themselves and to convince themselves rather than be infused with knowledge through 'dogmatic lectures'.[80] As Du Bois-Reymond made clear, the new facilities were not meant for the brilliant students, who hardly needed them. The main goal of the institute was to provide the 'less gifted' students with a 'thorough inductive training'.[81]

Intellectual and sensory restraint were not the only challenges facing the physiologist. If the chemist was confronted with bad smells and various dangers, the physiologist had to deal with other forms of unpleasantness, especially vivisection. The French physiologists Magendie and Bernard had given vivisection experiments a central place in physiology. Emotional control was therefore an additional prerequisite in the physiological laboratory. As Du Bois-Reymond confessed to his readers in 1848, while discussing his experiments in electrophysiology: 'The frogs in my experiments must have endured monstrous pain, as their behaviour, a horrible writhing and cooing, bore only too vivid witness.'[82] Even after the introduction of anaesthetic

79 R. Steven Turner, 'Hermann von Helmholtz and the Empiricist Vision', *Journal of the History of the Behavioral Sciences*, 13 (1977), 48–58; Scott Edgar, 'The Physiology of the Sense Organs and Early Neo-Kantian Conceptions of Objectivity: Helmholtz, Lange, Liebmann', in Flavia Padovani, Alan Richardson, and Jonathan Tsou, eds., *Objectivity in Science. New Perspectives from Science and Technology Studies* (Cham: Springer 2015), pp. 101–22.

80 Du Bois-Reymond, *Über die Grenzen*, pp. 26–7.

81 Du Bois-Reymond, *Über die Grenzen*, pp. 27–8.

82 Du Bois-Reymond, quoted in: Finkelstein, 'M. du Bois-Reymond Goes to Paris', pp. 274–5.

agents in the 1840s, public repugnance of animal experiments continued and even increased, culminating in the British Cruelty to Animals Act of 1876. Nevertheless, like his colleagues, Du Bois-Reymond emphatically defended the practice of vivisection. As he stressed at the opening of his new Berlin institute, a year after the Britons passed the new legislation, the suffering of frogs, rabbits, and dogs saved human lives and prevented human suffering.[83] Still, while students could watch vivisection demonstrations, vivisection experiments were generally excluded from the elementary practical exercises for medical students.[84] As Henning Schmidgen has suggested, this may well have been one of the reasons why physiologists like Ludwig and Du Bois-Reymond 'emphasized the predominance of the visual in the experimental work of physiology'.[85] We find the same predominance in Thomas Huxley's physiological laboratory, which focused its training program on microscopy.[86]

The Physical Laboratory: Measurement and the Primacy of Precision

Where physiology was making overtures to physics, the latter field was undergoing its own process of transformation. This process was at least partly connected to the rise of the physics laboratories in the second half of the nineteenth century. In the early nineteenth century, experimental physics and chemistry were in several instances connected to such an extent that it was often hard to distinguish between the fields. Leading 'experimental physicists', such as Louis Gay-Lussac and Henri Victor Regnault in Paris, Humphry Davy and Michael Faraday in London, James Prescott Joule in Manchester, Johann Christian Poggendorf and Heinrich Gustav Magnus in Berlin, and Hans Christian Ørsted in Copenhagen all shared a chemical background and all eschewed advanced mathematics. Tellingly, the leading German physics journal bore the title *Annalen der Physik und Chemie*.[87] Above all, experimental physics owed

83 Du Bois-Reymond, *Der Physiologische Unterricht sonst und jetzt*, pp. 22–3.

84 Henning Schmidgen, 'Pictures, Preparations, and Living Processes: The Production of Immediate Visual Perception (Anschauung) in Late-19th-Century Physiology', *Journal of the History of Biology*, 37 (2004), 477–513 (pp. 488–9).

85 Schmidgen, 'Pictures, Preparations, and Living Processes', pp. 489.

86 Gooday, '"Nature" in the Laboratory', pp. 307, 338–40.

87 Frans van Lunteren, '"Van meten tot weten". De opkomst van de experimentele fysica aan de Nederlandse universiteiten in de negentiende eeuw', *Gewina*, 18 (1995), 102–38 (p. 105); Rudolf Stichweh, *Zur Entstehung des modernen Systems wissenschaftlicher Diszplinen. Physik in Deutschland, 1740-1890* (Frankfurt a.M.: Suhrkamp, 1984), pp. 94–172.

its visibility to a series of remarkable discoveries, such as the polarization of light, the electromagnetic effect, thermoelectricity, the magnetic induction of currents, diamagnetism, and the magneto-optic effect.

Meanwhile, Franz Neumann in Königsberg and Wilhelm Weber in Göttingen were pioneering a very different kind of physics, in close collaboration with astronomers and mathematicians, above all Friedrich Bessel and Carl Friedrich Gauss. This new, so-called 'Gaussian physics' or 'measuring physics' was closely modelled on astronomical concepts and practices: it combined precision measurements, often with the use of a telescope, vibration-free foundations, research that partly focused on the instruments themselves, mathematical data-analysis, using the method of least squares, standardized instruments and units, and the use of potential functions and partial differential equations. Before the 1830s this approach was as marginal to physics as physiology was to medicine. The new kind of physics owed its visibility primarily to the collective effort, led by Gauss, to chart the local variations of terrestrial magnetism with the use of the magnetometer, developed by Gauss, according to a prescribed protocol. Ideally, the precise measurements were performed in a magnetic observatory that was totally free of iron and where the magnetometer and the telescope could be placed on a solid foundation.[88]

Both Neumann and Weber spread the new methods and approaches to their students, mainly through their seminars, an educational innovation that they had copied from the philologists. Lacking proper training facilities, both set up private laboratories for their students' use.[89] The laborious methods of Gaussian physics were quite demanding. One of Neumann's students, Gustav Kirchhoff, who was to become one of the main representatives of the 'Gaussian school', later admitted that he was originally taken aback by the drudgery of Neumann's seminar: 'boring observations and even more boring calculations'.[90] He was, however, soon won over to the Gaussian regime. Later, in Heidelberg, Kirchhoff would start his own seminar, where he would train students in a similar way. Among his students were the Dutch Heike Kamerlingh Onnes and the British Arthur Schuster.[91] Around

88 Frans van Lunteren, 'Astronomers and the Making of Modern Physics', in Ad Maas and Henriëtte Schatz, eds., *Physics as a Calling, Science for Society* (Leiden: Leiden University Press 2013), pp. 20–30.

89 For Neumann and his school, see: Olesko, *Physics as a Calling.* For Gauss and Weber, see: Christa Jungnickel and Russell McCormmach, *Intellectual Mastery of Nature: Theoretical Physics from Ohm to Einstein. Volume 1, The Torch of Mathematics, 1800–1870* (Chicago: University of Chicago Press, 1986), pp. 64–76.

90 Jungnickel and McCormmach, *The Torch of Mathematics,* p. 155.

91 Olesko, *Physics as a Calling*, pp. 215–19.

the turn of the century, both would oversee large and highly productive physics laboratories, in Leiden and Manchester.

When the number of students participating in Weber's laboratory exercises increased in the 1860s, he was finally allowed to hire a salaried assistant to direct the work in the seminar. The assistant position was filled by Weber's former student Friedrich Kohlrausch. Kohlrausch, who would be appointed as extraordinary professor in the following year, reorganized the laboratory exercises, which were now also open to chemists and pharmacists.[92] Onno Wiener later emphasized the pioneering nature of the Göttingen exercises that instilled a 'sharp criticism of the measurements' and a 'military disciplining of the observer'.[93] As many new German physics laboratories opened up, most would copy the Göttingen teaching methods. Particularly helpful was the laboratory manual, *Leitfaden der praktischen Physik*, that Kohlrausch published in 1870, the first of its kind in Germany. Starting with a general introduction to error calculus, it contained descriptions of experiments, experimental setups, and measuring techniques, as well as tables of physical quantities. Through its numerous editions, it would become the German bible of experimental physics.[94]

The success of the manual more or less secured the reform of physics that Neumann and Weber had started. From the 1870s onwards, experimental physics in Germany came to be identified with precision measurement. Precise measurement of known phenomena was in general preferred to exploratory work that aimed primarily at the discovery of unknown effects. Although smaller than Helmholtz's magnificent Berlin institute, August Kundt's Strasbourg institute, which opened its doors in 1882, came to be seen as the model institute for physics. The three-storey building featured numerous water, gas, and electrical connections, the latter for lighting and power purposes. It contained several rooms for undergraduates' laboratory exercises, the so-called *Practicum*, but most impressive were the research rooms for advanced students and staff. The experimental apparatus was set up on stone consoles that rested on stone pillars so as to exclude vibrations. The measuring apparatus was placed on another stone plate, kept at an appreciable distance from the experiment to avoid disturbances caused by the observer's motions and body heat. All rooms for magnetic and galvanic measurements were totally free of iron.[95]

92 Jungnickel and McCormmach, *The Torch of Mathematics*, pp. 104–5.

93 Onno Wiener (1906), quoted in: Stichweh, *Zur Entstehung des modernen Systems wissenschaftlicher Disziplinen*, p. 385.

94 David Cahan, 'The Institutional Revolution in German Physics, 1865–1914', *Historical Studies in the Physical Sciences*, 15 (1985), 1–65 (pp. 48–50).

95 Cahan, 'The Institutional Revolution in German Physics', pp. 23–33.

In 1888 Kohlrausch, now generally recognized as 'the master of measuring physics', would succeed Kundt in Strasbourg.[96] The latter had succeeded Helmholtz as director of the Berlin physics institute; Helmholtz himself had left the university to lead a new and prestigious research institute, the *Physikalisch Technische Reichsanstalt* in Berlin-Charlottenburg. The PTR, initiated by Werner Siemens and Helmholtz, aimed to link the interests of science, technology, and industry. Among the more pressing problems that plagued the electrotechnical industry was the lack of reliable methods for realizing electrical units, a problem that had occupied Weber for most of his life. The institute combined fundamental research with services for industry. It was the need for economic street lighting, and therefore for more precise standards of luminous intensity, that motivated research on black body radiation, which in turn resulted in the birth of quantum physics. More generally, perceived industrial interests played a dominant role in the founding of the large and costly physics institutes.[97]

In 1894 Kohlrausch had originally been invited to succeed Kundt in Berlin, an offer he had refused because he regarded the combined responsibilities for teaching and research as too demanding. Berlin then offered the chair to Emil Warburg, a former student of Kirchhoff who had also worked with Kundt in Strasbourg. Previously, Warburg had noted the large changes in the physics discipline, now well established, and the large importance attributed to physics in influential circles. Why, he asked, did the German states invest so much money in these costly facilities? Rather than pointing to industrial interests, Warburg emphasized modern ideas of *Bildung*. The teaching through practice of 'the precision methods of measurement', needed to understand 'the spirit of modern physics', required large institutes with extensive laboratories. Such exercises brought independent thinking and taught the student to appreciate the 'measure of accuracy', which was more important than the result itself:

> The young man will get used to considering carefully the claims that he makes. The feeling of scientific responsibility is awakened and strengthened. One can hardly find a better means for forming a serious, manly scientific character than these exercises.[98]

96 The expression is from Wilhelm Wien, 'Friedrich Kohlrausch', *Annalen der Physik* (4th series), 31 (1910), 449–54 (p. 452).
97 Cahan, 'The Institutional Revolution in German Physics', pp. 38–9.
98 Emil Warburg (1881), quoted and translated in: Cahan, 'The Institutional Revolution in German Physics', p. 41.

Figure 3.2 Students taking an elementary class at the Cavendish Laboratory, Cambridge.

The German physicists' predilection for precision measurements also found its way into leading physical laboratories abroad. We have already noted Maxwell's praise of the large magnetic project led by Gauss and Weber. Indeed, he made sure that the Cambridge Cavendish Laboratory contained a special room for magnetic measurements. It contained a pier resting on concrete foundations which were eighteen inches thick and made no contact with the walls. In fact, measuring the earth's magnetic field with the Kew magnetometer was Maxwell's favourite way of introducing students to the intricacies of physical measurement.[99] He discussed the instrument and its use in great detail in his *Treatise on Electricity and Magnetism*.[100] As Maxwell explained in his inaugural lecture,

> In experimental researches, strictly so called, the ultimate object is to measure something which we have already seen—to obtain a numerical

99 Malcolm Longair, *Maxwell's Enduring Legacy: A Scientific History of the Cavendish Laboratory* (Cambridge: Cambridge University Press, 2016), p. 63; Iwan Rhys Morus, *When Physics Became King* (Chicago: University of Chicago Press, 2005), p. 241.

100 J.C. Maxwell, *A Treatise on Electricity and Magnetism*, 2 vols. (New York: Dover, 1954), vol. 2, pp. 95–128.

> estimate of some magnitude. Experiments of this class—those in which measurement of some kind is involved–are the proper work of a Physical Laboratory.[101]

Maxwell also pointed to the link between physics and industry and their common need for precise electrical units:

> The new methods of measuring forces were successfully applied by Weber to the numerical determination of all the phenomena of electricity, and very soon afterwards the electric telegraph, by conferring a commercial value on exact numerical measurements, contributed largely to the advancement, as well as to the diffusion of scientific knowledge.[102]

Indeed, as Simon Schaffer has emphasized, much of the research in the Cavendish was centred on the determination of the unity of electrical resistance, the Ohm. Already in 1862, Maxwell had joined the committee, appointed by the British Association for the Advancement of Science (BAAS), to oversee the determination of fundamental standards in electricity.[103]

However, in his inaugural lecture Maxwell was careful not to dwell on these mundane motives. Like his German colleagues, he elevated the teaching of method above the establishment of results.

> Our principal work, however, in the Laboratory must be to acquaint ourselves with all kinds of scientific methods, to compare them, and to estimate their value. It will, I think, be a result worthy of our University, and more likely to be accomplished here than in any private laboratory, if, by the free and full discussion of the relative value of different scientific procedures, we succeed in forming a school of scientific criticism, and in assisting the development of the doctrine of method.[104]

In many respects, Maxwell followed in the footsteps of his former mentor William Thomson, the first physicist—or rather natural philosopher—in Britain to open his laboratory to his students. In 1857, to accommodate these students, Thomson managed to obtain several rooms beneath his lecture

101 Maxwell, 'Introductory Lecture', pp. 243–44 (see note 1).
102 Maxwell, 'Introductory Lecture', p. 246.
103 Schaffer, 'Late Victorian Metrology', pp. 23–56.
104 Maxwell, 'Introductory Lecture', p. 250.

hall, which, although located on the ground floor, became known as the 'cellar laboratory'.[105]

The opening of the laboratory coincided with the start of Thomson's involvement in Atlantic telegraphy and hence with the electric cable industry. Indeed, shortly before the opening, Thomson had been elected a Director of the Atlantic Telegraph Company. Thomson then dedicated his laboratory to precision measurements that aimed to overthrow the conventional wisdom in telegraphy, especially the use of narrow-cored, low-conductivity copper cables. Accordingly, Thomson and his students applied themselves to obtaining precise measurements of cable resistance. To this end, Thomson developed several new instruments, often by improving existing instruments. He increased the precision of a Helmholtz galvanometer already in 1857 by attaching a mirror to the rotatory magnet, using the reflected beam of light to indicate the current. Through the following decades, the new precision instruments, developed by Thomson and his students, rapidly found their way to other teaching laboratories in Britain and abroad.[106]

As Graeme Gooday has shown, Thomson's example was instrumental in the educational practices in later physics teaching laboratories in Britain.[107] Precision measurement became the core of practical physics teaching. Like their German colleagues, the Britons were eager to stress the pedagogical value of such training. In 1875 the London physicist George Carey Foster, another member of the BAAS committee, emphasized that

> in the study of physics we are obliged not only to learn a large number of new facts, but also to adopt new habits of learning ... [T]hese characteristics of the study of physics give it a value, as a means of training in habits of exact thinking which probably no other study possesses in the same degree.[108]

His Dublin colleague William Barrett could not agree more. The new measuring physics was particularly suitable for 'educating individual judgement by training the senses to habits of accurate observation and the mind to clear and precise modes of thought'.[109]

105 Graeme Gooday, 'Precision Measurement and the Genesis of Physics Teaching Laboratories in Victorian Britain', *British Journal for the History of Science*, 23 (1990), 25–51 (pp. 29–31).
106 Gooday, 'Precision Measurement', pp. 32–6.
107 Gooday, 'Precision Measurement', pp. 36–43.
108 G.C. Foster (1875), quoted in: Gooday, 'Precision Measurement', p. 43.
109 W. Barrett (1875), quoted in: Gooday, 'Precision Measurement', p. 43. For similar instances in the Netherlands, see: Van Lunteren, 'Astronomers and the Making of Modern Physics'.

Maxwell's Cambridge successor Lord Rayleigh immediately expanded and modernized the laboratory teaching. Among these changes was the introduction of an elementary practical course, consisting of a series of prescribed experiments, aimed at undergraduate students. Two so-called demonstrators, Richard Glazebrook and William Shaw, were hired to supervise the student work in the laboratory. Glazebrook and Shaw's book on *Practical Physics*, first published in 1884, became the standard practical manual in Britain. Meanwhile, the research in the Cavendish laboratory continued its former focus on electrical standards.[110] In 1900 Glazebrook would be appointed the first director of the National Physics Laboratory, the British equivalent of the Berlin *Physikalisch-Technische Reichsanstalt*. As the Prince of Wales explained at its opening in 1902, the aim of the new institute, much like that of its German model, was 'to bring scientific knowledge to bear practically upon our everyday industrial and commercial life'.[111]

Discussion

We have seen how university laboratories profited from and contributed to the gradual industrialization of Europe and how laboratory practices and values matched industrial needs and interests. We have also seen how these practices and values supported the emergence of disciplines and the scientist's quest for social legitimacy and social authority. We can even broaden the perspective somewhat further by connecting the broad changes in scientific practices and values to the nineteenth-century rise of the modern state.

Max Weber characterized the rise of the modern state as the introduction of bureaucratic arrangements such as specialization and a division of labour, hierarchical layers of authority, selection based on technical skills and competences acquired through training and experience, and, above all, a strong emphasis on impersonal rules and procedures. These changes were predicated on the rise of a new set of values and new type of personality, the *Berufsmensch*.[112] Late nineteenth-century laboratory life largely fits this description, notwithstanding the principles of academic freedom.

110 Longair, *Maxwell's Enduring Legacy*, pp. 82–91.

111 Quoted in: R. Moseley, 'The Origins and Early years of the National Physical Laboratory: A Chapter in the Pre-History of British Science Policy', *Minerva*, 16 (1978), 222–50 (p. 246).

112 As outlined in: Max Weber, *The Protestant Ethic and the Spirit of Capitalism* (London: Routledge, 1992).

From this perspective, modern laboratory values can, to a certain extent, be seen as bureaucratic values. As many of our protagonists argued, the essence of the new laboratory science was not to be found in its results but rather in its impersonal methods.[113] Science promoted clear, critical, and objective thinking, and therefore every educated member of a modern, liberal society could profit from scientific training. Regarding the rise of the modern bureaucratic state, it is, however, hard to say whether science was on the receiving end or on the giving side. The correct answer would probably be: both.

Seen in this way, it seems unlikely that the laboratory was the only or even the main vehicle for the mid-century change in academic values and associated virtues. Indeed, Ted Porter has analysed the role of quantification in the social sciences in the context of bureaucratic needs and public trust.[114] In the humanities we see a similar increasing emphasis on naked facts and impersonal procedures. Yet, there can be little doubt that in the natural sciences laboratory training and laboratory practices played a dominant role in ushering in the new scientific ethos, characterized by a stress on empiricism and such virtues as diligence, perseverance, meticulousness, and self-restraint. Moreover, as we have seen, these practices also determined the subtle differences in emphasis between the different sciences. Thus, physiologists tended to be specifically concerned about sensory information and objectivity, whereas physicists emphasized precision above all.

To prevent other potential misunderstandings, it must be emphasized that not all chemists, physiologists, and physicists conformed to the picture sketched above. Luminaries such as Liebig, Hofmann, and Helmholtz allowed and even encouraged more explorative kinds of research in their laboratories. There, the aim to make new discoveries was not frowned upon. In Hofmann's case this is highly understandable, given industry's constant need for new substances. In general, chemists seem to have adopted a less pronounced profile than physicists and physiologists. As the latter groups lacked the practical justification which chemists could offer, they appear to have developed a much stronger disciplinary identity and therefore to have emphasized pedagogical and disciplinary values to a much larger degree. Unlike chemists, physics students found their future employment in education rather than industry, and pedagogy hinges more on values

113 Paul Forman regards the 'primacy of procedure' (or a 'commitment to "methodism"') as a hallmark of modernity. See: Forman, 'The Primacy of Science in Modernity', p. 3.

114 Theodore M. Porter, *Trust in Numbers: The Pursuit of Objectivity in Science and Public Life* (Princeton, NJ: Princeton University Press, 1995).

than industrial productivity. In this regard, it is also telling that most late nineteenth-century physics students in Germany came from higher echelons of society and that the majority of physicists had attended a classical Gymnasium, whereas most chemists had attended a *Realschule* and shared a middle-class background.[115]

It is also important to emphasize that, already during the last decade of the nineteenth century or the *fin du siècle*, there were clear indications of a partial retreat from the restrictive ideologies of the previous decades. Under J.J. Thomson and Rutherford, precision measurement in the Cavendish gave way to chasing elusive and ill-understood radiation phenomena and to constructing quaint atomic models. In 1896 Thomson's Irish colleague George Francis FitzGerald dismissed his German colleague Wilhelm Ostwald's disenchanted view of science as 'worthy of a German who plods by habit and instinct. A Briton wants emotion.'[116] At the same time, many Germans, including Ostwald, turned against atoms and mechanical explanations, and some even advocated a revival of vitalism.[117] And even if this neo-romantic phase did not prove long-lasting, it does highlight the contingency and time-bound nature of strands of virtue discourse in science.

Finally, an interesting take from this exploration is the important role of newly developed instruments and related methods in the standardization of laboratory training for budding scientists. Where Liebig's combustion apparatus, the *Kaliapparat*, and similar instruments for the quantitative analysis of organic compounds played an important role in the formation of academic chemistry as an autonomous discipline,[118] something similar can be said of the achromatic microscope in large parts of medicine, such as pathology, histology, embryology; of the kymograph and its derivatives in physiology; and of the magnetometer and its offspring in physics. These examples recall Foucault's point about the important role of instruments in the process of disciplining.[119] Moreover, all these instruments practically embodied particular values, such as accuracy and reliability (the

115 Lewis Pyenson and Douglas Skopp, 'Educating Physicists in Germany circa 1900', *Social Studies of Science*, 7 (1977), 329–66 (pp. 338–43).

116 G.F. FitzGerald, 'Ostwald's Energetics', *Nature*, 53 (1896), pp. 441–2

117 For Ostwald's anti-materialism, see: R.J. Deltete, 'Wilhelm Ostwald's Energetics', *Foundations of Chemistry*, 9 (2007), 3–56, 256–313, and 10 (2008), 187–221. For German holism and vitalism see: Anne Harrington, *Reenchanted Science: Holism in German Culture from Wilhelm II to Hitler* (Princeton, NJ: Princeton University Press, 1999). For similar anti-modern tendencies in early twentieth-century science, see: Frans H. van Lunteren and Marijn J. Hollestelle, 'Paul Ehrenfest and the Dilemmas of Modernity', *Isis*, 104 (2013), 504–36.

118 See Chapter 2 in this volume.

119 I derived this point from Gooday, '"Nature" in the Laboratory', p. 337.

Kaliapparat), unadulterated seeing (the microscope), mechanical objectivity (the kymograph), and utmost precision (the magnetometer).[120] It seems likely, therefore, that these instrumental practices strongly informed the emergence of disciplinary identities.

About the Author

Frans van Lunteren is a historian of science at the Vrije Universiteit Amsterdam and Leiden University. His research interests include discipline formation, international collaboration, the laboratory, physics and related disciplines, and the changing concept of the 'laws of nature'. A recent publication on the nineteenth-century shifts in values and practices in experimental physics is 'Astronomers and the Making of Modern Physics', in A. Maas and H. Schatz, eds., *Physics as a Calling: Science for Society* (Leiden: Leiden University Press 2013), pp. 15–45.

120 For the accuracy and reliability of the *Kaliapparat*, see: Melvyn Usselman, Christina Reinhart, Kelly Foulser, and Alan J. Rocke, 'Restaging Liebig: A Study in the Replication of Experiments', *Annals of Science*, 62 (2005), 1–55 (p. 44). For a critical assessment of the role of the *Kaliapparat*, see: Ernst Homburg, 'Two Factions, One Profession: The Chemical Profession in German Society, 1780–1870', in David Knight and Helge Kragh, eds., *The Making of the Chemist: The Social History of Chemistry in Europe, 1789–1914* (Cambridge: Cambridge University Press, 1998), pp. 39–76 (pp. 54–5, 68–9).

Part II

Laboratory Networks

4 Chemistry in Zürich, 1833–1930: Developing the Teaching-Research Laboratory in the Swiss Context

Peter J. Ramberg

Abstract
This essay outlines the emergence of the chemical teaching-research laboratory at the University of Zürich, the Zürich Cantonal School, and the Swiss Federal Polytechnic during the nineteenth century. All three institutions were modelled after their German counterparts, but there were important differences that are reflective of the Swiss national and local cantonal contexts and involved a complex set of dual appointments and shared facilities that were absent at comparable chemical laboratories at German universities. This essay outlines the origins of these complex relationships and shows how this context, including the roles of cantonal and federal support, and the physical constraints created by shared laboratory facilities shaped chemical research and instruction in Zürich.

Keywords: Zürich, university, polytechnic, chemistry, Switzerland

Introduction

Although chemical laboratories certainly existed well before the nineteenth century, part of the Laboratory Revolution entailed a restructuring of the location and purpose of existing chemical laboratories after about 1840. The chemical laboratory moved from privately held spaces to universities, and group research replaced solitary work. Laboratories also became much larger spaces, used as 'factories' for the mass production of chemicals and chemists to meet the growing needs of industry. The template for this new kind of laboratory was created largely by Justus Liebig in Giessen, but as

Berkel, Klaas van, and Ernst Homburg (eds), *The Laboratory Revolution and the Creation of the Modern University, 1830-1940*. Taylor & Francis Group, 2023
DOI: 10.5117/9789463720434_CH04

Alan Rocke has outlined elsewhere in this volume, the creation of these new spaces for chemistry was by no means inevitable, and Liebig's success was the result of several contingent factors that happened to coincide during his early tenure in Giessen.[1]

Each new laboratory is the result of specific institutional, political, and personal factors, and if we are to get handle on the broader concept of a 'Laboratory Revolution', we need to understand these fine-grained details of the emergence of specific laboratories before we can step back to look at the broader causes and context of the revolution. In the case of chemistry, we have many good studies of the attempts—some successful, some not—to create research laboratories in France, Germany, the United States, and Great Britain, but fewer from elsewhere.[2] My own interest has been the development of chemistry in Zürich, which has seen little historical analysis despite its importance as a site of chemistry since the late nineteenth century. The emergence of academic chemistry in Zürich resembled the path followed in Germany but was also affected uniquely by the Swiss Federation and the local needs and finances of the Canton of Zürich. The result was a complex set of dual appointments and shared facilities that were absent at comparable chemical laboratories at German universities.[3]

1 See also: Alan J. Rocke, 'Organic Analysis in Comparative Perspective', in Frederic L. Holmes and Trevor H. Levere, eds., *Instruments and Experimentation in the History of Chemistry* (Cambridge, MA: MIT Press, 2000), pp. 273–310; Alan J. Rocke, 'Origins and Spread of the "Giessen Model" in University Science', *Ambix*, 50 (2003), 90–115 [also in this volume, Ch. 2].

2 Owen Hannaway, 'The German Model of Chemical Education in America: Ira Remsen at Johns Hopkins (1876–1913)', *Ambix*, 23 (1976), 145–64; Catherine M. Jackson, 'Re-Examining the Research School: August Wilhelm Hofmann and the Re-Creation of a Liebig Research School in London', *History of Science*, 44 (2006), 281–320; Yoshiyuki Kikuchi, *Anglo-American Connections in Japanese Chemistry: The Lab as Contact Zone* (New York: Palgrave, 2013); Gerrylynn K. Roberts, 'The Establishment of the Royal College of Chemistry: An Investigation of the Social Context of Early-Victorian Chemistry', *Historical Studies in the Physical Sciences*, 7 (1976), 437–85; Alan J. Rocke, *The Quiet Revolution: Hermann Kolbe and the Science of Organic Chemistry* (Berkeley, CA: University of California Press, 1993); Alan J. Rocke, *Nationalizing Science: Adolphe Wurtz and the Battle for French Chemistry* (Cambridge, MA: MIT Press, 2000); Margaret W. Rossiter, *The Emergence of Agricultural Science: Justus Liebig and the Americans, 1840–1880* (New Haven, CT: Yale University Press, 1975); John W. Servos, *Physical Chemistry from Ostwald to Pauling: The Making of a Science in America* (Princeton, NJ: Princeton University Press, 1990).

3 Peter J. Ramberg, 'Chemical Research and Instruction in Zürich, 1833–1872', *Annals of Science*, 72 (2015), 170–86. See also: George B. Kauffman, *Alfred Werner: Founder of Coordination Chemistry* (Berlin: Springer, 1966); Conrad Hans Eugster, '150 Jahre Chemie an der Universität Zürich', *Chimia*, 37 (1983), 1–44. An English translation of the latter also appeared as: Conrad Hans Eugster, '150 Years of Chemistry at the University of Zürich', *Chimia*, 62 (2008), 75–103.

Here I will recount the early development of chemistry in Zürich as a product of this complex system and compare it to what we know about the emergence of the teaching-research laboratory in Germany. While Carl Löwig established academic chemistry and the research laboratory in Zürich, the crucial period in the growth of chemistry in Zürich was the 1860s when Johannes Wislicenus revived chemical research after a relatively unproductive period under Georg Städeler. Wislicenus also untangled the complex relationship between the various institutions in Zürich responsible for teaching chemistry. By the late nineteenth century, Zürich was certainly the most productive site for chemistry in Switzerland and one of the most productive in the world, largely due to the close proximity and interaction of the University and Polytechnic.

Educational Reform in the Canton of Zürich

In the early nineteenth century, the city of Zürich had a number of educational institutions that had developed independently over several centuries. The oldest was the theology school known as the *Carolinum*, founded by Ulrich Zwingli in 1525. It was followed in the seventeenth century by the *Collegium Humanitatis* to improve the preparation of theology students, a medical institute in the eighteenth century, a 'political institute' with the aim of teaching law and producing civil servants, and, finally, the *Kunstschule* (school of arts and crafts) that emphasized practical study of mathematics, French, history, geography, and drawing.[4] All these institutions were independent with little coordination between them to produce a coherent educational system within the Canton. Students had opportunities to pursue non-theological subjects, but only at a low level, and courses in the natural sciences were rare.[5]

A coordinated, canton-wide educational system was implemented during the 1830s when the liberals came to power in Zürich and crafted a new cantonal constitution that explicitly called for the education of its citizens at all levels as a duty of the state.[6] The new liberal government worked quickly to establish a central educational council (*Erziehungsrath*) that

4 Fritz Hunziker, *Die Mittelschulen in Zürich und Winterthur, 1833-1933* (Zürich: Verlag der Erziehungsdirektion, 1933), pp. 14, 17.

5 Hunzieker, *Die Mittelschulen*, p. 26.

6 Gordon A. Craig, *The Triumph of Liberalism: Zürich in the Golden Age, 1830–1869* (New York: Scribner's, 1988), p. 127.

would oversee the educational needs of the Canton. The new law made school attendance obligatory from ages six to sixteen, and in the city of Zürich, pupils would attend a Cantonal School (*Kantonsschule*), which was divided into two separate but related institutions: a humanistic, classically oriented Gymnasium for the preparation of university students and a School of Industry (*Industrieschule*), resembling the German *Realschulen* (modern secondary schools) and consisting of an upper and a lower level, that would emphasize science and mathematics and train students for technical and commercial professions.[7] The Gymnasium and the *Industrieschule* had different curricula but shared faculty for common topics, especially mathematics and science.

The highest goal of the Zürich liberals was the creation of a new university in the city of Zürich, the first Swiss university since the founding of the University of Basel in 1460.[8] Liberals regarded the university as a symbol of modernity, and the new university was a conscious attempt to create an intellectual revival in Zürich. To fund the university, the council dissolved the foundation supporting the *Carolinum* and used the funds to combine the existing theological, medical, and legal schools with a new philosophical faculty. Following the model developed by Prussia, the new university would draw properly trained students from the Gymnasia, and the university in turn would train teachers for the various cantonal schools.

Both the Cantonal School and the University of Zürich officially opened in April 1833. The Cantonal School was located in the former *Carolinum* building, the site of the original theology school (fig. 4.1). The philosophical faculty of the University, including the chemical laboratory, would also be located with the Cantonal School in the *Carolinum* on the east bank of the Limmat river, and the rest of the University was located on the west bank of the Limmat. The University had an initial enrolment of 139 students and 46 faculty, mostly recruited from Germany, the most famous of which was the biologist Lorenz Oken.

Despite the lofty ambitions of the liberal founders, the University had a precarious existence during its first fifteen years, plagued by low enrolments

7 Bruno Quadri and Max Bandle, *Biografie einer Schule: von der Industrieschule über die Oberrealschule zum Mathematisch-Naturwissenschaftlichen Gymnasium: ein Kapitel Zürcher Schulgeschichte (1832-1992)* (Zürich: Schulthess, 1992), pp. 9–13.

8 Ernst Gagliardi, Hans Nabholz, and Jean Strohl, *Die Universität Zürich 1833-1933 und ihre Vorläufer* (Zürich: Verlag der Erziehungsdirektion, 1938); Gerold Meyer von Knonau, 'Die Universität Zürich in den Jahren 1833-1913', in *Universität Zürich Festschrift: Des Regierungsrates zur Einweihung der Neubauten 18. April 1914* (Zürich: Orell Füssli, 1914), pp. 11–100; Craig, *The Triumph of Liberalism*.

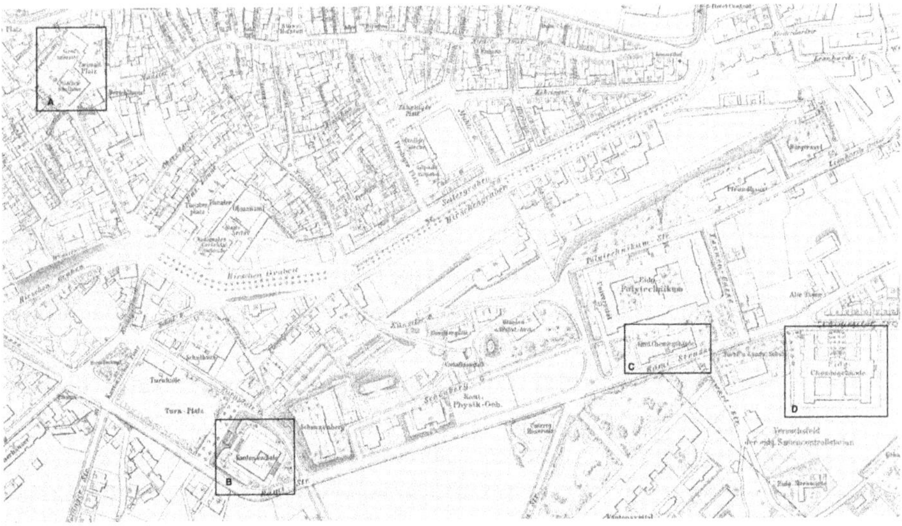

Figure 4.1 The locations of the different laboratories in nineteenth-century Zürich. A: The Carolinum, adjoining the Grossmünster. B: The 1842 Kantonsschule building. C: The 1861 Polytechnic Laboratory on Rämistrasse. D: The 1887 Polytechnic Laboratory. The Limmat river is just off the top of the map, and north is to the right. Scale is 1:3000. Source: *Plan der Stadt Zürich und Umgebung mit Angabe der Hausnummern* (Zürich: Hofer & Burger, 1887). ETH-Bibliothek Zürich, K 222050, https://doi.org/10.3931/e-rara-22990. Public Domain Mark.

and insufficient finances, and when the conservative opposition took power, they nearly closed it in 1839. When the liberals returned to power in 1846, on the eve of the *Sonderbund* war and the new Swiss constitution of 1848, the prestige of the University rose significantly in the hopes that it would become the national university of Switzerland. Beginning in 1846, the University received increased funding, but it was still not enough to make up for its initially low levels. The *Erziehungsrath* had initially decided in 1833 to emphasize the Cantonal School over the University. This came at the expense of the philosophical faculty, who consisted nearly entirely of extraordinary professors and *Privatdozenten* ('private lecturers'), whose paid appointments were at the Cantonal School but who also taught at the University. Faculty in the natural sciences were appointed to the *Industrieschule* within the Cantonal School. Oken was the only member of the philosophical faculty to hold a full professorship and receive a salary from the University.

In 1832 the *Erziehungsrath* called Carl Löwig (1803–1890) to teach at the *Industrieschule* and the University. Löwig was student of Leopold Gmelin in Heidelberg, where he had studied pharmacy and served as Gmelin's assistant. During this period Löwig determined that an orange-brown liquid he had obtained from brine in 1825 was the new element bromine.[9] During the next few years, Löwig studied the properties of bromine and its effect on organic compounds, especially alcohol. In 1829 he decided to practice pharmacy, but Gmelin convinced him to return to Heidelberg and complete his *Habilitation*, which he did in 1830.[10] Like many other new faculty in Zürich, Löwig accepted the paid position at the *Industrieschule* for sixteen weekly hours of instruction and received an unpaid appointment as extraordinary professor at the University, where he was expected to lecture five hours a week.[11] In 1836 he received an additional salary as an extraordinary professor and was eventually promoted to full professor.

As a sign of growth in chemistry instruction, Löwig was joined in the 1840s by Eduard Schweizer, a native Zürcher who had attended the *Industrieschule* on its opening in 1833 and was inspired by Löwig's lectures. Between 1836 and 1840 Schweizer served as Löwig's assistant and, in 1840, toured various prominent chemical laboratories and mining centres in Germany.[12] He returned to Zürich in 1841 and became *Privatdozent* in chemistry and mineralogy.[13] Schweizer became Löwig's assistant, directing the new laboratory of the relocated *Industrieschule* (see below) and lecturing in mineralogy at the University. In 1844 he began lecturing for Löwig at the *Industrieschule*, and in 1848 he took over completely as the instructor of chemistry, leaving Löwig responsible only for lecturing at the University (although Löwig formally retained both positions).[14] Schweizer published broadly in mineralogical, organic, and inorganic chemistry, but his most lasting contribution has since become known as 'Schweizer's reagent', a copper-ammonia solution that dissolves cellulose.[15] He also wrote two laboratory manuals and a textbook of inorganic chemistry for use at the *Industrieschule*.[16]

9 Hans Landolt, 'Carl Löwig', *Berichte der deutschen chemischen Gesellschaft*, 23 (1890), 905–09 (p. 906).

10 Armin Wankmüller, 'Löwig, Carl Jacob', in *Neue Deutsche Biographie*, 27 vols. (Berlin: Duncker, 1987), vol. 15, pp. 109–10.

11 Gagliardi, Nabholz, and Strohl, *Die Universität Zürich*, p. 295.

12 Anon., 'Mathias Eduard Schweizer', in *Programm der Kantonsschule in Zürich* (Zürich: Zürcher & Furrer, 1861), pp. 26–29; George B. Kauffman, 'Eduard Schweizer (1818–1860): The Unknown Chemist and His Well-Known Reagent', *Journal of Chemical Education*, 61 (1984), 1095–97.

13 Eduard Schweizer to Erziehungsrath, 26 June 1841, Erziehungswesen statement, 18 August 1841, Zürich, Staatsarchiv des Kantons Zürich (StAKZ), U110d.1 Einzelne Privatdozenten, Folder 9.

14 Anon., 'Schweizer', p. 27.

15 Kauffman, 'Schweizer'.

16 Eduard Schweizer, *Über den praktischen chemischen Unterricht an höheren technischen Lehranstalten* (Zürich: Orell and Füssli, 1851); Eduard Schweizer, *Praktische Anleitung zur Ausführung quantitativer chemischer Analysen*, 2nd ed. (Zürich: Orell and Füssli, 1853); Eduard

In 1843 Schweizer had also been hired as the first instructor of chemistry at the Cantonal veterinary school, established in 1820. Its existence was initially precarious, but in 1834, as part of the cantonal educational reforms, it was made a permanent institution (although separately administered from the Cantonal School and the University), and chemistry, zoology, and botany were added to the curriculum.[17] In 1843 it was decided that specialists should teach botany and chemistry, and Schweizer was hired to teach inorganic and organic chemistry.[18] The veterinary school was the least well-supported of the cantonal educational institutions in Zürich. The instructors in the ancillary sciences were always drawn from the *Industrieschule* and given temporary contracts, as the positions themselves were only provisional. By the mid-1850s, the existence of the school was again in doubt when Schweizer and the new director of the school, Rudolf Zangger, successfully convinced the cantonal educational council that the school should be supported more vigorously. They argued that it could be linked to the newly formed Polytechnic (more about this institution later) and that scientifically trained veterinarians were important for Swiss agriculture and the welfare of the Swiss cavalry.[19] As a result, the three faculty positions in chemistry, physics, and biology were made permanent and regularly funded, and chemistry instruction was expanded to three years in the curriculum.

In 1853 Löwig was called to the University of Breslau to succeed Robert Bunsen. To replace Löwig, the faculty called Georg Städeler, extraordinary professor of chemistry in Göttingen. Städeler had earned a doctorate under Friedrich Wöhler with a new synthesis of chloral in 1846 and became Wöhler's assistant.[20] His interests eventually turned to physiological chemistry, likely inspired by his friend and collaborator in the Wöhler laboratory Friedrich Theodor Frerichs (1819–1889), the future pathologist at the Charité hospital in Berlin. In 1850 and 1851, Städeler showed that urine contained carbolic acid and three previously unknown volatile acids.[21] On the basis of his work on physiological chemistry, he was appointed to the Göttingen faculty as

Schweizer, *Die unorganische Chemie. Ein Leitfaden für den ersten Unterricht* (Zürich: Schulthess, 1860).

17 Jörg Hohl, 'Die Entwicklung der Zürcher Tierarzneischule in den Jahren 1833 bis 1855' (Inaugural diss., University of Zürich, 1979), pp. 10–11; Peter Storck, *Die Anfänge der Tierarzneischule in Zürich* (Inaugural diss., University of Zürich, 1977).

18 Hohl, *Die Entwicklung der Zürcher Tierarzneischule*, pp. 38–39, 41.

19 Christian Senn, *Die Entwicklung der Zürcher Tierarzneischule in den Jahren 1856 bis 1882* (Inaugural diss., University of Zürich, 1981), pp. 6–7.

20 Georg Städeler, 'Über die Bildung des Chlorals aus Stärke und ein neues Zersetzungsprodukte desselben', *Annalen der Chemie*, 61 (1847), 101–21.

21 Georg Städeler, 'Über die flüchtigen Säuren des Harns', *Annalen der Chemie*, 77 (1851), 17–37.

extraordinary professor. Städeler had also written, but not yet published, a short introduction to analytical chemistry that would later become the long-lived *Leitfaden für die quantitative chemische Analyse unorganischer Körper.*[22]

The Zürich faculty was impressed by Städeler's lectures in a large variety of areas of chemistry, his experience as a laboratory supervisor, and his publications, which showed him to be a 'thoroughly educated, highly gifted, thoughtful and exact chemist'.[23] Städeler was hired in July of 1853 as ordinary professor of chemistry and director of the chemical laboratory. The transition between Löwig and Städeler is illustrative of the emerging relationship between University and *Industrieschule*. Schweizer now officially replaced Löwig as instructor at the *Industrieschule* and was also appointed extraordinary professor at the University. Städeler was appointed exclusively to the University faculty and had no teaching duties at the *Industrieschule*.[24]

The Swiss Federal Polytechnical Institute

The dream of Swiss liberals through the first half of the nineteenth century was the creation of a national university that would unite Switzerland. Indeed, the liberals who formed the University of Zürich had done so with the expectation that a national university would be created from an existing university. The new Swiss constitution of 1848 allowed for the creation of a national university and a polytechnical institute, and the attempt to establish the university was among the first actions of the new federal parliament. It proved to be a divisive issue, however, for linguistic, theological, and political reasons.[25] In February 1854 the university was soundly defeated, but parliament agreed to the formation of a polytechnical institute because training Swiss engineers in Switzerland was more practical than sending them to Paris, Karlsruhe, or Darmstadt and would meet the rapidly

22 Georg Städeler, *Leitfaden für die qualitative chemische Analyse unorganischer Körper*, 2nd ed. (Zürich: Orell and Füssli, 1857).

23 Philosophical Faculty to Erziehungswesen, 23 June 1853, StAKZ, U110b.1 Einzelnde Professoren, Folder 11.

24 Erziehungswesen to Regierungsrath, 30 March 1853, StAKZ, U110b.1, Folder 10.

25 An excellent overview of the deliberations leading to the formation of the Polytechnic is in: David Gugerli, Patrick Kupper, and Daniel Speich, *Transforming the Future: ETH Zürich and the Construction of Modern Switzerland, 1855–2005* (Zürich: Chronos Verlag, 2010), pp. 18–35. This is a translation of David Gugerli, Patrick Kupper, and Daniel Speich, *Die Zukunftmaschine: Konjunkturen der ETH Zürich* (Zürich: Chronos Verlag, 2005).

increasing need in Switzerland for industry and engineering projects such as dams and railroads.

The federal council in Bern commissioned a survey of existing polytechnical institutes in France, Germany, and Italy and appointed a commission to determine the best way to organize the new polytechnic in Zürich. The commission consisted of academics, including the Rector of the Zürich *Industrieschule* Joseph Wolfgang von Deschwanden and the Rector of the Aarau *Gewerbeschule* (school of crafts and industry), the chemist Pompejus Alexander Bolley, as well as politicians, including the industrialist Alfred Escher and Johann Kern. Under Escher's guidance, during the summer of 1854, the commission organized the new polytechnic after the Karlsruhe polytechnical school, which Escher had visited personally and knew well.[26] The new school would have five separate divisions (*Fachschulen*): architecture and construction, civil engineering, mechanical engineering, technical chemistry, and forestry. The commission also created a sixth division that contained the remnants of the liberal's dream of a national university—a small humanistic school consisting of modern languages, mathematics, natural science, general history and art history, Swiss law, and economics. Bolley succeeded in having pharmacy added to the technical chemistry division, arguing that it would increase enrolments enough to justify the needed laboratory space and equipment.

After the new Swiss federal parliament chose Bern as the capital of Switzerland, they would choose Zürich as the home of the new polytechnic. Local officials initially sniffed at the prospect of hosting a polytechnical institute instead of a university, but Alfred Escher convinced them of the advantages, noting that the institute would otherwise go to Basel, and it was a good compensation for losing the capital to Bern.[27] The new institute was designed by the authorities from the top down to serve a specific national purpose, and it was not organized like a university. The administrative head of the Polytechnic would be the director, who would be chosen from the faculty by the Swiss School Council (*Schweizerische Schulrat*, founded in 1854 specifically to supervise the Polytechnic), not by the faculty, for a two-year, renewable term. Each division also had a Division Head (*Abteilungsvorstand*) chosen by the School Council.[28] Faculty had relative *Lehrfreiheit* (freedom to teach topics of choice) but did not have the power to recommend or evaluate

26 Wilhelm Oechsli, *Geschichte der Gründung des eidgenossische Polytechnikum mit einer Übersicht seiner Entwicklung 1855-1905* (Frauenfeld: Huber, 1905), p. 75.

27 Gugerli, Kupper, and Speich, *Transforming the Future*, pp. 34–35.

28 Oechsli, *Die Geschichte der Gründung*, p. 157.

new faculty, who would be chosen by the School Council led by its president. Only the sixth division, containing the liberal arts, would be allowed to grant the *venia docendi* for the rank of *Privatdozent*. The young men attending the Polytechnic would not be students (*Studenten*) but pupils (*Schüler*) and would have no *Lernfreiheit* (freedom to follow courses of choice), as they were required to follow a specific curriculum and pass exams at each level. The administration had a variety of means for monitoring academic progress with regular obligatory tutorials and exams and for disciplining pupils for disobedience or delinquency.[29]

Faculty at the Polytechnic would be better paid and have better facilities than those at the university, but their positions, particularly those in engineering, did not carry the same cachet as a university appointment. Teaching at the Polytechnic, especially in the general philosophical division, nearly always involved a conflict between the ideals of scholarship and the academy, on one hand, and practical applications, on the other. In the first appointments, the School Council attempted to bridge this gap by making fifteen joint appointments between the Polytechnic and the university between 1855 and 1857.[30] Many of these joint appointments would be in the natural sciences, for example Albert Mousson in experimental physics, Oswald Heer in taxonomic botany, Arnold Escher von der Linth in geology, and Gustav Kenngott in mineralogy. The Polytechnic also created a position in mathematical physics filled by Rudolf Clausius, who was also appointed to the university. Hiring established professors from the university granted the Polytechnic some prestige and saved money for the cash-strapped university.

Escher and the first president of the School Council, Johann Kern, also had another motive for these joint appointments. Escher and Kern had been strong proponents of a national university, and they had hoped that by integrating the Polytechnic closely with the university, the two could in effect be merged into a single institution. Recruiting faculty from the university and offering them double appointments began to fulfil this wish. But when the second, extremely effective, and long-serving president of the School Council, Karl Kappeler, began his duties in 1859, he saw this vision in conflict with the purpose of the Polytechnic. He perceived that the professors with university appointments continued to present their lectures with the needs of the university in mind rather than the Polytechnic, and he thought that the university students attending the Polytechnic were a bad influence on the discipline of the Polytechnic students. He began to slowly break ties to the university

29 Gugerli, Kupper, and Speich, *Transforming the Future*, pp. 92–93.
30 Gugerli, Kupper, and Speich, *Transforming the Future*, p. 68.

by leaving joint professorships unfilled and by not allowing faculty to hold appointments elsewhere without permission of the School Council. In 1863 university students were no longer allowed to enrol in Polytechnic lectures.

For the chemistry division, the planning commission originally envisioned three chairs, one in technical chemistry, one in general and theoretical chemistry, and a position explicitly for a French-speaking chemist that went unfilled when they attempted to call Charles Gerhardt but could not meet his demands.[31] The planning commission advertised the faculty openings all over Europe, but the council was impressed by Städeler's success after two years as a lecturer and laboratory director in Zürich and chose to appoint him to the position. In accordance with Escher's and Kern's initial desire to merge the two institutions, Städeler would be paid by the Polytechnic but hold a joint appointment between the Polytechnic and the university, holding lectures and supervising the laboratory for both, and continue to supervise and grant doctoral degrees.

For the position in technical chemistry, the Polytechnic chose Bolley, an accomplished chemist at the Cantonal School in Aarau and a member of the initial committee establishing the composition of the faculty and the curriculum at the Polytechnic. Bolley studied mineralogy and metallurgy in Heidelberg, where he was an assistant to Gmelin and became interested in technical chemistry. In 1838 he moved to the Cantonal School in Aarau.[32] While at the Cantonal School, Bolley was an extremely productive chemist with an interest in theoretical chemistry (the chemistry of tin salts, zinc, sodium borate) and analytical chemistry, including methods for detecting ammonia and nitric acid and improving techniques for the purification of hydrochloric acid, but his specialty was the chemical composition of fibres, especially silk, and the chemistry of dyes. As Wislicenus later described it, Bolley was a master at applying the principles of theoretical chemistry to improve the techniques of applied chemistry.[33] Bolley was also a prolific editor and author, contributing extensively to the *Handwörterbuch der reinen und angewandten Chemie* and publishing the *Handbuch der technisch-chemischen Untersuchungen* (1853). By 1854 Bolley's excellent reputation in technical chemistry and his Swiss citizenship made him a natural choice as the first professor of technical chemistry at the Polytechnic.

31 Oechsli, *Die Geschichte der Gründung*, p. 185.

32 Emil Kopp and Johannes Scherr, 'Alexander Pompejus Bolley', *Berichte der deutschen chemischen Gesellschaft*, 3 (1870), 813–20; Johannes Wislicenus, *Gedächtnisrede auf P.A. Bolley* (Zürich: Schabelitz, 1871).

33 Wislicenus, *Gedächtnisrede*, pp. 10–11.

The Zürich Chemical Laboratories

The Carolinum Laboratory

Having reviewed the complicated origins of the four Zürich institutions that included chemistry in their curriculum—Cantonal School, University, Polytechnic, and Veterinary School—we can now turn more explicitly to the development of instructional chemical laboratories in Zürich. Until Löwig arrived in Zürich in 1833, there had been no tradition of chemical research in Zürich, and there were no formal instructional or research laboratories for chemists. Chemistry for both the *Industrieschule* and the university was located in the old *Carolinum* building (fig. 4.1), on the ground floor where the ovens (*Backöfen*) once were. There are no surviving plans to indicate the size and arrangement of rooms, but the space included an instructional laboratory, a lecture hall with auditorium style seating, and a 'depository' for apparatus and chemicals.[34] There does not appear to be any dedicated research space, and there was only one unqualified building manager for assistance in lectures and laboratories.[35] Löwig coped with the inadequate conditions in the *Carolinum* by choosing projects that required relatively little equipment: he analysed Swiss mineral waters and wrote a landmark text, *Chemie der organischen Verbindungen* (1839), the first comprehensive systematic presentation of carbon compounds that served 'as the "Beilstein" of the time, and was found in every chemists hand'.[36]

The New Cantonal School Laboratory on Rämistrasse

In 1842 Löwig moved into a new laboratory located in the new Cantonal School building on Rämistrasse, just northeast of the *Carolinum* (fig. 4.1). This new laboratory would serve the students of the *Industrieschule*, the University, and eventually the Veterinary School. A drawing of the main laboratory (fig. 4.2) from the early 1850s, done in the same style as the more famous drawing of Liebig's Giessen laboratory from the 1840s, shows that it consisted of a central fume hood surrounded by workbenches with a capacity of about twenty students.[37] The exact size of this laboratory is unknown, but the drawing of the main laboratory space itself suggests that it might have been about 100 square meters. Floor plans have not survived, but in

34 Gagliardi, Nabholz, and Strohl, *Die Universität Zürich*, p. 295; Antrag der Bau Dept. betreffend die Einrichtung eines chemischen Laboratoriums im Chorherrngebäude, 10 July 1833, StAZH, U118.1.

35 Gagliardi, Nabholz, and Strohl, *Die Universität Zürich*, p. 295.

36 Landolt, 'Carl Löwig', pp. 906–07.

37 StAKZ U118.1, Chemisches Laboratorium.

Kantonsschule laboratory, c.1850. Löwig is on the left, with the cigarette, and Schweizer is near the fume hood, wearing the cap. Courtesy: Archives, Department of Chemistry, University of Zürich.

addition to this laboratory space, there was almost certainly additional space for a lecture hall, offices, and storage.

This new laboratory significantly improved Löwig's material conditions, allowing him to resume research in organic chemistry, focusing on organic compounds that contained the metals arsenic, lead, antimony, and bismuth. At first, he began research on his own, but by the late 1840s he had a group of at least eight students working on different organometallic compounds.[38] His long-time assistant Eduard Schweizer, for example, prepared ethyl antimony, and the young future chemist Hans Landolt (a native of Zürich) prepared tetramethyl antimony and ethyl arsenic. As Löwig himself noted, his work in organometallic chemistry actually predated the far better known research by Edward Frankland, who is generally credited with the first preparation of organometallic compounds in 1849.[39]

Löwig's inspiration for assigning small projects to his students as part of a larger research programme, a hallmark of Liebig's model, is not clear

38 Carl Löwig, 'Zur Geschichte der organischen Metallverbindungen', *Journal für praktische Chemie*, 60 (1853), 348–52; Carl Löwig, 'Zur Geschichte der organischen Metallverbindungen', *Journal für praktische Chemie*, 65 (1855), 55–62.

39 Colin A. Russell, *Edward Frankland: Chemistry, Controversy and Conspiracy in Victorian England* (Cambridge: Cambridge University Press, 1996), p. 106.

from the surviving records. Löwig was exactly the same age as Liebig and not his student, but by the mid-1840s, just as Löwig had obtained the means for creating a more robust research programme, he must have been well aware of Liebig's successful research model. Whatever the inspiration, Löwig certainly created the first teaching-research laboratory in Zürich and his laboratory was one of the earliest versions of Liebig's teaching-research laboratory outside of Germany, along with August Hofmann's laboratory at the newly established Royal College of Chemistry in London and Gerrit Jan Mulder's laboratory in Utrecht.[40]

Research productivity at the university laboratory under Städeler is more difficult to ascertain, as fewer surviving records exist. Städeler did continue his own productive research programme on physiological chemistry, but the known dissertations produced under Städeler's direction do not indicate that his students' research followed a coherent research programme.[41] The teaching-research laboratory had a good start in Zürich with Löwig, but it entered a relative lull during Städeler's tenure.

The Polytechnic Laboratory on Ramistrasse

When the Polytechnic opened in 1855, Bolley's technical chemistry laboratory was located in the *Carolinum*, but Städeler's analytical chemistry laboratory was located in the Cantonal School laboratory, creating additional strains on an already cramped and inadequate space. At the Cantonal School laboratory, the primary demand for laboratory space came from the University and Polytechnic, and Schweizer and his *Industrieschule* and veterinary pupils were relegated to the dark and humid 'lower laboratory' in the cellar of the Cantonal School building.[42]

The overcrowding of the laboratories in the Cantonal School building would be relieved in 1861 when the Polytechnic completed construction of its

40 On Hofmann's laboratory, see: Jackson, 'Re-Examining the Research School'; Roberts, 'The Establishment of the Royal College of Chemistry'. For Mulder, see: H.A.M. Snelders, 'Het chemisch laboratorium "De Leeuwenbergh" te Utrecht', *Tijdschrift voor de geschiedenis der geneeskunde, natuurwetenschappen, wiskunde en techniek*, 7 (1984), 129–40. I thank Ernst Homburg for bringing Mulder's laboratory to my attention.

41 Paul Liechti, *Beiträge zur Kenntniss der aromatischen Säuren* (Zürich: Zürcher and Furrer, 1869); Victor Merz, *Untersuchungen über das Titan, Silicium und Boron* (Inaugural diss., University of Zürich, 1864); Jakob Neukomm, *Ueber das Vorkommen van Leucin, Tyrosin und andere Umsatzstoffe im menschlichen Körper bei Krankheiten* (PhD diss., University of Zürich, 1859); Arnold Rossel, *Beiträge zur Kenntniss des Anisaldehyds* (Zürich: Zürcher and Furrer, 1868); Emil Ruge, *Beiträge zur Kenntniss des Wismuthverbindungen* (Inaugural diss., University of Zürich, 1860).

42 Anon., 'Schweizer', pp. 27–28.

Figure 4.3 The exterior of the Rämistrasse laboratory. The people standing at the far end of the building give an idea of the scale. The main Polytechnic building is immediately behind the laboratory. Courtesy: ETH Bildarchiv.

own chemistry laboratory as the first phase of its new building programme.[43] The new chemistry laboratory was the Polytechnic's first building, funded jointly by the Canton of Zürich and the Swiss Federation. It was completed for the summer semester of 1861 on Rämistrasse just north of the *Kantonsschule* building (fig. 4.1). Designed under the supervision of Bolley and Städeler, it consisted of two wings laid along a north-south axis along Rämistrasse, with Städeler's analytical portion occupying the northern half and Bolley's technical portion on the southern half. Each wing contained a lecture hall and laboratories for research and instruction. There were 30 workplaces for technical chemistry and 36 places for analytical, which was almost immediately expanded to 40.[44] The total floor space of the ground floor was about 1,474 square meters, with additional floor space available in the basement (fig. 4.3).[45] The chemical laboratory would be followed by an observatory in 1863 and by the main building for the Polytechnic in 1864. This latter was designed by the professor of architecture Gottfried Semper and was located immediately to the west of the chemistry building

43 Bericht des eidgenössischen Polytechnikum über das Jahr 1872, p. 16. ETH Bibliothek, Archive.
44 Bericht des eidgenössischen Polytechnikum über das Jahr 1872, p. 8. ETH Bibliothek, Archive.
45 StAKZ, 103a.1, Universität-Polytechnikum Beziehungen.

overlooking Zürich in the position that would come to dominate the city's appearance from the west. The Semper building was financed by the Canton and would be shared by the university and the Polytechnic until 1894 when the university completed construction of a separate building on Rämistrasse immediately south of the Semper building.

Johannes Wislicenus in Zürich, 1860–1872

Group research would revive during the 1860s when Wislicenus arrived and began climbing the academic ladder. Wislicenus had come to Zürich from Halle in late 1859, and in February 1860 he became *Privatdozent* at both the University and the Polytechnic and began to accumulate various positions in the next two years. He began teaching in the summer of 1860, but in the following fall, Schweizer became gravely ill and died unexpectedly. Wislicenus was the only suitable candidate in town and agreed to take over Schweizer's courses at the *Industrieschule*.[46] By the middle of November, Wislicenus also began co-teaching chemistry at the veterinary school, and in February 1861 he was appointed as the instructor at the veterinary school.[47] In January of 1861 Städeler had moved to the new Polytechnic laboratory building, and Wislicenus was appointed director of the chemical collection for the *Industrieschule* and University. In April 1862 he was formally appointed as the full-time instructor of chemistry at the *Industrieschule*.[48]

In 1864, as part of Kappeler's move to separate instruction at the Polytechnic from the university, Städeler's position was made exclusive to the Polytechnic, and his courses would no longer be available to university students. Wislicenus was appointed as Städeler's successor at the university, but as extraordinary professor. He was granted a release from teaching at the veterinary school but remained instructor at the *Industrieschule*. Although officially an extraordinary professor, Wislicenus had become the *de facto* ordinary professor of chemistry, responsible for both lectures and the laboratory at the university.

46 Gustav Zschetzche to Erziehungswesen, 17 October 1860, StAKZ, U87.3, Industrieschule Lehrerschaft.

47 Erziehungswesen statement, 13 February 1861, StAKZ, U112.b.1 Tierarzneischule. Lehrerschaft im Allgemein.

48 From the surviving documentation in the Zürich Cantonal Archive, the appointment as director of the chemical collection was done separately from the appointment as instructor. It is unclear why this was a separate appointment or whether the curator of the chemical collection received a salary in addition to the instructor's salary.

Wislicenus requested and was granted an assistant and an intern, but the winter semester of 1864–65 proved difficult; when he accepted the professorship without giving up the teaching position at the *Industrieschule*, Wislicenus had essentially recombined the positions that Städeler and Schweizer had divided in 1853. The next few years remained arduous for Wislicenus, although he managed have his assistant Armin Baltzer appointed to teach mineralogy and then general chemistry at the *Industrieschule*. He was still responsible for teaching advanced chemistry until a formal promotion could occur. In December 1867 he was finally promoted to full professor and fully released from his position at the *Industrieschule*, to be replaced by Baltzer.[49]

Throughout the 1860s there was also a significant increase in attendance in both lecture and laboratory. Städeler had averaged about 25 students each semester, but during Wislicenus' tenure, enrolment steadily increased, peaking in the summer of 1870 with 67 students in organic chemistry.[50] Attendance in the laboratory peaked at 58 students during 1867–68, dropped for two years, and increased significantly in the summer of 1869 when Wislicenus introduced a 'half-laboratory' course (*Halbpraktikum*). This course was designed for medical students and required only half the time in the laboratory as chemistry students to relieve the overcrowding in the laboratory.[51]

In addition to the increased general enrolment, there was an increase in the number of advanced students in Wislicenus' laboratory. In his annual report for 1865, Wislicenus noted that aside from Baltzer and his intern, there were 'five men occupied with independent work' in the laboratory.[52] In the summer of 1866 he was also joined by two *Privatdozenten*, Victor Merz (1839–1904) and Wilhelm Weith (1846–1881), both graduates of the Polytechnic. Merz had studied chemistry at the *Industrieschule* and was inspired by Städeler's lectures at the Polytechnic. Weith had also been a student at the Polytechnic and worked with Wislicenus at the *Industrieschule*. He obtained his doctorate in Heidelberg in early 1865 and returned to Zürich, earning his Habilitation in May of 1866.[53] Merz and Weith became close friends and collaborators, and between 1868 and 1881 the two were inseparable in

49 Regierungsrath to Erziehungswesen, 14 December 1867, StAKZ, U110.b.1, Einzelnen Professoren, Folder 17.

50 StAKZ, U99.1 Allgemeines Vorlesungsverzeichnisse. Bd. 1, 1838-1910, Bd II SS1835-WS1863/64.

51 Wislicenus, Vorschlag, 12 January 1869, StAKZ, U118.1 Chemisches Laboratorium.

52 Wislicenus to Erziehungswesen, 31 January 1866, StAKZ U118.1, Chemisches Laboratorium.

53 Kenngott to Erziehungswesen, 19 July 1866, and Erziehungswesen Statement, 7 August 1866, StAKZ, U110d.1, Folder 32.

the laboratory, co-authoring 31 papers and adding a significant number of courses to the University and the Polytechnic.

During the 1860s Wislicenus gradually began to eclipse Städeler in research productivity and the size of his group. In 1860, under Städeler, there was one *Privatdozent* (Wislicenus), no assistants for the university laboratory, and two assistants for the Polytechnic laboratory. This meant that there was a total of four advanced workers in the laboratory in 1861, not including *Doktoranden* ('PhD students'). Städeler did not average more than one *Doktorand* per year, so the total number of advanced members of the laboratory was probably no more than five or six.

In 1865, when Wislicenus reported five chemists in addition to himself and two assistants, he had already surpassed Städeler in the number of advanced students. In 1866, when Merz and Weith joined the laboratory, the number had likely increased to seven, including Wislicenus but not *Doktoranden*. If we add *Doktoranden* and other known members of the laboratory—nine students named as co-authors in papers—there is a total of nineteen known advanced students in the university laboratory between approximately 1862 and 1872. This averages about two advanced students per year over ten years, making an annual average of nine advanced members of the laboratory, including Wislicenus. It is likely there were other students who stayed at the laboratory and left without publication or completing a dissertation, making the average number nine or ten members at any one time, or approximately double the number working under Städeler. Wislicenus' students authored eighteen papers in the *Annalen der Chemie* in some form, and Wislicenus' total of 29 published papers was nearly twice Städeler's output. Merz and Weith were also extremely productive, producing five solo papers each, and fourteen co-written papers between 1868 and 1872.

All this increased productivity occurred despite the constraints of the laboratory in the Cantonal School building that had been built for only twenty students and still served students at three different institutions. Wislicenus continuously lobbied the educational council for additional assistants and funds for necessary apparatus and for remodelling and enlarging the laboratory space. As we saw above, Wislicenus was successful in acquiring funds for more assistants and dividing up the teaching duties with Baltzer, Merz, and Weith, but improving the material support of the laboratory for three different institutions proved to be more difficult. He did get funding for some minor renovations, but the central problem of having the same laboratory space serve three different institutions remained, especially in light of increasing enrolment. In 1866 Wislicenus wrote to the authorities that 'This rush of new students to the university laboratory exceeds the boldest expectations that I harboured when I assumed my current position,' and the current laboratory budget was far from sufficient to equip the laboratory with equipment and chemicals.[54] Until 1869 the laboratory ran continuous deficits.

Wislicenus' struggle with the laboratory budget during the 1860s was primarily a continuation of the traditionally low budgets endured by Löwig and Städeler that were reflective of the relatively poorly funded university itself. Wislicenus noted several times to the authorities that the facilities and funding of the laboratory were inadequate, and it was a struggle to catch up with even smaller German universities.[55]

The need for more space and funding finally reached a head in October of 1868 when Wislicenus began to see the University as the least important of the three institutions sharing the laboratory space. He urgently requested either an expansion of the Cantonal School building or at least moving the *Industrieschule* and veterinary school laboratories to a separate space in the same building.[56] This request went unanswered, but in December of 1868 Wislicenus finally gained some much-needed leverage when he received a call to replace Christoph Schönbein at the University of Basel, and after a month of negotiations, he decided to remain in Zürich, with a promise from the *Erziehungswesen* to completely separate, financially and physically, the Cantonal School and university laboratories. The agreement allowed Wislicenus to equip the university laboratory with much-needed new instruments, apparatus, and chemicals. It also helped him remodel the existing laboratory to accommodate more students (principally by removing the central fume hood; see fig. 4.2) and create separate spaces in the Cantonal School building for the university laboratory and the *Industrieschule* and veterinary school laboratories. He soon thereafter also managed to have Merz appointed as extraordinary professor, finally restoring the arrangement in 1860 when Städeler and Schweizer were ordinary and extraordinary professors.[57]

In spring of 1871 Wislicenus would succeed Städeler at the Polytechnic. Städeler had been forced to retire early in the fall of 1870 after nearly a decade of poor health, and the Swiss School Council President, Karl Kappeler, offered Wislicenus the chair of theoretical chemistry, which Wislicenus accepted

54 Wislicenus to Erziehungswesen, 21 April 1866, StAKZ U118.1, Chemisches Laboratorium.

55 'All recently formed institutes at their opening were completely outfitted, while our institute in 1864 can only be furnished with the absolutely necessary [items].' Wislicenus to Erziehungswesen, 31 January 1866, StAKZ U118.1, Chemisches Laboratorium.

56 Wislicenus to Erziehungswesen, 16 October 1868, StAKZ, U118.1, Chemisches Laboratorium.

57 Wislicenus to Erziehungswesen, 15 May 1869, and 29 October 1869, StAKZ, U110b.1, Folder 18.

after some deliberation.[58] In spring of 1871 Wislicenus made the move down Rämistrasse from the *Kantonsschule* building to the Polytechnic laboratory but retained a courtesy position of professor at the University with the ability to train university students. In some respects, the move to the Polytechnic made little difference in Wislicenus' teaching, as he followed the same teaching schedule and would not need to repeat the struggle to reduce his teaching load. Städeler already had funds for two laboratory assistants with better compensation than the sum offered by the University. Funding for the laboratory, however, had not kept up with the increased enrolment. Even in 1861, the new laboratory was too small, with 36 workspaces, and Städeler ran a deficit for at least six of the years between 1859 and 1869 to meet the demands of increasing enrolment.[59]

Wislicenus also found the laboratory of the Polytechnic deficient in several other respects. Städeler's private workroom was extremely small, there was an insufficient number of gas lines for the number of students, and the workplaces were not consistently stocked with equipment. The most urgent problem, he noted, was the severely disorganized state of the chemical collection. All would need to be properly stored in stoppered jars, a task that had taken him eight years to complete at the University.[60] In 1871 laboratory enrolment ballooned again to over 60 students, and Wislicenus scrambled to outfit twenty more workplaces in the cellar.[61] In July of 1871 Wislicenus requested an increase in the state subsidy for the laboratory, noting that the Polytechnic laboratory had long been insufficiently funded and that the demands on the laboratory would only increase with a pending requirement that both chemistry pupils and the pupils in the science education division have an additional semester of chemistry with an increase of required weekly hours, causing a surge in *Praktikanten* even without a general increase in attendance.[62] His request was honoured. Wislicenus' laboratory budget was nearly doubled and laboratory fees increased, and in 1872 the Swiss educational council reported that under the 'excellent' (*treffliche*) direction of Wislicenus, 'the outfitting of the laboratory has greatly improved', with new gas lines and a newly organized, bottled, and labelled chemical collection, with over 500 new samples.[63] Unfortunately

58 Karl Kraut, 'Georg Städeler', *Berichte der deutschen chemischen Gesellschaft*, 4 (1871), 425–28.

59 Städeler-Schulrat correspondence, 1860–66, ETH Bibliothek, Archive, SR1.

60 Wislicenus to Kappeler, 6 February 1871, ETH Bibliothek, Archive, SR3:1870.53a; Kappeler to Wislicenus, 29 March 1871, ETH Bibliothek, Archive, SR1:1871.26.

61 Wislicenus to Kappeler, 15 March 1872, ETH Bibliothek, Archive, SR3:1872.76.

62 Wislicenus to Kappeler, 21 July 1871, ETH Bibliothek, Archive, SR3:1871.275.

63 Bericht des eidgenössischen Polytechnikum über das Jahr 1872, p. 11. ETH Bibliothek, Archive.

for the Polytechnic, at the time of this report, Wislicenus had already left Zürich for Würzburg in his native Germany.

Comparisons and Conclusions

The institutional structure of chemical research and instruction in Zürich bears some similarity to German institutions. The Cantonal School and University were explicitly founded on the German model, with the University emphasizing academic freedom and incorporating a similar faculty structure. The Polytechnic was modelled after Karlsruhe with its divisional structure and more restricted academic freedom for both faculty and students. The federally funded Polytechnic did attempt to accommodate French-speaking Switzerland, but like the University, it remained within the sphere of German-speaking universities, with many faculty recruited from Germany itself.

In their correspondence with cantonal authorities, both Löwig and Wislicenus explicitly saw German universities as their point of comparison and competition. Wislicenus specifically wanted to make Zürich's laboratory one of the best among the 'German' universities, and he repeatedly argued that comparable German laboratories were better funded and equipped.[64] Like the original chemical laboratories in Marburg, Giessen, Göttingen, and Heidelberg, the first Zürich chemical laboratory in the *Carolinum* was also a repurposed existing building. The growth in enrolment in the Zürich laboratories, the construction of the Polytechnic laboratory in 1861, and the renovations of the university laboratory in 1869 all closely parallel the rapid growth of chemical institutes in Germany during the 1860s.[65]

The University chair in chemistry was initially placed in the philosophical rather than the medical faculty. In this respect it resembles some German universities but certainly not all, as one of the trends of the first half of the nineteenth century was the movement of chemistry from the medical to the philosophical faculty, as it became recognized as a part of natural philosophy. The committee establishing the Polytechnic also recognized the theoretical aspects of chemistry by creating separate chairs for analytical/theoretical and applied chemistry. While pharmacy was taught at both

64 Gustav Zschetzsche to Erziehungsdirektor, 27 March 1869, StAKZ, U118.1, Chemisches Laboratorium.

65 Jeffrey A. Johnson, 'Academic Chemistry in Imperial Germany', *Isis*, 76 (1985), 500–24; Rocke, *The Quiet Revolution*; Rocke, *Nationalizing Science*.

schools, neither the University nor the Polytechnic was developed with the explicit aim of training pharmacists, as was the case for many German universities. Löwig was also hired to teach exclusively chemistry, whereas some of the smaller German universities the same size as Zürich (Halle and Würzburg, for example) would have a combined chair for lecturing both chemistry and physics well into the nineteenth century.

But most importantly, academic chemistry in Zürich also developed under the unique constraints of the University and Polytechnic as cantonal and federal institutions. The physical proximity of the University and Polytechnic in the same city, with dual appointments and a shared chemical laboratory until 1861, resulted in arrangements that occurred rarely, if at all, in the German system. Cantonal officials took advantage of this proximity to save money, as the finances of the University were comparable to smaller provincial universities in Germany. In 1868 the educational council asked whether Wislicenus was 'worth the sacrifice' of the funds necessary to keep him in Zürich, and Wislicenus complained about the council's indecisiveness and vacillation. In 1871 the educational council asked the Zürich faculty to justify replacing Wislicenus after his move to the Polytechnic because the Polytechnic offered a similar array of courses. The faculty had to remind the council that the Polytechnic courses were not available to university students. When he succeeded Wislicenus, Merz was paid a far lower salary than Wislicenus.[66]

During the 1860s Wislicenus played a key role in the development of chemical research and instruction in Zürich. He facilitated a lasting separation between chemical instruction at the *Industrieschule*, the Veterinary School, and the University and addressed some of the problems of chronic insufficient funding and Städeler's neglect to reinvigorate the teaching-research model in Zürich. He resolved similar problems, although on a smaller scale, at the Polytechnic.

As his successor at the University, Merz was competent but had an undistinguished tenure, and he continued to direct the university laboratory until his retirement in 1893. He was succeeded by the extremely productive Alfred Werner, rightly celebrated as one of Zürich's most famous and productive chemists. Werner would go on to create a very successful 25-year teaching-research programme in inorganic chemistry involving dozens of students.[67] After Werner's premature death in 1919, he was succeeded by Paul Karrer, who created a forty-year research programme on plant pigments

66 Erziehungswesen statements, 15 February and 29 March 1871, StAKZ, U110b.1, Folder 18.
67 Kauffman, *Alfred Werner*.

and vitamins. Significantly, the appointments of Merz, Werner, and Karrer were essentially all internal. Merz and Karrer were native Swiss, whereas Werner was from Mulhouse in Alsace, culturally and linguistically related to northern Switzerland, and he had studied under Arthur Hantzsch at the Polytechnic. Merz, Karrer, and Werner all remained in Zürich for their entire careers.

Chemistry at the Polytechnic did not follow this pattern. Wislicenus lasted only eighteen months before leaving for Würzburg. In the same period in which the University had three professors of chemistry (Merz, Werner, and Karrer), the Polytechnic had six professors of theoretical chemistry: Victor Meyer (1872–85), Arthur Hantzsch (1885–92), Eugen Bamberger (1892–1905), Richard Willstätter (1905–12), Herman Staudinger (1912–26), Richard Kuhn (1926–28), and Leopold Ružička (1928–57).[68] Until Ruzcika, all of them were Germans, the longest-serving with a term of fourteen years. Of these six, five returned to Germany (Bamberger resigned due to ill health). Perhaps it is not coincidental that Ruzcika, a Croatian, remained in Zürich for thirty years, finally giving the Polytechnic greater stability. Victor Meyer foresaw this revolving door at the Polytechnic, calling it a 'first class waiting room'.[69]

Yet, in the long run, both strategies of looking internally (the University) and externally (the Polytechnic) paid off. Werner, Karrer, Willstätter, Staudinger, Kuhn, and Ruzcika would all receive Nobel Prizes, although Willstätter and Kuhn did not do their prize-winning work at the Polytechnic.

The fate of the chemical laboratory buildings at the University and Polytechnic is also instructive and reflects the differences in cantonal and federal funding. In 1887, 26 years after the opening of the original Rämistrasse laboratory, Arthur Hantzsch moved into a palatial laboratory a short distance away on Universitätsstrasse (fig. 4.1). Merz inherited the vacated Rämistrasse laboratory from the Polytechnic in 1887, finally moving chemistry out of the Cantonal School building after 45 years. Werner would inherit the Rämistrasse building, and was extraordinarily successful despite its extremely cramped and unhealthy conditions.[70] The Polytechnic and University would finally achieve some parity in infrastructure in 1909 when Werner oversaw the construction of a large, well-equipped chemistry building, the first building after 86 years that the University would construct

68 The sequence of famous theoretical chemists at the Polytechnic overshadows the later expansion of chemistry at the Polytechnic. For example, in 1893 Frederick Treadwell (1857–1918) was appointed as the first professor of analytical chemistry.

69 Gugerli, Kupper, and Speich, *Transforming the Future*, p. 70.

70 Kauffman, *Alfred Werner*, p. 44.

for chemistry.[71] Werner's successor, Paul Karrer, inherited this building in 1919, and with modifications, it served chemistry at the University until 1978. In 1909 yet another kind of parity between the institutions was achieved when after some deliberation, the Polytechnic decided to grant doctoral degrees and allow the faculty and students more autonomy.

Despite the relatively neglected facilities at the University, by the late nineteenth century, Zürich had certainly become one of the leading centres for chemical research in the world. What made chemistry in Zürich so successful? A complete answer to this question would require more thorough study of the institutional history of late-nineteenth century Zürich, but here I will briefly consider four general factors. First, the German teaching-research model found fertile ground in Zürich, as it was deliberately transplanted there by Germans familiar with it (especially Löwig). Löwig's laboratory was one of the earliest successful teaching-research laboratories in chemistry outside of Germany, emerging at the same time as Hofmann's laboratory in London and Mulder's laboratory in Utrecht. Second, chemistry itself saw explosive growth in the mid to late nineteenth century. The emerging chemical industries, including those in Switzerland, had a need for trained chemists, with or without a doctorate. The presence of the Polytechnic provided students, including many foreigners, with a route into chemical industry without the equivalent of the *Abitur* (German final qualification at the end of secondary education), and these students could also eventually work towards a doctorate; Merz, Baltzer, and Werner for example, all followed the route from Polytechnic to the University. Third, the presence of the Polytechnic, with its extensive federal support, also certainly made Zürich unique among Swiss cantons with a university and made a larger chemical community possible. Chemistry in Basel, Bern, and Geneva, supported exclusively by cantonal funds, languished relative to Zürich. Fourth, we cannot ignore the important role various laboratory directors played. Löwig had the temperament and a topic with great potential for multiple related smaller projects for students. Wislicenus and later Werner would attract a significant number of students despite the relatively underfunded facilities. Before 1909 faculty at the Polytechnic could funnel students to the University for doctorates, as faculty at the Polytechnic retained the power to grant them through the University. In short, this mixture of the Liebig model, external economic conditions, flexible routes into industry or academia, state-of-the-art space and equipment, and outstanding faculty combined to create one of the most successful sites of chemistry in the world.

71 Kauffman, *Alfred Werner*, p. 44.

About the Author

Peter J. Ramberg is Professor of History of Science at Truman State University in Kirksville, Missouri. His research area is the conceptual and institutional history of chemistry in the nineteenth century, particularly in Germany and Switzerland. The essay in this volume is a revised version of his article 'Chemical Research and Instruction in Zürich, 1833–1872', *Annals of Science*, 76 (2015), 170–86.

5 Island Kingdoms in the Making: The New Laboratories and the Fragmentation of Dutch Universities c.1900

Klaas van Berkel

Abstract

New laboratories not only reflect current ideas about science and science pedagogy; they also shape ideas and behaviour. The rise of the modern research laboratory in the second half of the nineteenth century contributed to the emergence of a new self-image for universities in Europe. The need to house the new laboratories often resulted in laboratories being scattered across the city, thus fragmenting university life. Instead of a close-knit community, the university became an archipelago of institutes, laboratories, lecture halls, and administrative buildings, where life for professors and students centred on their own institute rather than the entire university. The situation in the Netherlands, where the state-funded universities expanded tremendously around 1900, offers a key case study of these developments.

Keywords: university expansion, the Netherlands, *commune vinculum*, fragmentation of university life, unity of science

Introduction

Britain's Prime Minister Winston Churchill once famously said, 'We shape our buildings, but afterwards they shape us.' He said this in October 1943 during a debate in the House of Commons on restoring the Houses of Parliament in London, following the destruction of the House of Commons chamber in a German air strike in 1941. Churchill meant that the buildings

Berkel, Klaas van, and Ernst Homburg (eds), *The Laboratory Revolution and the Creation of the Modern University, 1830-1940*. Taylor & Francis Group, 2023
DOI: 10.5117/9789463720434_CH05

we live and work in somehow direct our behaviour, our thoughts, and our perception of reality. Thus, the House of Commons, with its characteristic layout of two opposing sets of bench rows, had shaped political culture in Great Britain, and since Churchill felt comfortable with this way of doing politics, he advocated the complete restoration of the chamber.[1]

The history of laboratories offers manifold examples of the same phenomenon. Sophie Forgan has convincingly demonstrated that the way laboratories and lecture halls were constructed in the nineteenth century not only helped to create scientific disciplines but in many ways also shaped or 'disciplined' the students.[2] She points to the practical chemistry classroom at Liverpool University College around 1890, where benches were arranged in curved rows, raking gently upwards from the lecturer's desk. This enabled the lecturer to maintain eye contact with all the students and hold the attention of each, thus maintaining control and discipline. Forgan also points to the change in the spatial arrangement of laboratories—which she attributes to Justus Liebig—characterized by placing free-standing benches where students could do their work. This meant that the room and its occupants could be seen at a glance, and the professor could easily walk around the laboratory to oversee students' work. The element of control, which is essential to the modern laboratory, is evident here.[3]

However, Churchill's maxim is relevant for understanding not only individual buildings and spaces but also a set of buildings, such as the laboratories of a particular university. Most researchers of university buildings focus on individual buildings—the halls, the laboratories, the library, or the student residences—while the urban planning aspect, whether or not there was one, is usually left out of consideration. In this chapter, I would like to focus on this specific issue of the expansion of universities and their laboratories in the late nineteenth century. I would like to concentrate on the intended or unintended and often unrecognized impact of the spatial arrangement of university buildings on ideas and perceptions of science, its

1 Churchill's quote was brought to my attention by: Remieg Aerts, 'Architectuur en representatie. De cultuurgeschiedenis van politiek en ruimte', in Remieg Aerts, Klaas van Berkel, and Babette Hellemans, eds., *Alles is cultuur. Vensters op moderne cultuurgeschiedenis* (Hilversum: Verloren, 2018), pp. 95–109 (p. 95).

2 Sophie Forgan, 'The Architecture of Science and the Idea of a University', *Studies in the History and Philosophy of Science*, 20 (1989), 405–34 (pp. 424–8). See also Frans van Lunteren's chapter in this volume.

3 Nineteenth-century university laboratories are almost completely absent from: Peter Galison and Emily Thompson, eds., *The Architecture of Science* (Cambridge, MA: MIT Press, 1999). The contribution by Sophie Forgan, 'Bricks and Bones: Architecture and Science in Victorian Britain', pp. 181–208, deals exclusively with museums.

practitioners, and their place within the university. The material I will focus on derives from the history of the three state-funded Dutch universities in the late nineteenth and early twentieth centuries: Utrecht, Leiden, and Groningen. This was a time of remarkable expansion within the Dutch university system in general and experimental science, medicine, and psychology in particular.[4] The new buildings were necessary to accommodate the new science of the times, but at the same time they changed the self-image of the scientists who worked and lived in them, not through their individual architecture but because of their distribution across the city. In doing so, these laboratories also changed the nature of the university of which they were a part.

The *Commune Vinculum*

In 1875 Utrecht University in the Netherlands was discussing where to build a new laboratory for the physical sciences. Until then, students had attended their lectures in physics, meteorology, and astronomy either at the professor's home or in the Physiological Laboratory of the renowned Franciscus Cornelis Donders in the northern part of the city.[5] The physics professor, the equally eminent Christophorus (Chris) Buys Ballot, had a strong preference for building the projected new laboratory close to the Physiological Laboratory, and Donders, always ready to lend a helping hand, had already privately bought part of this site to prevent others from interfering with this plan. Yet, to Donders's annoyance, the Board of the University (the curators), supported by both the government in The Hague and the city of Utrecht, decided to locate the new physics laboratory elsewhere, at the opposite end of the city and at the southernmost tip of its former defensive works, which had been in disuse since 1830. The architect saw this as the most convenient location.

Donders was not a man to be easily daunted, however. He believed that what was at stake here was not a matter of convenience but one of principle.

4 See: Klaas van Berkel, Albert van Helden, and Lodewijk Palm, eds., *A History of Science in The Netherlands. Survey, Themes, and Reference* (Leiden: Brill, 1999), pp. 130–69. See also: Klaas van Berkel, 'Low Countries', in Hughes Richard Slotten, Ronald L. Numbers, and David N. Livingstone, eds., *Modern Science in National, Transnational, and Global Context. The Cambridge History of Science, vol. 8*, 8 vols. (Cambridge: Cambridge University Press, 2020), pp. 305–24.

5 On this laboratory, built in 1867: F.C. Donders, 'Het physiologisch laboratorium der Utrechtsche Hoogeschool', *Onderzoekingen gedaan in het physiologisch laboratorium der Utrechtsche Hoogeschool*, 3/1 (1872), 1–12. On the importance of periodicals like these *Onderzoekingen* (an in-house scientific journal), see the chapter by Dorien Daling in this volume.

On his initiative, the University Senate sent a letter to the Board of the University in June 1875, stating the following:

> We deceive ourselves in thinking that a number of scattered buildings constitute a university. The nature of a university requires the realization of the 'commune vinculum', the common bond, the reciprocal penetration and interaction of all branches of human knowledge.[6]

The concentration and proximity of all university buildings and facilities in one spot were essential, according to the Senate, for a university to flourish. The Senate referred to the situation in Leipzig, Bonn, and Kiel, places that Donders had visited a few years earlier. Even the situation in Amsterdam and Delft would be far preferable to the one in Utrecht, should the plan of the Board of the University be realized. The Senate therefore urged the Board to reconsider its decision and to relocate the new physics laboratory to the site next to Donders's Physiological Laboratory. But the appeal to the curators was of no avail, and the construction of the new laboratory proceeded as planned.

The Physics Laboratory was only the second purpose-built laboratory at Utrecht University in the nineteenth century. In 1845, when Gerrit Jan Mulder, recently appointed as professor of chemistry, was given a chemistry laboratory where his students could also do their practicals, he was assigned De Leeuwenbergh, a renovated church (A in fig. 5.1).[7] This practice had been quite common in the past: universities were mostly housed in buildings that had been built for other purposes, such as churches and convents. It was only in 1867 that Donders moved into his new, purpose-built Physiological Laboratory, situated—in fact, tucked away—on the slope of an ancient bulwark in the northern part of the city (B in fig. 5.1). The site of the new Physics Laboratory could not be farther away from Donders's laboratory (C in fig 5.1).

Over the next few decades, more than ten other new laboratories and clinics were built at different sites in Utrecht, apparently without a master

6 'Men bedriegt zich, naar het oordeel van den Senaat, wanneer men van meening is, dat tal van verspreide inrichtingen een universiteit vormen. Het wezen eener universiteit onderstelt de verwezelijking van het commune vinculum, het wederzijds elkander doordringen en op elkander inwerken van alle takken van menschelijke kennis.' Quoted in: J.A. Schuur, 'Naar de stadsrand. Academische gebouwen uit de periode 1815-1915', in A.W. Reinink and J.A. Schuur, eds., *Bouwen voor Utrechts universiteit. Architectuur en stedebouw binnen de stad* (Utrecht: Matrijs, 1985), pp. 84–105 (p. 85) (quote translated by the author).

7 H.A.M. Snelders, 'Het chemisch laboratorium "De Leeuwenbergh" te Utrecht', *Tijdschrift voor de Geschiedenis der Geneeskunde, Natuurwetenschappen, Wiskunde en Techniek*, 7 (1984), 129–40.

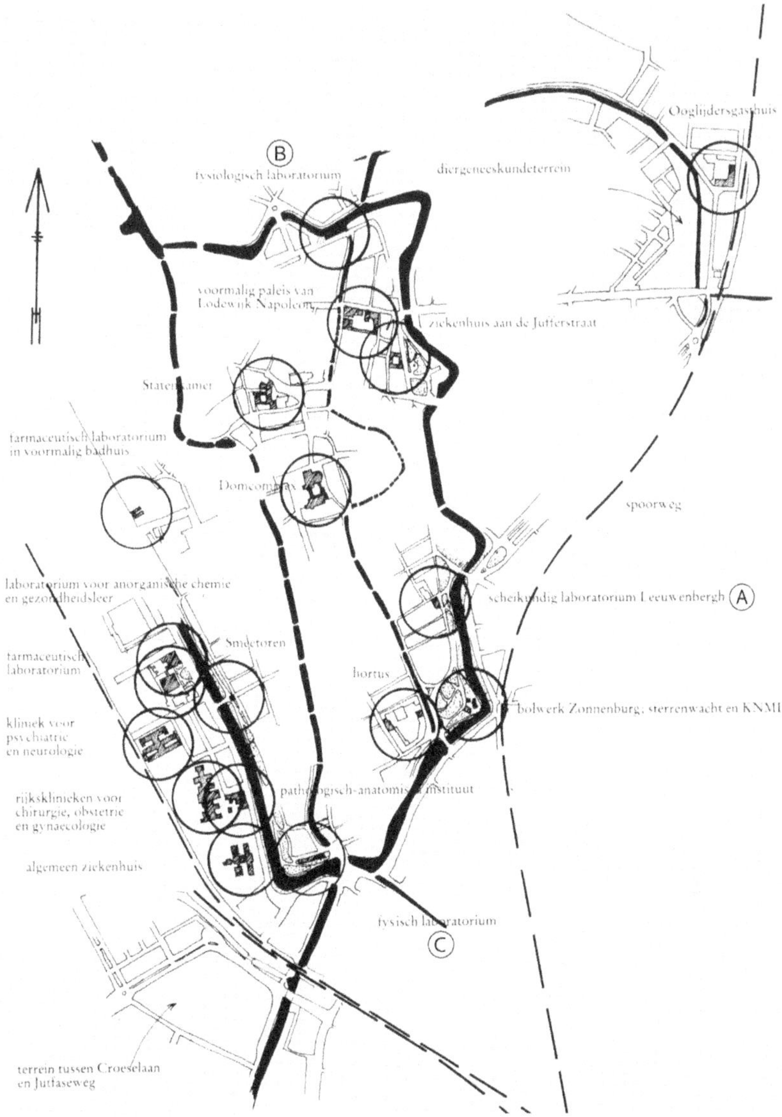

Figure 5.1 Map of the city of Utrecht, showing the laboratories and clinics established in the nineteenth and early twentieth centuries. Source: Schuur, 'Naar de stadsrand', p. 84. Courtesy: Universiteitsmuseum Utrecht.

plan and in different architectural styles. (The Physiological Laboratory was neo-gothic, at least on the outside, whereas the Physics Laboratory was classicist and monumental.) By 1900 Utrecht University had been transformed from a close-knit, compact university located near the Academy Building to an archipelago of lecture halls, laboratories, hospitals, a botanical garden, and an astronomical observatory scattered across the city. What did this mean for the University as a whole? Was this spatial fragmentation indeed detrimental to the University, as Donders claimed, or did it have no impact at all? And if the latter were the case, was this because it really does not matter where the sites of knowledge are located, or were specific countermeasures taken to neutralize the negative impact of this spatial fragmentation?

Safety Concerns in Leiden

The situation in Leiden was slightly different.[8] Until the end of the century, new laboratories and institutes were built in the old city, where the University was concentrated along one of the canals in the southwestern quarter of the city, the curving Rapenburg. The old Academy Building, the chief site of the University since its foundation in 1575, was located halfway along this canal, with the Hortus Botanicus and the Observatory (newly built in 1860) on the piece of land between the Academy Building and the canal surrounding the city. To the north of this canal zone were the academic hospital (1867) and the adjoining Pathological Laboratory (1885), both located on the site of a former bastion in the city's defensive works. (These works had lost their function in the early nineteenth century, but the city wall and the canal encircling the city still marked the limits of the municipality of Leiden.) At the eastern end of the canal zone was the Physical, Chemical, and Anatomical Laboratory (1859). This multi-purpose laboratory was located at a site called the Little Ruins (Kleine Ruïne), an open space on the northern side of the canal (called Het Steenschuur here), which had been cleared, unintentionally, when a ship full of gunpowder exploded in 1807. This explosion had cleared an even larger space on the other side of

8 For the situation at Leiden, see: David Geneste, *Werkplaatsen van de wetenschap. Functie- en vormbetekenis van natuurwetenschappelijke laboratoriumgebouwen van de Leidse universiteit in de tweede helft van de negentiende eeuw* (MA thesis, Leiden University, 2005). A summary of Geneste's findings was published in a series of articles in the Leiden-based journal *Stielz*: David Geneste, 'Laboratoriumarchitectuur van de Leidse universiteit, 1669-1859', *Stielz*, 2 (2004), 13–14; 'Van "Monstrum horribili visu" naar "Oud-Hollandsche stijl". De architectuur van universitaire laboratoriumgebouwen in Leiden van 1859 tot 1885', *Stielz*, 3 (2004), 11–14; 'Leidse laboratoriumgebouwen rond 1900. Werkplaatsen van de neogotiek', *Stielz*, 4 (2004), 12–14; 'Van cité medicale naar wetenschapspolder. Leidse laboratoriumgebouwen in de twintigste eeuw', *Stielz*, 1 (2005), 14–18. Geneste extensively discusses the architecture of the individual buildings, but he ignores the aspect of urban planning. So too does Willem Otterspeer, *Groepsportret met dame, IV. De strategie van de aanpassing. De Leidse universiteit, 1876-1975* (Amsterdam: Prometheus, 2021), pp. 157–60.

the canal, the Great Ruins (Grote Ruïne), which was also still unoccupied at the end of the century. These two sites in the old city, the Little and Great Ruins, offered ample space for the huge, new laboratories that had become necessary by the end of the nineteenth century. The new laboratories had to provide accommodation for a growing number of students and professors as well as reflect the newly acquired status of the sciences. More space was needed. As the open spaces of the Great and Little Ruins could accommodate the new science buildings, there was no immediate need for the University to spread its laboratories across the surrounding areas. The explosion of 1807 meant that Leiden University could hold on longer to the idea of a *commune vinculum* and build its facilities close to one another.

This changed at the end of the nineteenth century when Leiden University could no longer postpone the building of a modern laboratory for the pharmaceutical sciences. Pharmacy had become an academic discipline in 1876 and had been assigned a rather primitive laboratory in the Papengracht, a small street near the northern end of the Rapenburg. Professor Hendrik Paulus Wijsman had campaigned for an up-to-date laboratory from day one of his appointment in 1890, but the Board of the University and the central government proved uncooperative. Wijsman suggested several locations for this laboratory, including the Great Ruins and tracts of land outside the city, but things did not begin to move until Wijsman teamed up with the professors of inorganic and organic chemistry, who were still housed in the chemical, physical, and anatomical laboratory on the Little Ruins. Their neighbour, the ambitious professor of physics Heike Kamerlingh Onnes, was claiming more and more space for his research on low temperatures and was keen to see his chemistry colleagues go. Eventually, in 1894, both the Board of the University and the central government agreed to move pharmacy and the two branches of chemistry to a location outside the city, the Vreewijk estate. Located south of the city, it was officially in another municipality (Zoeterwoude) but within sight of Leiden. It was the first time a university facility was built outside the city, but it would certainly not be the last. The Pharmaceutical Laboratory, a rather luxuriously decorated and ornamented Gothic 'citadel of science', was opened in 1898 and was soon followed by the laboratories for organic and inorganic chemistry.[9]

9 According to Geneste, *Werkplaatsen van de wetenschap*, pp. 86–95, there was more than one reason why the architect Jacob van Lokhorst had chosen the neo-gothic style. It allowed for a more flexible and functional arrangement of the spaces compared to the rigid layout of the classicist buildings that had been fashionable earlier. In addition, whereas the Gothic style was formerly associated with Roman Catholicism, the archenemy of science and the Enlightenment, a secular interpretation of the Gothic style was promoted towards the end of the century because

A major reason for moving to a site outside the city, thereby loosening the ties between the various university departments, may have been concerns about safety in modern laboratories. In 1875 the *Fabriekswet* ('Factory Law') had prescribed safety measures for industrial sites, but these regulations (including the need to obtain permission to store dangerous materials and carry out dangerous operations) were soon extended to a wide range of facilities, including university laboratories.[10] From 1890 on, complaints were heard in Leiden about dangerous situations in Kamerlingh Onnes's Physics Laboratory, where gas cylinders were stored, it was said, without adequate safety measures being taken. Even though the risk of the gas cylinders exploding was exaggerated, the municipal authorities were particularly nervous about the fact that a nearby school was potentially in danger. These complaints in 1895 evolved into a famous dispute between Kamerlingh Onnes and the city of Leiden and led to a serious delay in his efforts to be the first to liquify gases such as oxygen and helium.[11] The municipal authorities were very much in favour of moving Kamerlingh Onnes's laboratory to somewhere outside the city, but they were unable to prevent the central government from issuing a permit that allowed Onnes to continue his research, provided that he introduced additional safety measures. It is highly likely that the decision to relocate the pharmaceutical laboratory, together with the chemical laboratories, to the Vreewijk site was made with this battle over the Physics Laboratory in mind.[12]

it spoke to the heightened self-respect of the scientists, who saw themselves as priests of the new science. The University Museum of Science in Oxford (1860), which bore the mark of the English art critic John Ruskin, is supposed to have led the way in this re-evaluation of the Gothic style. Geneste does not, however, present clear indications that this was indeed the case.

10 On the effect of the Factory Law, see: J. Lintsen sr., 'De werking van de Hinderwet tijdens de industrialisatie van Nederland (1890-1019)', *Jaarboek voor de Geschiedenis van Bedrijf en Techniek*, 4 (1987), 190–210.

11 Dirk van Delft, *Freezing Physics. Heike Kamerlingh Onnes and the Quest for Cold* (Amsterdam: KNAW, 2007), pp. 261–86.

12 Geneste already pointed out that the safety issue may have been a reason for moving to a site outside the city: 'The low population density of this area apparently offered fewer safety risks in the event of fire or other laboratory-related dangers than the overcrowded inner city.' ([De] relatief lage bevolkingsdichtheid van dit gebied [bood] blijkbaar minder veiligheidsrisico's bij brand of ander laboratoriumgerelateerd gevaar dan de overbevolkte binnenstad.) Geneste, *Werkplaatsen van de wetenschap*, p. 74. He does not relate the decision to move to a place outside the city to the discussion about laboratory safety that erupted in the mid-1890s and attracted nationwide attention. The Royal Academy of Arts and Sciences became involved and in 1896 issued a detailed report in favour of Kamerlingh Onnes, a prominent member of the Academy. The Academy pointed out in its report that a chemical laboratory was much more dangerous than the Leiden physics laboratory. Otterspeer, *Groepsportret met Dame IV*, p. 157 also mentions

A Special Case: The University of Groningen

For several reasons, the University of Groningen is an even better case study for research into the dispersal of university laboratories across the city. Firstly, the number of laboratories that were established between 1865 and 1915 surpasses that of Leiden or Utrecht. (Groningen acquired its first purpose-built laboratory building in 1866, also a physiological laboratory as in Utrecht.) This was partly because Groningen had been severely neglected by central government and had some catching up to do. Moreover, a major fire completely destroyed the central Academy Building in 1906, thereby speeding up the programme to rebuild the laboratories that had been housed there. Secondly, in the case of Groningen, we have at our disposal rich documentation about the new laboratories thanks to the University's memorial volume that was published in 1914 to mark the University's 300th anniversary. Apart from a bulky 300-page history of the University during the nineteenth century by the famous historian Johan Huizinga, the memorial volume also contains extensive and detailed descriptions (with a large number of photographs and ground plans) of all the laboratories and the other academic buildings in Groningen.[13] This makes it worthwhile to study Groningen in more detail.[14] I will not say much about the interior of these buildings,[15] nor will I dwell on their architectural style, interesting though it might be.[16] Instead, I will concentrate on the issue of urban planning

the concerns about safety but again does not relate them to the discussion about Kamerlingh Onnes's Physics Laboratory. R.M. Elskamp, ed., *Farmacie in Leiden* (Leiden: Gist Brocades, 1985) is silent on the issue.

13 *Academia Groningana MDCXIV-MCMXIV. Gedenkboek ter gelegenheid van het derde eeuwfeest der universiteit te Groningen* (Groningen: Wolters, 1914).

14 Much of the following is taken from: Klaas van Berkel, *Universiteit van het Noorden. Vier eeuwen academisch leven in Groningen*, 3 vols. (Hilversum: Verloren, 2014–22), vol. 2: *De klassieke universiteit, 1876-1945*. See also: Klaas van Berkel and Guus Termeer, *The University of Groningen in the World. A Concise History* (Amsterdam: Pallas Publications, 2021).

15 Sieger Vreeling, *Geen Stijl. Een rijkere architectuurgeschiedenis* (Hilversum: Verloren, 2022) rightly points out that focusing solely on the outward appearances of a building gives a distorted image of the work of the architect, who was more concerned with such trivial issues as foundations, water conduits, electric cables, choice of material, and the sewage infrastructure. He partly illustrates his ideas with a detailed reconstruction of the building process of some university laboratories in Groningen.

16 A curious fact about Dutch university laboratories built between 1876 and 1910 is that the architecture was the work of only two architects, Jacob van Lokhorst and his chief assistant and successor Jan Vrijman, the official architects of the Department of Education in The Hague. This resulted in a rather uniform choice of style (from neo-gothic to neo-renaissance, with some hints of classicism) and the recurring re-use of designs. The result is that the laboratories built

(or the lack of such planning) between 1866, when a new Physiological laboratory was erected, and 1914, the year in which the memorial volume was published. Besides, the outbreak of the First World War in 1914 and the ensuing budget cuts by the central government put an end to the building programme that had started a couple of decades earlier.

Figure 5.2 offers a map of the city of Groningen around 1910. It shows the city not long after the ramparts that once protected the city from foreign invaders were torn down. Some of these defensive works had been transformed into a park (in the northwest), while others were used to build a hospital (in the northeast) or to lay out stately boulevards (in the south). I have marked all the laboratories and other university buildings on the map, both the old ones and the new ones erected between 1877 and 1914.

The heart of the University of Groningen was located right in the heart of the city, as it still is today (no. 1). It is the Academy Building, the administrative and representative centre of the university, where new professors gave their inaugural lectures, where Senate meetings were held, where students took their exams, and where the Board of the University convened.

The Academy Building was also home to the small administrative office of the University. It was built in 1906–9 on the site of the building that had burnt down in 1906. In the mid-nineteenth century, this older building had housed several laboratories, such as those for chemistry and physics, but these had already partly moved out before the great fire of 1906. After 1909, however, the new Academy Building accommodated a Psychological Laboratory, the first university laboratory for psychology in the Netherlands, especially created for the professor of philosophy and psychology Gerardus Heymans.

Across the street (or actually behind the neo-gothic church that faced the Academy Building) was the University Library (2), which is still located there today. In the early modern period, the University Library had actually resided in the church (or, to be more precise, in the medieval church that had stood there) until it had to move out when the church was given back to the small Catholic community in Groningen in the beginning of the nineteenth century. Next to the Academy is the Laboratory for Hygiene

in Leiden looked very much like the laboratories in Utrecht and Groningen, all three of them universities that were financed and controlled by the central government. There is a remarkable resemblance, for instance, between the Physics Laboratory in Groningen (1892) and the Pharmaceutical Laboratory in Leiden (1898). The situation is different in the building programmes for laboratories at the new University of Amsterdam (financed by the city of Amsterdam), even though other architects also long preferred a 'neo-style' over more contemporary styles.

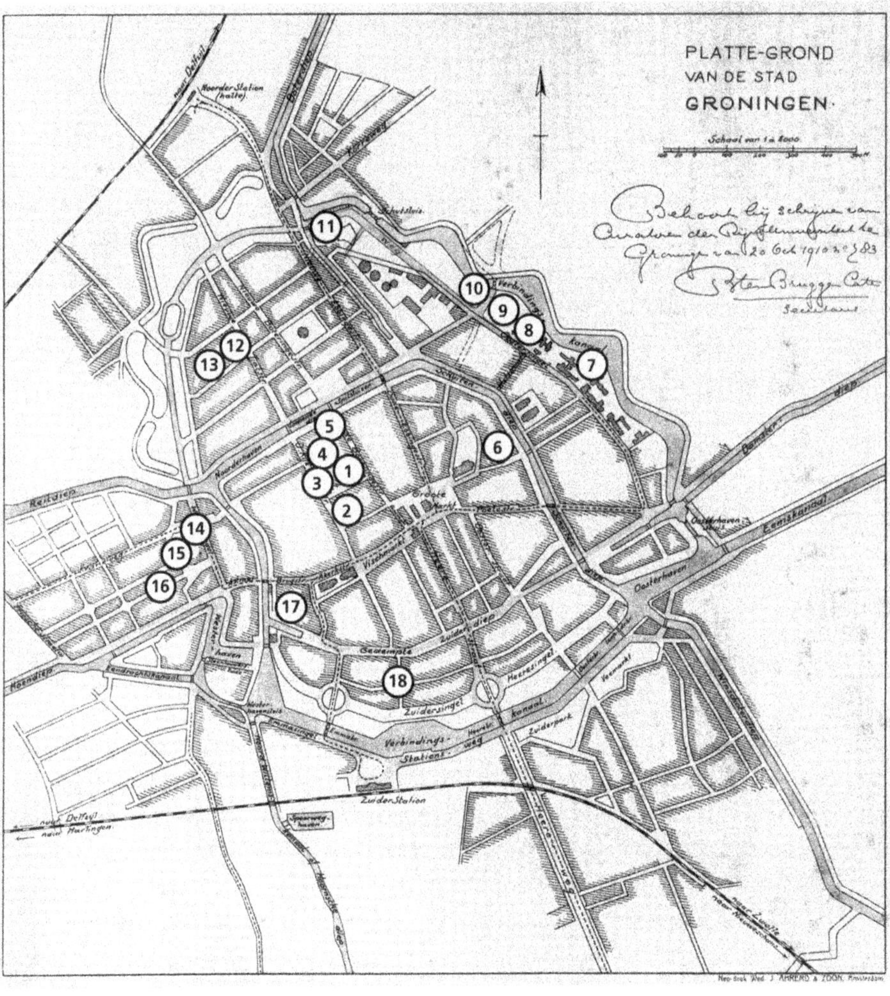

Figure 5.2 Map of the city of Groningen in 1910, with the numbered laboratories and institutes of the university. Source: Van Berkel, *Universiteit van het Noorden*, vol. 2, p. 177.

and Public Health (3), built in 1884, and behind the Academy Building we find the Laboratory for Physiology (4). When it was constructed in 1866 for Izaac van Deen, the professor of physiology, this laboratory was quite modern, but by the end of the century it had become too cramped, and in 1911 the then professor of physiology, Hartog Jacob Hamburger, moved to a new building on the outskirts of the city. In 1913 the former Physiological Laboratory was transformed into Jacobus Cornelius Kapteyn's Astronomical Laboratory. Close to the Academy Building is the Corps de Garde (5), a seventeenth-century guardhouse used first by the professor of agricultural science, next by the professor of mineralogy, and then, until 1913, by the professor of astronomy (before he moved to 4).

At number 6 we leave the oldest parts of the city and move outwards. In numerical order we encounter the Children's Hospital (6), affiliated to the University, the General University Hospital (7), with its individual clinics and laboratories, and the Anatomical and Pathological Laboratories (8 and 9). Number 10 is Hamburger's new Physiological Laboratory, constructed in 1911, and 11 is the Laboratory of Inorganic Chemistry, completed in 1912. We then move on to the site of the old botanical garden (12 and 13), the location of both the Pharmaceutical Laboratory (12), still occupied by the professor of organic chemistry, and Jan Willem Moll's Botanical Laboratory (13).[17] To the west of the inner city, we have the Physics Laboratory (14), built in 1892, with the Mineralogical and Geological Institute right behind it (15).[18] Number 16 is a private home that was fitted out as the Pharmaceutical Laboratory in 1906, while 17 is the Zoological Laboratory and 18 the Ophthalmological or Eye Clinic, also affiliated to the University.

Building on the Edge of Town

There were two major factors that determined this specific spatial layout of the university buildings. The first is that the city lost its key role in defending the country against foreign—meaning German—attacks in 1874; the second was the great fire in the Academy Building in 1906. Both factors are very specific to Groningen, since Utrecht and Leiden had lost their defensive function before the middle of the nineteenth century and did not experience any event akin to the destruction of its Academy Building.

After the Franco-Prussian War of 1870, which the Germans so easily won, the Dutch government decided to alter its defence strategy against possible attacks from the new German Reich. Before 1870 the Dutch had put their trust in a string of heavily fortified cities in the east of the country, such as Groningen, Zwolle, Deventer, and Arnhem. All these cities were surrounded by extensive fortifications consisting of an intricate system

17 On the construction of this laboratory: Jaline de Groot, *'Doelmatig maar tevens schoon'. Architectuur & wetenschap in het Botanisch Laboratorium in Groningen, 1892-1899* (MA thesis, University of Groningen, 2016). A unique source for analysing the construction of this particular laboratory was the extensive correspondence between professor of botany Jan Willem Moll and the architect Jacob van Lokhorst, who resided in The Hague.

18 For this institute, see: Sieger Vreeling, *Kasteelheer. F.J.P. van Calker (1841-1913), hoogleraar mineralogie en geologie* (Groningen: De Kleine Uil, 2021).

of defensive walls, canals, ramparts, and a surrounding countryside that could be inundated. This had worked in the seventeenth century, when the city of Groningen had been besieged by the bishop of Munster and had withstood weeks of bombardments and attacks issuing from encroaching trenches. Then, in 1870, the Germans had proved that such fortifications were no longer of any avail. So, in 1874, the cities just mentioned lost their defence function and were permitted to demolish their defensive works if they wished. The municipal authorities of Groningen certainly wanted to do so because the city was so densely populated by the mid-nineteenth century that it needed more space for new housing and industry outside the city. Parts of the former defensive works were also reserved for healthful parks surrounding the inner city.[19]

The city of Groningen did not own these sites, however. The former defensive works were the possession of the central government, and the state did not want to hand over these precious sites to the city for free. The city could buy the land or let the government decide what to do with it. This resulted in protracted legal battles that the city ultimately lost. Some plots were indeed sold to the city, both for parks and new houses, whereas others were kept in reserve for buildings that the central government went on to build itself.

These buildings included new university buildings. In 1876 Parliament had passed a new higher education act stipulating that, from then on, a number of new disciplines had to be taught at Dutch universities by separate professors. It was also implied, given the experimental turn in several of these new disciplines, that each of the new professors should have their own laboratory. At the same time, growing student enrolments, in part the result of the modernization of secondary education, forced the government to expand the existing laboratories or to build completely new ones for the disciplines involved, such as botany and physics. The first new laboratory was the Laboratory for Hygiene and Public Health, right beside the Academy Building (3 on the map, built in 1884). In the meantime, preparations had begun for the construction of a new Physics Laboratory, which up to that time was still housed in the Academy Building. However, not just any site would do for this new laboratory, given the special field of interest of Herman Haga, the new professor of physics whose career was devoted to electrical and magnetic precision measurements. Haga needed a firm foundation for his precision instruments, and the only place he deemed adequate were

19 See: Pim Kooij, *Groningen 1870-1914. Sociale verandering en economische ontwikkeling in een regionaal centrum* (Assen: Van Gorcum, 1987).

the former bastions of the defensive works surrounding the city. The heavy load of bricks and stones in these bastions had compressed the substratum to such an extent that Haga would be able to use his instruments without fear of disturbances caused by passing trucks. He found his favourite spot to the west of the old city (14 on the map). In 1892 he could officially put his new laboratory into use. It was notable not only for its architectural beauty but also for its lack of iron and steel. (The architect had replaced iron with other building materials to avoid disrupting Haga's magnetic and electrical measurements.) It was the first laboratory to be built on the former defensive works, but it was soon followed by the Mineralogical and Geological Institute, the General University Hospital (in 1903), and the Anatomical, Pathological, Physiological, and Chemical Laboratories.

In the meantime, disaster had struck the University. In August 1906 the Academy Building caught fire and burnt down. The magnificent Zoological Museum on the top floor was completely lost, and the rest of the building was also ruined. The professors in the faculties of Arts, Theology, and Law and their colleagues in Chemistry and Pharmacy had to find somewhere else to teach and to carry out their research. The University found them new accommodation in churches, the provincial archives, a deserted hospital, and some residential houses, but—more than ever—new, efficient, and modern laboratories needed to be built for the displaced professors. This not just concerned the chemistry and pharmacy professors but also the professors in modern languages and history, who were eager to have their *Seminare*, as most German universities now had. This resulted in the construction of a separate Laboratory for Experimental Psychology on the ground floor and in the basement of the new Academy Building, a new Chemical Laboratory in the northern quarters of the city, and linguistic and historical seminars in the Corps de Garde, the former institute of the professor of astronomy (see 5, 11, and 16 on the map).[20] The professor of zoology found refuge in some old buildings that had been abandoned by the General University Hospital many years ago. The diaspora of the laboratories and institutes, already well underway before 1906, was thus given added momentum by the Academy Building fire of that year.

20 Heymans's Psychological Laboratory in the new Academy Building is one of the best-studied laboratories in Groningen in the early twentieth century. See: Douwe Draaisma, Bikram Lalbahadoersing, and Eric Haas, 'Een laboratorium voor de ziel. Heymans' Laboratorium voor Experimentele Psychologie 1892-1927', in Douwe Draaisma, ed., *Een laboratorium voor de ziel. Gerard Heymans en het begin van de experimentele psychologie* (Groningen: Historische Uitgeverij, 1992), pp. 12–26.

A Collection of Semi-Independent Communities

This branching-out of the University had all manner of unintended consequences. Professors who had given their lectures in the Academy Building used to meet each other regularly in the professorial waiting room, a separate room where they could have coffee together before reading their lectures and where theologians, historians, and natural scientists mingled. Now, after the fire, they no longer saw each other so often. Of course, their conversations could continue during the monthly Senate meetings that professors were expected to attend. And there was still the monthly meeting of the 'professorial circle'—an informal opportunity to meet on a Saturday night before and after listening to a talk given by one of their number. But professors who valued the cross-disciplinary contacts that, in their eyes, defined a university, deplored the disappearance of the professorial waiting room.

Professors who already had their own laboratory had not taken part in the small talk in the professorial waiting room. They lectured at their laboratory, institute, or clinic, where none of their colleagues came. For them, their professional life revolved around their laboratory. Unlike the professors in the Faculty of Law or the Faculty of Arts, they no longer lectured in their homes. Even the intermediate examinations (*tentamina*) were held in the laboratory. (In the Law and Arts faculties, these intermediate examinations continued to be held in some professors' homes until the 1960s.) It was now the other way around. Elements of domestic life were transferred to and grafted onto the life of the laboratory. Some professors moved part of their private library—especially their collections of off-prints—to the laboratory library, where more people could make use of it. Furthermore, the professor's office in the laboratory, where he received visitors and students, was often called the 'sitting room' or 'living room', as if he were at home (see fig. 5.3). Apart from the desk where he worked, the sitting room was furnished with easy chairs and a coffee table, all rather cosy. For the professors, it was something like 'a home away from home'.[21] The University, as represented

21 *Academia Groningana*, pp. 539–47 (see note 9). With the exception of the professor of botany, it was not customary in the Netherlands for a professor to live in or next to his laboratory (as was the case in Germany). And in 1880 even the Groningen professor of botany Petrus de Boer had to move away from the Hortus because his house was going to be demolished to make way for the new Pharmaceutical Laboratory. De Boer protested, arguing that it was essential for all professors to live in, or at least very close to, their laboratories. The Royal Academy of Sciences supported him and claimed that this was indeed widely acknowledged all over the country. In the end De Boer did not have to find a new home elsewhere in the city; a new house was built for him just around the corner, allowing him to enter the Hortus Botanicus from his backyard.

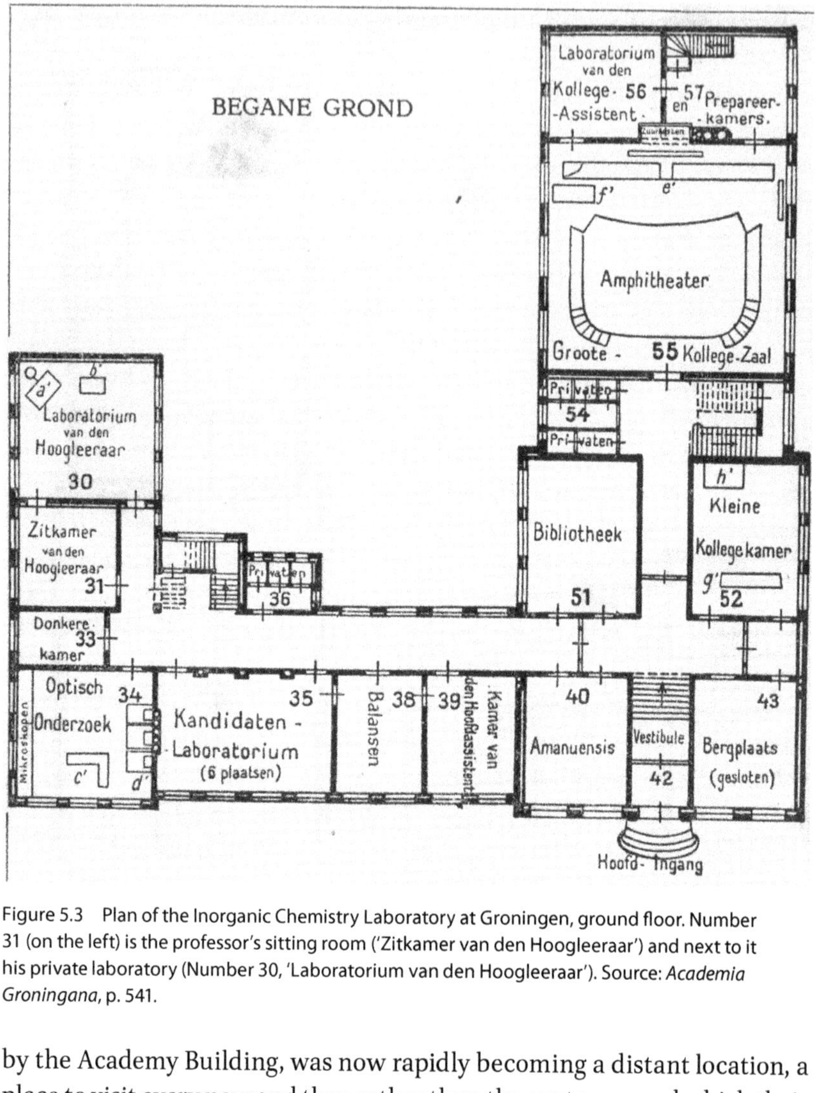

Figure 5.3 Plan of the Inorganic Chemistry Laboratory at Groningen, ground floor. Number 31 (on the left) is the professor's sitting room ('Zitkamer van den Hoogleeraar') and next to it his private laboratory (Number 30, 'Laboratorium van den Hoogleeraar'). Source: *Academia Groningana*, p. 541.

by the Academy Building, was now rapidly becoming a distant location, a place to visit every now and then rather than the centre around which their professional lives revolved.

Much the same can be said about the assistants and auxiliary personnel at the laboratory, the instrument makers, apprentices, and errand boys. The growing number of students and the rising demands of scientific

Klaas van Berkel, 'Het huis van de professor. De Koninklijke Akademie van Wetenschappen en de scheiding tussen werk en privé in de negentiende eeuw', in Esther van Gelder, Eric Jorink, Ilja Nieuwland, Marlise Rijks, and Alice Spruit, eds., *'Dingen die ergens toe dienen'. Verhalen over materiële cultuur van wetenschap* (Hilversum: Verloren, 2017), pp. 167–9.

research meant that the professor was surrounded by an increasing number of people who supported him in his research and his teaching. In the old days, the professor of chemistry in Groningen had only one amanuensis, who tended to the fire in the stove, arranged the instruments for the experiments the professor wanted to demonstrate during his lectures, introduced new students—if there were any—to the practicals, and cleaned the rooms. By 1900 all laboratories had specialized staff for all these duties: the caretaker or concierge, who often lived in the laboratory or in a house next door, along with an instrument maker, errand boy, cleaning lady, and assistants (mostly older students or PhD students). The university staff was thus rapidly expanding, especially at the lower echelons.[22] These people formed a small community of their own, with little contact with other 'tribes' in other laboratories. Christmas was celebrated at the laboratory, with a Christmas tree and small presents, and when Saint Nicholas came to the Netherlands in early December, he also managed to pay a visit to every laboratory in Groningen—as was the case in Leiden and Utrecht. The professor was the head of this small community and insofar as the university was breaking apart and becoming an island nation, the professor was the king of these separate principalities, officially part of one empire but in reality semi-independent. A professor was of course still officially accountable to the Board of the University (and to the Minister of Education in The Hague), but within the walls of his own laboratory his authority was undisputed.

In this context the community created like this at a laboratory is often compared to a family, with the professor described as its head. This family included not only the assistants and the technical staff but also the more advanced students, the cleaning ladies, and the young errand boy. The family feeling generated a sense of loyalty to the professor that knew no bounds. It emerged not just from informal rituals, such as collective coffee breaks or rivalry with other laboratories, but was also actively stimulated by commemorative practices. These included both informal gatherings (staff members' birthdays, Christmas parties) and formal celebrations of prizes that the professor had won or the 25th anniversary of his PhD defence, or his appointment as professor. Truus van Bosstraeten offers an excellent analysis of these commemorative practices at the Heymans Institute for Pharmacology at Ghent University in Belgium in the first half of the twentieth century, something she was able to do because so many pictures were taken—and preserved—of these commemorative

22 See also the chapters by Peter Morris and Bas Nugteren in this volume.

events.[23] It is also hard to think of any professors who made more effort to create a family feeling than Jan Frans Heymans and his son and successor Corneel Heymans, both heads of the Ghent laboratory. But the idea that a laboratory was a family home was widespread in the decades around 1900. There were differences in this regard between the various countries in Europe, reflecting the family structure that predominated in each country—authoritarian in France and Germany and a more persuasive approach in the Netherlands, for instance—but what has been called the 'domestication of science' was a widespread phenomenon.[24] It may simply have been the continuation of a long tradition. In the early modern period, professors had considered themselves to be not only teachers but also the mentors of students, *in loco parentis*—taking the place of the students' parents. The idea of the professor as the 'good father' of a laboratory community thus was perhaps simply the modern equivalent of this age-old image that they had of themselves.

The Fragmentation of Student Life

Student life was also deeply affected—and fragmented—by the rise of the laboratory.[25] In the middle of the nineteenth century, science students (there weren't many at that time) attended their lectures and their few practicals in the Academy Building. By 1914, however, they almost never visited this building anymore and instead spent most of their working hours in the laboratories on the outskirts of town. In the mid-nineteenth century almost all students were members of the still existing Student Association (Vindicat atque Polit, as it is called, Vindicat for short), but this was changing in the last decades of the century. There were several reasons for this: the recruitment of students from non-academic circles, the growing dissatisfaction with the bizarre and sometimes rather cruel

23 Truus van Bosstraeten, 'Dogs and Coca-Cola: Commemorative Practices as Part of Laboratory Culture at the Heymans Institute Ghent, 1902–1970', *Centaurus*, 53 (2011), 1–30. This article was one of the results of the large-scale research project 'Laboratory Culture in Belgium, 1870–1950' at the University of Leuven.

24 It remains an open question whether the paternalistic style of managing a laboratory fostered or inhibited creativity. For some insightful remarks, see: Jeffrey A. Johnson, 'Hierarchy and Creativity in Chemistry, 1871–1914', in Kathryn M. Olesko, ed., *Science in Germany: The Intersection of Institutional and Intellectual Issues*, *Osiris*, 5(1) (1989), 214–40. Johnson deals exclusively with chemical laboratories.

25 For the following paragraph, see: Van Berkel, *Universiteit van het Noorden*, vol. 2, pp. 315–40, esp. pp. 334–9.

rituals practised in the student association, the high cost of membership, and the loss of contact with other students because of the University's spatial fragmentation. This caused a new kind of student organization to blossom: the faculty association (*Faculteitsvereniging* in Dutch). Five of these faculty associations were established in Groningen in 1907, one for each faculty. Those of the faculties of Medicine and Science were the most active. As these faculties by now had the largest number of students, these faculty associations, which indeed comprised practically all the students in the faculties involved, had more members than the others. The faculty associations not only looked after students' material needs but also organized lectures by famous scientists and physicians. These lectures were not held in the Academy Building but mainly in the laboratories or clinics, where the professor was so kind as to allow the students to use his lecture hall for the event. The professor would usually lend a helping hand when it came to inviting a well-known foreign confrere, and it was not uncommon for this foreign guest to stay at the professor's home during his visit to Groningen. Finally, the faculty associations also organized social events. In 1928 the faculty association of the Faculty of Science even acquired its own clubhouse when a small engine room, which had fallen out of use after the laboratory was renovated, was handed over to the faculty association.[26] The General Student Association, Vindicat, was not amused. It saw this as a direct threat to its position, but could do nothing about it. The inauguration of the new clubhouse for science students symbolized the growing fragmentation of the University at the student level.

Nostalgia

The General Student Association's protest was not the only or the first instance of growing dissatisfaction with the new reality of a university that was breaking up into semi-independent units. At the inauguration of the new Academy Building in 1909, the acting rector magnificus, the philosopher and psychologist Gerardus Heymans, gave a speech in which he commented on the fact that the University was falling apart. With some nostalgia, he called to mind the former building, which had burnt down in 1906. In many ways, this building had been obsolete, with its cramped quarters, its out-of-date laboratories, and its gloomy lecture halls. Yet the building had symbolized the unity of science—something that Heymans

26 Van Berkel, *Universiteit van het Noorden*, vol. 2, p. 689.

valued highly. It had been the place where all the professors met on a regular basis, both in the Senate meeting room and the professorial waiting room, and it was therefore especially painful to lose it.

> The dismay we felt was of a different order than what we would have felt after the most beautiful laboratory burned down; this dismay stood in no proportion to the material and even moral damage that was done.[27]

For all its defects, the old Academy Building had still been the 'mother house' of the University. Now that virtually all the professors had found a place outside the Academy Building, these days were over. Heymans conceded that the University's rapid expansion was a sign of unmistakable progress and growth, something to be proud of, but it was also a threat to the very idea of a university, which he felt was intimately tied up with the unity of the sciences.

The Physical Society as a Counterweight

Heymans was not the only professor to deplore the breaking apart of the university. In fact, even before the turn of the century, some professors had taken steps to remedy this situation and to restore some kind of unity in academic life. One of the countermeasures taken involved revitalizing an old scientific society in Groningen, the Physical Society (nowadays the Royal Physical Society of Groningen, *Koninklijk Natuurkundig Genootschap*). Although the Society as such was not part of the University, it has always played an important role in academic life. The history of the University of Groningen would be incomplete without due attention being paid to this society.[28]

Originally, in 1801, the Society was founded by a couple of wealthy, enterprising students who wanted to carry out the experiments that their professors only talked about. In the first half of the nineteenth century, the Physical Society provided laboratory space and instruments for professors to conduct scientific research that the University offered no opportunity for. After the middle of the century, however, when the University was finally

27 As quoted in: *Jaarboek Rijksuniversiteit te Groningen* (1908–9), 29–30 (translated by the author).

28 C. Wiese, ed., *Een spiegel der wetenschap. 200 jaar Koninklijk Natuurkundig Genootschap te Groningen* (Bedum: Profiel, 2001).

acquiring the means to set up decent laboratories and to hire technical staff to support the professors, the role of the Society as an auxiliary institution of the University was more or less over. Like so many other scientific societies, it gradually evolved into an institution for the popularization of science. The lectures were intended for the general public, and the instruments were sold to the new, modern high school that was established in 1864. Membership dropped and the Society would have withered away if nothing had happened. By the end of the nineteenth century, however, a few professors from the Faculty of Science established a 'Scientific Department' within the Society with the full support of its board, now presided over by Florentius Groneman, the head of the aforementioned high school and a physicist himself. Professor of pharmacy Johan Frederik Eijkman, who had been a member of the highly influential Amsterdam Society for the Promotion of Science, Medicine, and Surgery and had seen the importance of that society's sessions for the coherence of the Faculties of Science and Medicine at the University of Amsterdam, suggested establishing a separate scientific department within the general Physical Society in order to offer a meeting place for researchers in natural science, medicine, and related disciplines. Eijkman became the first president, but not for long. He was soon replaced by the professor of chemistry Arnold Frederik Holleman and then by the eminent astronomer Kapteyn, who was liked by everyone. From 1900 on, the new Scientific Department met monthly, first at the 'Concerthuis', a former concert hall in the city centre, which was the home of the Physical Society (located just below no. 6 on the map of Groningen, fig. 5.2), and then in its members' laboratories. There, the professors of science and medicine mingled with the highly skilled teachers from the local high school and the gymnasium, thus making up in some way for the loss of unity within the University itself.

Conclusion

The students in Groningen also took measures, albeit a little later than the professors, to counter the disruptive forces at play in the modernizing and disintegrating University.[29] In 1924, for instance, they set up an independent students' weekly called *Der Clercke Cronicke* ('Student Chronicle'), in which all the student organizations could advertise their meetings and where students could discuss any issues relating to science, scholarship, student

29 For the following paragraph, see: Van Berkel, *Universiteit van het Noorden*, vol. 2, pp. 660–84.

life, and social or even political life in general. The weekly was also a vehicle for the University to announce when professors would start teaching again after the holidays, who had taken their exams, and who had defended their dissertations. All students, either through their organizations or by subscribing to the newspaper directly, were readers of *Der Clercke Cronicke*. Thus, the newspaper re-established, at least on paper, the unity of the University that had been undermined, unintentionally of course, by building so many laboratories scattered throughout the city.

The results of measures like these were disappointing. Neither the establishment of the Scientific Department of the Physical Society nor the publication of the student weekly *Der Clercke Cronike* proved effective at stopping the further disintegration of the University. The success of the Scientific Department of the Physical Society—and it was a success—was not matched by comparable arrangements in other faculties. And the student weekly, while bringing together students from very different backgrounds and with widely diverse interests, also highlighted and deepened the very fundamental antagonisms within the student body.

It was not until the twentieth century that real measures were taken to counter the disruptive impact of the local diaspora of academic institutions in Groningen. After the Second World War, as the University again rapidly expanded and was in dire need of new buildings, a decision was taken to concentrate the laboratories and institutes of one particular faculty in one place.[30] Utrecht University had led the way with the creation of the Uithof, a campus in the rural area east of the city. In Groningen beginning in the late 1960s, the idea of a campus, modelled on the American university, led to the creation of a whole new series of science laboratories on a site on the outskirts of the city, at the time called Paddepoel. (It was renamed the Zernike Complex much later, after the physicist Frits Zernike, who won the Nobel Prize for Physics in 1953, Groningen's first Nobel Prize winner.) Over the course of time, the Faculties of Economics and Spatial Sciences (Geography) also moved to that location. Yet, the Academy Building, with its administrative and representative functions, remained in the inner city, as did the University Library and the Faculties of Arts, Law, Social

30 The idea of concentrating the University of Groningen at three locations was developed by the secretary of the Board of the University, J.L.H. Cluysenaer, in 1958, but in 1959 the decision to locate the science departments in the meadows north of the city was taken when the municipality of Groningen offered this location. See: Jaap Bruintjes and Marijke Martin, 'Het ruimtelijk beleid van de universiteit', in Auke van der Woud, ed., *De innige betrekking tussen stad en hogeschool. Architectuur en stedebouw aan de Rijksuniversiteit Groningen 1950-1984* (Groningen: Rijksuniversiteit Groningen, 1984), pp. 6–24.

Sciences, Philosophy, and Religious Studies. (The University Hospital and the Medical Faculty also remained at their former location, to the east of the inner city.) The concentration and re-establishment of links between the separate faculties by creating a campus was thus only partially successful.

The story is not very different at other universities, such as Leiden, where the natural sciences and a Science Park were created in a polder to the west of the city (Leeuwenhoek), or in Utrecht, where the Uithof was developed in the Johannapolder to the east, also for the natural sciences and some other departments. A real campus, where all departments were located, was only realized in some of the younger universities, such as the Vrije Universiteit Amsterdam (Free University Amsterdam), Radboud University in Nijmegen, and the technical universities of Delft, Eindhoven, and Twente. Spatial dispersal therefore remains a characteristic of many of the Dutch universities.

Acknowledgements

Earlier versions of this chapter were read at the conference 'The Laboratory Revolution. The Rise of the Laboratory and the Changing Nature of the University', University of Groningen, 26–27 October 2017, and the workshop 'The Architecture of Science and the Humanities', NIAS, Amsterdam 15–16 May 2019. I wish to thank Fabian Kraemer for inviting me to this workshop. I would also like to thank the participants at both meetings for their comments.

About the Author

Klaas van Berkel recently retired as Rudolf Agricola Professor of History at the University of Groningen. In his research he focuses on cultural history and the history of science, especially the history of scientific institutions (academies, universities). In 2021 he published, with Guus Termeer, *The University of Groningen in the World. A Concise History* (Amsterdam: Pallas Publications).

6 A Fertile Ecosystem: University Chemical Laboratories and their Suppliers in Fin-de-Siècle Paris

Pierre Laszlo

Abstract

In Fin-de-Siècle Paris, the laboratory revolution was powered by an ideology comparable to the Dreyfus case: a determination for military revenge against Germany. Hence the need to beat the Germans at their own game, excellence in chemical science and technology. While the French state instituted faculties of sciences in both Paris and the provinces, commercial suppliers sprouted in Paris to provide laboratory equipment, glassware, and chemicals. They also fed upon the contemporary rise of photography. The French government established the *Laboratoire municipal de chimie* in Paris with the mission of making food and drink safe for Parisians. This chapter focuses on some leading commercial suppliers with shops in the Latin Quarter, conveniently close to their laboratory clients at the Sorbonne.

Keywords: fin-de-siècle Paris, Sorbonne, pharmacists, laboratory suppliers, photography

Introduction

This chapter is set in Paris during the *fin-de-siècle* period 1880–1910. This was a time when Paris was increasingly recognized as the cultural capital of Europe.[1] Moreover, Parisian institutions of higher learning had something of a fresh start of after the Franco-Prussian War, and this setting offers excellent opportunities for studying the Laboratory Revolution.

1 Christophe Charle, *Paris, fin de siècle. Culture et politique* (Paris: Le Seuil, 2016).

Berkel, Klaas van, and Ernst Homburg (eds), *The Laboratory Revolution and the Creation of the Modern University, 1830-1940*. Taylor & Francis Group, 2023
DOI: 10.5117/9789463720434_CH06

In the aftermath of the Franco-Prussian War of 1870, the French government and the elite discussed reasons for their defeat.[2] The consensus was that Prussia had won primarily due to its superiority in science and technology.[3] Accordingly, it was decided to upgrade the French educational system, especially in the sciences. The necessary resources were allocated, and money was poured into this national emergency. Existing institutions of higher education were revamped, especially with respect to research and science education. Novel institutions, meant to occupy the leading edge, were built.[4] In short, public opinion was intent upon revenge and beating the Germans at their own game.

Easier said than done! In chemistry especially, Germany at that time led the world. And whereas French intellectuals boasted of their excellence in abstract thought and theory, French experimental science was weaker. The ruling elite were keenly aware of this blind spot and strove to remedy it, an instance of the Laboratory Revolution in a protracted mode.

Some Milestones in French Chemistry

The deliberate post-1870 effort by the French State to further higher education and boost the sciences paid off. Novel institutions reflected the liveliness of French chemistry. The *Société chimique de Paris*, founded in 1857, blossomed at the end of the century and sprouted provincial offshoots in Nancy (1895), Lyon (1898), Lille and Toulouse (1902), and Marseille and Montpellier (1905).[5]

In the provinces faculties of sciences were endowed with new, modern laboratories.[6] Albin Haller (1849–1925) was the first director of the *Institut de chimie* in Nancy upon its creation in 1887. The following year he synthesized nitric acid. In Toulouse Paul Sabatier (1854–1941) studied heterogenous catalysis by metals and, together with Jean-Baptiste Senderens (1856–1937), devised the nickel hydrogenation catalyst in 1897. At the turn of the century in Lyon, Victor Grignard (1871–1935) discovered the eponymous reaction

2 Christophe Charle, 'Elite Formation in Late Nineteenth Century: France Compared to Britain and Germany', *Historical Social Research/Historische Sozialforschung*, 33 (2008), 249–61.

3 Pierre Laszlo, *La leçon de choses* (Paris: Austral, 1995), p. 160.

4 Jean-François Picard, 'L'Organisation de la science en France depuis 1870: un tour des recherches actuelles', *French Historical Studies*, 17 (1991), 249–68.

5 [Société chimique de France], *Centenaire de la Société Chimique de France, 1857–1957* (Paris: Masson, 1957).

6 Mary Jo Nye, *Science in the Provinces: Scientific Communities and Provincial Leadership in France, 1860–1930* (Berkeley/Los Angeles: University of California Press, 1986).

and reagents. In Grenoble François-Marie Raoult (1830–1901) established the laws for cryoscopy and ebullioscopy, which Robert Lespieau (1864–1947) would popularize in Paris during the early 1890s.[7]

In Paris Marcellin Berthelot (1827–1907) had synthesized acetylene from the elements with his *œuf électrique* ('electric egg'; an oven) in 1862.[8] Alsatian in origin, Charles Adolphe Wurtz (1817–1884) gradually built up his laboratory in Paris, not at the Sorbonne but in the medical faculty. By contrast to eminent chemists such as Marcellin Berthelot and Henri Sainte-Claire Deville (1818–1881), he was a prominent advocate of the atomic theory. He devised the Wurtz coupling reaction, which formed carbon–carbon bonds from reacting alkyl halides with sodium. A gifted educator, he published an influential textbook of organic chemistry. But most importantly for the topic of this book, he also induced his co-workers to publish the annual *Agenda du chimiste* starting in 1877, a resource with all kinds of data useful in the chemical laboratory, a kind of forerunner of the *Handbook of Chemistry and Physics*.[9]

In 1882 some of the Alsatian refugees who were influential in their new, Parisian base set up the *École de physique et de chimie industrielles* (*de la Ville de Paris*).[10] Charles Friedel (1832–1899) and James Crafts (1839–1917) devised their all-important alkylation or acylation reaction in 1877.[11] In 1896 Henri Becquerel (1852–1908) discovered radioactivity. In 1898 the Curies, Pierre (1859–1906) and Marie (1867–1934), discovered radium and polonium. Their laboratory then consisted of a mere shed in the courtyard of the *École de physique and chimie industrielles*.

Parisian chemistry in the 1880s and 1890s was split between the establishment and non-conformists. The latter were defenders of atomic theory, which had yet to receive the approval of officialdom, and they sheltered in

7 Richard Williams, 'François-Marie Raoult et la loi de Raoult', *Reflets de la physique*, 29 (2012), 22–23.

8 See also: Marcellin Berthelot, *Chimie organique fondée sur la synthèse* (Paris: Mallet-Bachelier, 1860).

9 Collective: a group of Wurtz's co-workers *Agenda du chimiste* (Paris: Hachette et Cie., 1896). It started in 1877 and carried on into the turn of the century. Most suppliers advertised in its pages.

10 Ana Carneiro and Natalie Pigeard, 'Chimistes Alsaciens à Paris au 19ème siècle: un réseau, une école?', *Annals of Science*, 54 (1997), 533–46; Terry Shinn, 'Des sciences industrielles aux sciences fondamentales: La mutation de l'Ecole supérieure de physique et de chimie (1882-1970)', *Revue Française de Sociologie*, 22 (1981), 167–82.

11 Xavier Bataille and Georges Bram, 'La découverte de la réaction de Friedel et Crafts', *Comptes Rendus de l'Académie Des Sciences – Series IIC – Chemistry*, 1(4) (1998), 293–96.

Adolphe Wurtz's laboratory at the *École de médecine*.[12] A professor at the *École de pharmacie*, Auguste Béhal (1859–1941), seceded: while Berthelot was officially teaching organic chemistry at the *École de pharmacie* using the formalism of equivalents, Béhal taught there—in the very same institution— an informal parallel course in organic chemistry, based on atomic theory, from 1890 to 1897.[13] As an offshoot from his rebellion, Ernest Fourneau (1872–1949) and his brother-in-law Marc Tiffeneau (1873–1945) started a research group in 1903.[14]

Parisian Laboratory Suppliers

The growth of chemical research and the construction of several new laboratories in Paris naturally also led to a growing demand for laboratory instruments and chemicals. In this chapter we shed an unexpected light on the Laboratory Revolution by investigating the role of commercial laboratory suppliers. It provides an overview of what laboratory equipment then consisted of. It identifies and locates the main Parisian laboratories and points to the relative influences of the German and the British suppliers. It depicts the interplay of scientists and suppliers in what was, to use an inescapable metaphor, a rich ecosystem.

The commercial laboratory suppliers numbered between ten and twenty during the period 1880–1910.[15] Their number however pales in comparison to the suppliers to artists, specifically painters, whose number during the same

12 Alan J. Rocke, 'Atoms and Equivalents: The Early Development of the Chemical Atomic Theory', *Historical Studies in the Physical Sciences*, 9 (1978), 225–63; Charles Adolphe Wurtz and Armand Gautier, 'Laboratoire de chimie biologique de la Faculté de Médecine', *Rapport sur l'École pratique des hautes études*, 1 (1873), 37–39.

13 Marcel Delépine, 'Souvenirs: Auguste Béhal: I. son Agrégation; II. la notation atomique', *Revue d'histoire de la pharmacie*, 48 (1960), 249–54.

14 Known as *La Molécule*, it had monthly meetings, each of which featured a presentation of recent work. It gathered about twenty former béhalian students, most of them graduates from the school of pharmacy. Béhal had been chief pharmacist at the Cochin Hospital, where he was seconded by Amand Valeur (1870–1927) and Edmond Blaise (1872–1939), two of the founding members of La Molécule. Other members included Gabriel Bertrand (1867–1962), Albert Buisson (1881–1961), Raymond Delange (1887–1948), Marcel Delépine (1871–1965), Justin Dupont (1869–1943), Louis Givaudan (1875–1936), Nicolas Grillet (1871–1947), and Marcel Sommelet (1867–1952). See: Viviane Quirke, *Collaboration in the Pharmaceutical Industry* (London: Routledge, 2012), p. 56.

15 A short list of the most important suppliers to laboratories would include Adnet, Alvergniat Frères, Brewer Frères, Chenal & Douilhet, Darrasse, Exupère, Georges Fontaine, Lelièvre, Levieil, Pellin, Poulenc Frères, Emile Rousseau, and Paul Rousseau.

period ranged 50–100.[16] This is not the place for a detailed comparative analysis, though. A brief summary will suffice. Both types of suppliers catered to creativity, artistic or scientific. The former was the province of individuals, highly idiosyncratic as a rule. Moreover, each artist had favourite suppliers, no more than half a dozen, to whom he—very seldom she—was loyal.[17] The latter catered to small groups in laboratories. Both types kept substantial inventories in their shops and accordingly needed large stockrooms. Both types were family businesses, passed on from father to son and often from husband to widow. A contrasting feature was the geography: suppliers to painters—*marchands de couleurs*, i.e. dealers in paints—were located in most of Paris *intramuros*, not only in the Montparnasse and Montmartre areas where many of the artists lived. By contrast, suppliers to laboratories set up shop predominantly in the Latin Quarter, in close proximity to the labs. They supplied not only academic laboratories but also hospital laboratories. With the steady increase of the Parisian population during the second half of the nineteenth century, from 1.5 million in 1850 to nearly 4 million in 1900, and with the heightened concern about epidemics, hygiene, and health, hospitals needed laboratories to analyse their patients' bodily fluids.[18] Typically each such laboratory was headed by a chief pharmacist, most likely to have received training at the *École de pharmacie*.[19] Along with those hospital laboratories, major new chemistry laboratories dedicated to public health and hygiene were established during the period.

16 Manuel Charpy, 'Les ateliers d'artistes et leurs voisinages', *Histoire urbaine*, 26(3) (2009), 43–68.

17 A short list of the most important would include Blanchet, Bourgeois Aîné, Briault, Contet, de Rivière, Dubus, Gay, Haro, Louis Latouche, Lefranc et Cie, Sennelier, Tanguy, and Tasset & Lhote; Clothilde Roth-Meyer, 'Les marchands de couleurs à Paris au XIXe siècle' (PhD diss., Paris 4, 2004).

18 Jack D. Ellis, *The Physician-Legislators of France: Medicine and Politics in the Early Third Republic, 1870–1914* (Cambridge: Cambridge University Press, 1990).

19 Let me mention in passing some such chief pharmacists: Émile Bourquelot (1887–1919) at Hôpital Laënnec; Joseph Bougault (1870–1955) at Hôpital Trousseau, followed by the hospitals Hérold and Tenon; H. Carrion at Hôpital Saint-Antoine; Paul-Louis Chastaing (1847–1907) at La Pitié hospital; Henri-Charles Cousin (1863–1936) at Broussais hospital; Marcel Delépine at Bretonneau hospital in the 1890s (he would end his career as a professor at Collège de France); Jean-Victor Gasselin (1861–1916) at Hôpital Tenon; Léon Grimbert (1860–1931) at La Pitié from 1882 and Hôpital Cochin from 1897 until 1906; Marcel Guerbet (1861–1938), chief-pharmacist at Hôpital Bichat, replaced by Maurice François; Henri Hérissey (1873–1959) at Hôpital Bretonneau; E. Léger at Hôpital Beaujon; Patein at Hôpital Lariboisière; Ludovic Portes (1846–1923) at Saint-Louis hospital; the aforementioned Marc Tiffeneau at Hôpital Boucicaut (1897–1902); Eugène Villejean at Trousseau, Lariboisière, and the Hôtel-Dieu (1884–1916). All were scientists who, in addition to their hospital duties and their analytical work assisting physicians, followed their curiosity to explore the natural world.

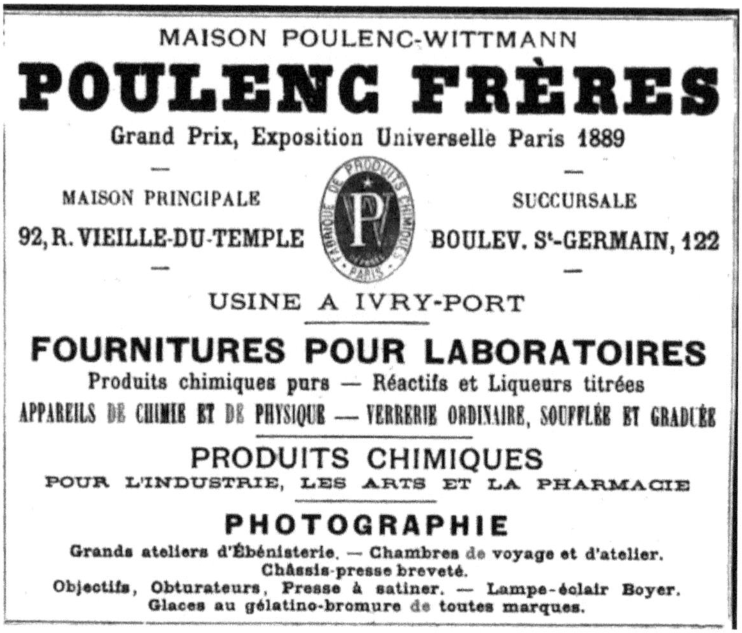

Figure 6.1 Advertisement for Poulenc Frères. Source: A. Wurtz, *Dictionnaire de chimie* (Paris, 1894).

A New Government Laboratory

The *Laboratoire municipal de Paris* (hereafter: *Laboratoire municipal*) was a major client for laboratory suppliers in Paris at the end of the nineteenth century; it was admirably staffed and well-funded. This state-of-the-art chemistry laboratory monitored the quality of drinking water and the authenticity of food products and wine and thus closely monitored the health of city dwellers.[20] Those were its core missions. Adolphe Wurtz, originally from the Alsace and someone whose early career had benefited from the contributions of German chemistry, was behind the creation of that laboratory.

The *Laboratoire municipal* was set up by the authorities in 1879 in part because of demographics—the Parisian population had grown to a few million—and in part because of the lurking epidemics—cholera (1854, 1862), typhus, typhoid, yellow fever in Saint-Nazaire (1857), tuberculosis, and flu.[21] The laboratory opened in 1881.

20 *Deuxième congrès de chimie pure et appliquée*, 5 vols. (Paris: 1897), vol 3, p. 175; Henry Huet-Deaunay, *Le Laboratoire municipal et les falsifications ou Recueil des lois et circulaires* (Paris: F. Pichon, 1890).
21 J. Bertillon, 'Epidemic Influenza in France', *Transactions of the Epidemiological Society of London*, n.s. 9 (1891), 103; Patrice Bourdelais and Jean-Yves Raulot, *Une peur bleue: histoire du choléra en France 1832-1854* (Paris: Payot, 1987); Allan Mitchell, '"Obsessive Questions and Faint

The budget of the laboratory, in addition to the salaries, was 200,000 francs per year, which translates, as an order of magnitude, to about one million euros today. This was a very large amount at the time.[22] The City of Paris located the new laboratory in a central building, near the police headquarters and Notre Dame cathedral on the La Cité island, in former barracks that were renovated into brand-new laboratories. The renovation was directed by Charles Girard (1837–1918), the appointed head of the new laboratory. He had been Wurtz's lieutenant at the *École de medecine*.[23] To help him run the new laboratory, Girard had another former Wurtz co-worker, Anatole Dupré, as the assistant director. The staff numbered 67, viz. the director, two assistant directors, three principal chemists, 29 chemists, 26 expert-inspectors, six secretaries, one photographer, five laboratory assistants, and a coach driver.[24]

Note the hierarchical organization, typical of both France and the period. The role of the inspectors was to visit food stores and collect samples for analysis, both chemical and bacteriological. Foremost were analyses of wine, milk, and drinking water.[25] Wine was emphasized primarily because quite a few adulterated wines were being sold following the devastation of French vineyards by the phylloxera pest (an insect).[26]

Answers: The French Response to Tuberculosis in the Belle Epoque', *Bulletin of the History of Medicine*, 62 (1988), 215–35; Scarlett Beauvalet-Boutouyrie and Pierre Boutouyrie, 'Du geste qui tue au geste qui sauve. Épidémies et procédures médicales invasives à Paris au XIXe siècle: l'exemple de la maternité de Port-Royal', *Annales de démographie historique*, 1997, no. 1 (1998), 135–55; Jack D. Ellis and David A. Barnes, 'The Making of a Social Disease: Tuberculosis in Nineteenth-Century France', *Journal of Interdisciplinary History*, 27 (1997), 519–20; Elizabeth T. Hurren, 'Poor Law versus Public Health: Diphtheria, Sanitary Reform, and the "Crusade" against Outdoor Relief, 1870–1900', *Social History of Medicine*, 18 (2005), 399–418.

22 Léon Martin, ed., *Encyclopédie municipale de la ville de Paris*, 2 vols. (Paris: Georges Roustan, 1902–04), vol. 2, p. 1868.

23 Initially a student of Pelouze, he had earlier been involved in the budding dyes industry at Société La Fuchsine: Henk van den Belt, 'Why Monopoly Failed: The Rise and Fall of Société La Fuchsine', *The British Journal for the History of Science*, 25 (1992), 45–63 (pp. 45–46); Willem J. Hornix, 'From Process to Plant: Innovation in the Early Artificial Dye Industry', *The British Journal for the History of Science*, 25 (1992), 65–90; Ernst Homburg, 'The Emergence of Research Laboratories in the Dyestuffs Industry, 1870-1900', *The British Journal for the History of Science*, 25 (1992), 91–111.

24 Martin, ed., *Encyclopédie municipale*, vol. 2, p. 1867.

25 Alfred Picard, ed., *Rapports du jury international. Exposition Universelle de 1889 à Paris.* Group VI, Part 6 (Paris: Imprimerie nationale, 1892), p. 294; Jean-Pierre Goubert, *The Conquest of Water: The Advent of Health in the Industrial Age* (Cambridge/Oxford: Polity Press in association with B. Blackwell, 1989).

26 Roger Pouget, *Histoire de la lutte contre le phylloxéra de la vigne en France: 1868-1895* (Paris: INRA/ Quae, 1990); Allesandro Stanziani, 'La falsification du vin en France, 1880-1905: un cas de fraude agro-alimentaire', *Revue d'histoire moderne & contemporaine*, 50 (2003), 154–86.

Figure 6.2 *Le laboratoire municipal de chimie de Paris.* A painting by Ferdinand Gueldry (1858–1945), presented to the Salon of 1887. Notice the glassware and the earthenware. Some chemists wore a large apron but no lab coats. Courtesy: Carnavalet Museum of the City of Paris.

Two rooms in the basement were allocated to chemical and bacteriological analyses of water samples[27] and to wine distillation.[28] On that lower level were also two dark rooms, one for photometry and spectroscopy and another for photography, a stockroom for supplies, a meeting room for the inspectors, a room for gas analyses, and a room for machinery such as vacuum pumps. On the upper level were the offices of the director and his assistants, two dark rooms, three large laboratories with a maximum capacity of 37 workers, an office for the secretaries, and a reception lobby.

The *Laboratoire municipal* prided itself upon owning a Jobin spectroscope.[29] At the official opening ceremony, guests were thus presented with absorption spectra of wine, fuchsin (the red dye patented by the Renard

27 Léon Dufour, 'Notice sur le laboratoire municipal de Paris', in *Petit dictionnaire des falsifications* (Paris: Félix Alcan, 1902), p. 188.

28 Pouget, *Histoire de la lutte contre le phylloxéra*; Stanziani, 'La falsification du vin en France'.

29 François Dupont, *Deuxième Congrès International de Chimie Pure et Appliquée. Comptes-Rendus*, 5 vols. (Paris: Association des chimistes de sucrerie et de distillerie, 1897), vol. 3, p. 179;

brothers in 1859), blood, and sodium, along with emission spectra of sodium, potassium, silver, and copper.

In his speech, Jules Cambon (1845–1935), then general secretary, i.e. executive director of Préfecture de Police, mentioned a then nagging concern among Parisians: the stench. Paris was stinking to the high heavens due to the lack of a network for connecting toilets in houses and buildings to the sewage system. Between late July and early October 1880, the horrible smell peaked. It happened again in the summer of 1881.[30]

The *Laboratoire municipal*, as shown by the number of the analyses it performed, continually increased production during the end of the century. The activity of the *Laboratoire municipal* testified to the Laboratory Revolution in a progressive way. Setting it up was an important step in the build-up of the future welfare state. By investigating the quality of food and drink, the Third Republic was paternalistic; the French government monitored the health of its citizens in a Big Brotherly-manner.

Let us turn now to the second-in-command at the *Laboratoire municipal*, a most interesting figure in his own right.

Dupré: a Designer of Instruments

Anatole (Pierre Victor Désiré) Dupré springs out of the all-too-few sources I was able to locate, like a character from a novel by Jules Verne (1828–1905). He was a representative of the Age of Science at its heyday in nineteenth-century France: he was both an inventor and a pioneer glaciologist. Born about 1850, he was a displaced Alsatian following the 1870 defeat France suffered against Prussia. Accordingly, in like manner to a great number of fellow-Alsatians, the young Dupré—then in his late teens or early twenties—left his hometown of Strasbourg for Nancy and in December 1871 was offered a position there as a laboratory technician in chemistry at the local university.[31] He subsequently moved to Paris and in 1878 became a member of Adolphe Wurtz's group, working in his laboratory at the *École de*

Conseil municipal de Paris, *Rapports et Documents*, deuxième partie, de 94 à 166 (Paris: Imprimerie Municipal, 1901), p. 465.

30 David S. Barnes, 'Scents and Sensibilities: Disgust and the Meanings of Odors in Late Nineteenth-Century Paris', *Historical Reflections / Réflexions Historiques*, 28 (2002), 21–49; David S. Barnes, *The Great Stink of Paris and the Nineteenth-Century Struggle Against Filth and Germs* (Baltimore, MD: Johns Hopkins University Press, 2006); Donald Reid, *Paris Sewers and Sewermen: Realities and Representations* (Cambridge, MA: Harvard University Press, 1991).

31 Isaure Triby, *Manifestations of Cultural Change: Alsatian Identity between 1871 and the Interwar Period. Three Case Studies* (PhD diss., University of Exeter, 2012).

médecine in Paris. Wurtz, as mentioned above, also came from the Alsace. Some years earlier, during the summers from 1866 to 1869, Dupré would accompany a friend, an older Alsatian named Charles Grad (1842–1890), on mountaineering trips through the Alps, genuine scientific explorations monitoring various glaciers.[32]

Dupré stayed only a few years in Wurtz's laboratory. In 1881 he joined the newly created *Laboratoire municipal* as its associate director. The director, Charles Girard, was a fellow chemist and a fellow Alsatian from Wurtz's laboratory. In the *Laboratoire municipal* Dupré proved himself to be a tinkerer of genius. It is a very long list indeed of the scientific apparatus he either modified (and improved) or designed. He worked on: an apparatus for solvent depletion of samples, determination of CO_2 concentrations, determination of the densities of gases from their flow rates, an azotometer, a dialyser, drying enclosures, an ebullioscope, an eudiometer, a fractional still, an incineration oven, a refractometer, a thermostat, an ureometer, etc.[33]

Many of those instruments devised by Dupré would be manufactured and sold by Alvergniat Frères, to whom we now turn. The story of this supplier is typical in many ways: a shop located in the Latin Quarter, near the Sorbonne; gradual diversification of the articles offered for sale; their increased sophistication over time; and, parallel to the powerful move of the country into its second industrialization at the end of the century, vastly increased technicity on the part of its owners.

32 A. Charles Grad, *Observations sur les glaciers de la Viège et le massif du Monte-Rosa, etc.* (Paris: Challamel, 1868); A. Charles Grad, 'Observations sur les glaciers du Grindelwald', *Bulletin de la Société des sciences naturelles de Strasbourg*, 2 (1869), 75–82; A. Charles Grad, 'Fragments d'un voyage aux Alpes', *Revue d'Alsace*, 3 (1869), 409–14; A. Charles Grad, 'Une Campagne sur le glacier d'Aletsch. Août et Septembre 1869', *Annales des voyages, de la géogaphie, de l'histoire et de l'archéologie* (July 1870), 6–19; A. Charles Grad, 'L'ozône', *Bulletin de La Société d'histoire Naturelle de Colmar*, 11 (1870), 195–210; A. Charles Grad, 'Théorie du mouvement des glaciers', in *Association Française pour l'avancement des sciences. Compte rendu de la 3me session. Lille 1874*. (Paris: Secretariat de l'Association, 1875), pp. 279–86.

33 Ph. Pellin, 'Réfractomètre de M. A. Dupré. Appareil pour mesurer les indices de réfraction des liquides ou des gaz, construit pour le laboratoire municipal de Paris', *Journal de Physique Théorique et Appliquée*, 8 (1889), pp. 411–15; Alfred Picard, ed., *Exposition universelle de 1889. Rapport du jury international*. Groupe VI, Part 6 (Paris: Imprimerie nationale, 1892), pp. 292–98; Raoul Neveu, *Catalogue général illustré: verrerie soufflée et graduée, porcelaine, appareils et ustensiles pour laboratoires scientifiques et industriels* (Paris: published by the author, 1910), available online at http://cnum.cnam.fr/redir?M9932; Claude Viel, 'Le laboratoire et les instruments de chimie, du XVIIe à la seconde moitié du XIXe siècle', *Revue d'histoire de la pharmacie*, 90 (2002), 7–30.

The Alvergniat Frères Suppliers

This is a rags-to-riches story in the American style. However, not only is it a true story, but it occurred in nineteenth-century France as a prelude to the *Belle Époque*, a period thus named for large wealth and middle-class prosperity.

Charles-Désiré Alvergniat (1831–1883) was born in the village of Droué in the Loire area. His brother Auguste-Adrien (1834–1898) was also born there. Their father, Joseph Alvergniat, was a carpenter in that village; he was originally from the Auvergne in central France, hence the family name. The family came to Paris in 1838. Charles was apprentice to a glassblower who worked for the Fastré company, and one of their clients was the physicist Victor Regnault (1810–1878), a professor at Collège de France. As soon as Charles had learned the craft, he established himself up as a glassblower on what became rue Séguier.

Adrien, after other rather menial jobs, had his brother show him how to blow glass. Soon, he became proficient in the art and outskilled Charles in this vocation. He also set up a company, which was located near the Sorbonne in 1862–3. Because he was so good and so reliable at glassblowing, his business became a big success. Adrien was the first Frenchman able to build Geissler tubes, so the French laboratories would no longer have to import them from Germany. From Geissler tubes, he would go on to make Crookes tubes and radiometers.[34]

Such was his success with scientists of the Sorbonne and with other scientific laboratories in the Latin Quarter that his elder brother joined him towards the end of the 1860s: the Alvergniat Frères company was born. It moved in 1874 from Passage de la Sorbonne to more ample premises at 10 rue de la Sorbonne—just across from the university. From glassblowing, Alvergniat Frères diversified not only into all kinds of glassware, including Erlenmeyer vials, but also into artefacts made of china porcelain, earthenware, and stoneware, such as crucibles, jugs, funnels, and Büchner funnels for filtration under partial vacuum. Indeed, Adrien Alvergniat became famous also for building and selling water pumps that the chemical engineer Henri Lasne, who was mostly active in the 1880s and 1890s, had devised. From water pumps, he went on to mercury pumps and to other apparatus for establishing, maintaining, and measuring a vacuum.[35] In 1892 those

34 C.V., 'Nécrologie', *La Nature, revue des sciences*, 26 (1898), p. 270; Nestor Gréhant, *Manuel de Physique Médicale* (Paris: Baillière, 1869), p. 84.

35 Henri Becquerel, *Phosphorography of the Infra-Red Region of the Solar Spectrum: Wave-Length of the Principal Lines* (London: Taylor & Francis, 1883).

included the high-precision Spengel mercury pump. Alvergniat frères also made and sold thermometers. Alvergniat Frères thus underwent a very successful development period in the 1870s and 1880s. By the end of the century, they were a well-established supplier across from the Sorbonne, right next to the new laboratories under construction.

Starting in 1879, Victor Chabaud headed the company. In 1890 he purchased it and only added his name to that of the Alvergniat brothers. Under his management, the company continued to thrive and to innovate, particularly in the area of X-ray devices. At the turn of the century, Jules Thurneyssen (1868–1929), a graduate from the *École polytechnique*, purchased the company from Chabaud. Thus, the company started by an enterprising craftsman of rural stock came to be led by an engineer with the highest credentials and from the most prestigious engineering school in the country.

I turn now to another supplier of chemicals, glassware, and other apparatus, a rival to Alvergniat Frères with a stockroom and shop in the same neighbourhood.

Chenal & Douilhet

Georges Chenal and Douilhet—whose first name I was unable to find—took over an existing supplier, Billault, who had himself succeeded Fontaine. Their shop had a highly desirable address, 22 rue de la Sorbonne, at the corner of place de la Sorbonne: in the very centre of the Latin Quarter, facing the main building of the Sorbonne.[36] Billault had succeeded Fontaine in the aftermath of the Tuesday, 16 March 1869 explosion that killed six persons, including Fontaine's son Véron, who was preparing shipment of a potassium picrate-filled torpedo to the Navy yard in Toulon.[37]

Chenal and Douillet were both pharmacists in training. Douilhet obtained his degree from the *École de pharmacie* in Paris in 1889, and Chenal obtained his in 1893, also from the *École de pharmacie*. It was a prominent supplier firm, illustrated by the fact that Billault received a gold medal for his exhibit in the 1889 Paris World's Fair. Douilhet and Chenal took

36 Ministère du commerce, de l'industrie, des postes et télégraphes, *Rapport du Jury International Exposition Universelle Internationale à Paris de 1900* (Paris: Imprimerie nationale, 1902), vol. 14, p. 316; 'Enquête sur les industries chimiques françaises', *Revue Scientifique*, 3 (1905), 118.

37 Sacha Tomic, 'La gestion du risque chimique en milieu urbain: les conséquences de l'explosion du magasin Fontaine à Paris en 1869', in Thomas LeRoux, *Risques industriels. Savoirs, Régulations, Politiques d'assistance. Fin XVIIe-XXe siècles* (Rennes: Presses universitaires de Rennes, 2016), pp. 279–303.

Figure 6.3 Laboratory oven, Chenal & Douilhet make, date unknown. Courtesy: Musée de Bretagne, Rennes, France.

over from him as soon as Georges Chenal had graduated from the *École de pharmacie*. Already in 1894 their firm advertised in the *Bulletin de la Société Chimique de France.*

To direct their factory at Billancourt (a suburb down the river), they recruited Léon Séquard, a chemist trained at the *École de physique et de chimie Industrielles* (PC for short) in the class of 1882. Existing products in their catalogue, which they had inherited from their predecessor Billault, included broths similar to the so-called *bouillie bordelaise*—a liquid mix with copper salts to be sprayed on vine plants—from the aftermath of the phylloxera epidemic. This disaster started in the 1860s, and by the beginning of the 1880s, three-quarters of the French vineyards had been wiped out. French chemists hastened to develop chemical weapons to prevent a renewal of the disease once the vineyards had been replanted with resistant stock imported from California.[38]

By continuing to market Billault's treatment for vines, they ensured sustained income. Another wave they rode successfully was also a treatment for ailments, but this time of people rather than of plants: peptones.[39] What are, or rather were, peptones? These were polypeptides as hydrolysed from a protein source—such as meat or blood—using pepsin. Starting in the 1850s, they became highly popular with physicians, who prescribed them to prevent or cure a variety of conditions. The belief was that their ingestion

38 Bernard Ginestet, *La Bouillie Bordelaise* (Paris: Flammarion, 1975).

39 Charles Guyotjeannin, 'La pharmacie française aux expositions universelles de 1878, 1889 et 1900', *Revue d'histoire de la pharmacie*, 81 (1993), 313–30 (p. 319).

Figure 6.4 Filtering at the Chenal & Douilhet factory. From *Enzymes et produits physiologiques*. Source: Chenal, Douilhet & Cie.: Paris, 1904.

or injection would desensitize the body. They enjoyed great popularity until roughly the Great War.[40]

In addition to pepsin and peptones, Chenal & Douilhet advertised other biochemical products, in particular surrenin—we know it as adrenalin—that was isolated from pigs in their factory in Billancourt.[41]

Laboratory suppliers sometimes contributed to a scholarly journal. This was the case in 1897 with a paper in the *Bulletin de la Société Chimique de France* (pp. 392–95): Messrs. Tixier, Chenal, Ferron, and Douilhet described a novel fractioning apparatus, which they termed very powerful and sensitive. Made entirely of glass, it had two parts, an analyser and a condenser. The former was a bulbed tube, the latter a small, subsidiary condenser.

A late nineteenth-century event shows the porosity of the interface between laboratory suppliers and their scientific clients. At the 1900 Paris World's Fair, Chenal & Douilhet presented an exhibit that drew rave

40 P. Chapoteaut, *Comparative Researches on the Preparation and Use of Different Peptones* (Paris: Berger-Levrault & Cie., 1883).

41 *Enzymes et Produits Physiologiques* (Paris: Chenal Douilhet & Cie, 1904).

reviews in addition to winning them a gold medal. It consisted of 'large samples' of the pure 'rare earth' elements in a showcase. Note the apparent polar opposites in the preceding sentence: 'rare' and 'large'. Indeed, they boasted kilogram sizes; part of the display was a twenty-kilogram sample of a crystalline mixed nitrate of lanthanum and ammonium.[42] Another sample shown was a crystalline samarium oxide Sm_2O_3, a rarity at the time and presumably a first. In terms of advertising for the Chenal & Douilhet company, this was the whole point, viz. to impress the crowds that flocked to the *Exposition Universelle* ('World's Fair') with their ability to turn rarity into abundance.

How did they do it? They drew upon a brand-new technique devised by Eugène Demarçay (1852–1903), a former professor at the *École polytechnique* who had resigned his position and opened a private laboratory in Paris. He was the first to confirm the discovery of radium by Pierre and Marie Curie in 1898, and he discovered the element europium.[43]

The Demarçay modus operandi consisted of an electric arc set between metallic electrodes, coupled with a spectroscopic apparatus.[44] Demarçay would melt and vaporize samples and then monitor the presence of an element from its characteristic emission lines. Demarçay subsequently crystallized that rare earth element as the double nitrate of the trivalent metal and magnesium.[45]

By using the Demarçay methodology and applying it to hefty (several-hundred-kilogram) quantities of monazite, a sandy mineral imported from North Carolina, Léon Séquard (mentioned above) prepared large amounts of several lanthanides—gadolinium, lanthanum, neodymium, praseodymium, samarium, and yttrium.[46] The import of sands rich in monazite from North

42 Georges Urbain, *Recherches sur la séparation des terres rares* (Paris: Gauthier-Villars, 1899); O. Baudouard, 'Les Terres Rares', *Le Radium (Paris)*, 1 (1904), 106–11.

43 Abel Rey, 'La découverte de la radioactivité et le mouvement des idées scientifiques', *Revue Philosophique de la France et de l'Étranger*, 82 (1916), 340–73.

44 Robert K. DeKosky, 'Spectroscopy and the Elements in the Late Nineteenth Century: The Work of Sir William Crookes', *The British Journal for the History of Science*, 6 (1973), 400–23.

45 Eugène Demarçay, 'Sur un nouveau mode de fractionnement de quelques terres rares', *Comptes Rendus de l'Académie des Sciences*, 130 (1900), 1019–22; Eugène Demarçay, 'Sur le samarium', *Comptes Rendus de l'Académie Des Sciences*, 130 (1900), 1185; Eugène Demarçay, 'Sur le gadolinium', *Comptes Rendus de l'Académie des Sciences*, 131 (1900), 343; Eugène Demarçay, 'Sur un nouvel élément, l'Europium', *Comptes Rendus de l'Académie Des Sciences*, 132 (1901), 1414; A. Etard, 'Notice sur la vie et les travaux d'Eugène Demarçay', *Bulletin de la Société Chimique de Paris* (1904), I–VIII.

46 [Ecole supérieure de physique et de chimie (Paris)], *Rapport général sur l'historique et le fonctionnement de l'Ecole municipale de physique et de chimie industrielles* (Paris: Imprimerie générale Lahure, 1900), p. 115; Henri Moissan and François Dupont, *IVe Congrès International de*

Carolina to Europe served to manufacture Auer gas lamps, using the bright light from incandescent thorium oxide. Séquard worked on monazite samples already depleted in thorium.[47]

In brief, the display at the 1900 Paris World's Fair showed the Chenal & Douilhet company as an equal of the best academic laboratories; it could therefore be relied upon to supply top-of-the-line chemicals and laboratory equipment.[48] One of the scientists who published a raving review of their exhibit at the 1900 Paris World's Fair was Justin Dupont, a young chemist who in 1903 joined the group *La Molécule*, started by Fourneau and Tiffeneau.[49]

I turn now to yet another family-owned and run company from the same area in Paris, adjacent to the Sorbonne.

The Poulenc Frères Suppliers

The Poulenc Brothers owned a leading Parisian supplier. Who were they? Three brothers: Gaston, Emile, and Camille. In 1900 Gaston was 48 years old, Emile 45, and Camille 36. All three had studied pharmacy in Paris. The eldest, Gaston, and the youngest, Camille, both married into the same Marcilhacy family. Another significant family detail is that Gaston served as tutor to the orphaned future novelist Paul Vialar (1898–1896).[50]

Their father Etienne Poulenc (1823–1878) left his native Espalion in the Aveyron department to complete his training as a pharmacist in Paris.[51] He married Pauline Wittmann (1828–1910), the daughter of a druggist. In Paris he founded a company, Wittmann and Poulenc, which would soon afterwards become the Poulenc Company of laboratory suppliers. Following

Chimie Appliquée, tenu à Paris du 23 au 28 juillet 1900: Compte-Rendu in Extenso (Paris: Association des chimistes de sucrerie et de distillerie, 1902), p. 309.

47 [State Board of Agriculture (Raleigh)], *North Carolina and Its Resources* (Winston, NC: M.I. & J.C. Stewart, 1896), p. 101; anon., 'The Radio-Activity of Thorium', *The Journal of Gas Lighting, Water Supply & Sanitary Improvement*, 91 (1905), 174.

48 Jean-Christophe Mabire, *L'Exposition Universelle de 1900* (Paris: L'Harmattan, 2000).

49 Baudouard, 'Les Terres Rares'.

50 Luc Antonini, 'Les Poulenc, entre musique et industrie', *Généalogie Magazine*, 2004; 'Notices Biographiques – Vialar Paul', *Robert Denoël* [website], <https://www.thyssens.com/03notices-bio/vialar_p.php> (accessed 3 January 2019).

51 R. Béteille, 'Émigration et propriété foncière dans l'Aveyron', *Revue géographique des Pyrénées et du Sud-Ouest*, 37 (1966), 389–406; Jean-Claude Farcy and Alain Faure, *La mobilité d'une génération de français: recherche sur les migrations et les déménagements vers et dans Paris à la fin du XIXe siècle* (Paris: INED, 2003).

his death, his widow took over the business before handing it over to their three sons.[52]

Poulenc Frères thus supplied laboratory ware, chemicals, glassware, and instruments. Theirs was a family business in the aveyronnais mould, still very much in force nowadays: an offshoot of the nuclear family, with each family member contributing according to his or her ability and taste. This translates into rather formidable strength.

Which directions did the Poulenc brothers give their company? Each left a somewhat distinct imprint. After receiving its responsibility from their mother, Gaston the eldest kept the company going and thriving. Emile, the second son and a fervent Catholic, was fond of the arts. He expanded the line of photographic products, which became very successful. As for Camille, the youngest of the three, he was the most scientifically minded. He went through a second training, in chemistry, and prepared a D.Sc. doctoral dissertation under the direction of Henri Moissan (1852–1907), also a pharmacist-turned-chemist. Camille's doctoral work dealt with fluorides, including the first preparation of PF_5, phosphorus pentafluoride.[53] His chemical bent prompted Camille to serve as a go-between to bridge the manufacture of chemicals and their applications in the laboratory. Moreover, Camille Poulenc had an easy pen: the numerous notes he wrote in the annual catalogues of Poulenc Frères about the elements and compounds for sale were worthy of scientific journals.[54]

The Photography Line

Often part of a chemical laboratory, a darkroom is itself a small laboratory. The late nineteenth century brought those twin features to the fore. Photography was one of the numerous technologies—an art, too—spawned by the Laboratory Revolution.

The main chemical for capturing images was gelatinobromide of silver, coated on glass plates; it was introduced in 1871 and widespread from 1878 onwards. Poulenc Frères started supplying it early on once Etienne Poulenc's widow had passed on the business to their sons. Emile greatly developed that

52 Pierre Cayez, *Rhône-Poulenc, 1895-1975: contribution à l'étude d'un groupe industriel* (Paris: Armand Colin/Masson, 1988), p. 13.

53 Camille Poulenc, 'Sur un nouveau corps gazeux, le pentachlorofluorure de phosphore', *Annales de Chimie et de Physique*, 6 (1891), 548.

54 Camille Poulenc, *Les produits chimiques purs en photographie, leur nécessité, leur contrôle* (Paris: Charles Mendel, 1908).

novel and most lucrative aspect of the business. Not only chemicals but also cameras—Kodak cameras in due time—and other hardware were offered. The company acquired a high reputation among amateur photographers.[55]

To uphold that public image, in the 1890s the Poulenc brothers brought out a novel, highly sophisticated line of photographic paper for prints using platinum salts. Those platinum papers were very expensive, but they provided sumptuous rendering of blacks and shades of grey. To prepare such supports for photographic prints, a solution of ferric oxalate and potassium chloroplatinate was spread on a quality paper. After exposure to intense light, it was immersed in a potassium oxalate developer, followed by a dilute hydrochloric acid fixative.[56]

Of course, given the popularity of photography, other Parisian suppliers vied for the same segment of the market. One such competitor was Paul Rousseau. In the early 1890s his shop was at 17 rue Soufflot, the street running down from the Panthéon to the Luxembourg gardens. There was no factory to back it up, only a combination laboratory-workshop-shipping on rue Laromiguière, a small street in the same general area as the *École polytechnique* and the *École normale supérieure*. By 1895, with the store on rue Soufflot far too small, Rousseau moved a short distance to rue des Fossés Saint-Jacques.[57]

During the second half of the nineteenth century, photography enjoyed a close relationship with the sciences. It had a reputation for accuracy. So much so that, at the Préfecture de Police, Alphonse Bertillon (1853–1914) pioneered the use of criminals' mugshots—full-face and profile—together with anthropometric measurements and fingerprints in order to catch serial offenders.[58]

Photography also enjoyed multiple applications in laboratories. One of those was the discovery of radioactivity (1896) by Henri Becquerel, a

55 Victor Roux, *Manuel opératoire pour l'emploi du procédé au gélatino-bromure d'argent* (Paris: Gauthier-Villars, 1881); anon., 'Nouveautés Photographiques: La Lampe-Éclair Poulenc', *L'Amateur Photographe* (1890), 259–60; Charles Gravier, 'La photographie et ses applications industrielles', *L'Amateur Photographe* (1890), 222, 363, 369, 384–90; Ph.C., 'Des épreuves positives', *Journal mensuel du photo-club Toulousain*, (15 May 1898), 1–2; Jacques Boudet, *Le monde des affaires en France de 1830 à nos jours* (Paris: Société d'édition de dictionnaires et encyclopédies, 1952), p. 194.

56 A. Courrèges, *Ce qu'il faut savoir pour réussir en photographie* (Paris: Gauthier-Villars et fils, 1894); Frédéric Dillaye, *Les nouveautés photographiques* (Paris: Librairie illustrée, 1896), pp. 79–80.

57 Anon. (advert), 'Nouveautés Scientifiques', *La Nature: revue des sciences et de leurs applications aux arts et à l'industrie*, 1154 (1895), 28

58 Piazza Pierre, *Aux origines de la police scientifique. Alphonse Bertillon, précurseur de la science du crime* (Paris: Karthala Editions, 2011).

supremely gifted experimental physicist who had the genius to develop otherwise virgin photographic plates which had been stored next to uranium salts.[59] There were quite a few other uses of photography for scientists in fin-de-siècle Paris, such as micrography, images of celestial bodies, high-speed pictures, photometry, anthropology, Kirlian photography, portraits of madmen and sick people, stereographs, and X-rays.[60]

An Addiction that Launched an Industry

At the turn of the twentieth century, cocaine was a highly popular over-the-counter drug. This was primarily due to Sigmund Freud (1856–1939) and to his book on cocaine.[61] Two commercial drinks based on extracts of coca were very popular then: since 1863 in France, a tonic wine made by Angelo Mariani (1838–1914); and, of course, Coca-Cola in the US. In France the Mariani wine, with cocaine included, entered the pharmaceutical Codex in 1884. It remained legal until 1910.[62]

We now turn again to the Poulenc Frères company. Camille Poulenc, who had been classmates with Ernest Fourneau during their joint internship in pharmacy, recruited him as director of their firm's fabrications. In like manner to Camille Poulenc, the pharmacy graduate Fourneau gave himself additional professional training as a chemist. He went to Germany, where, in succession, he joined the laboratories of Theodor Curtius (1857–1928) and Ludwig Gattermann (1860–1920) in Heidelberg, of Emil Fischer (1852–1919) in Berlin for a year (he there synthesized glycylglycine), and of Richard Willstätter (1872–1942) in Munich.[63] Cocaine was a foremost research interest for Willstätter.

59 Abel Rey, 'La découverte de la radioactivité et le mouvement des idées scientifiques', *Revue Philosophique de la France et de l'Etranger*, 82 (1916), 340–73; Jean-Louis Basdevant, *Henri Becquerel à l'aube du XXe siècle: 1896-1996, centenaire de la découverte de la radioactivité* (Paris: Éditions de l'École polytechnique, 1996).

60 *Paris Exposition 1900. Guide pratique du visiteur de Paris et de l'Exposition* (Paris: Hachette, 1900), p. 305; Gilbert Beaugé, 'De l'apparence des caractères au caractère des apparences. Photographie et anthropologie: 1839-1912', *Le Monde alpin et rhodanien. Revue régionale d'ethnologie*, 23 (1995), 81–144; Caroline Fieschi, 'L'illustration photographique des thèses de science en France (1880-1909)', *Bibliothèque de l'école des chartes*, 158 (2000), 223–45.

61 Sigmund Freud, *Über Coca* (Vienna: Verlag von Moritz Perles, 1885).

62 H.W. Maier, *La Cocaïne. Historique, Pathologie, Clinique, Thérapeutique, Défense Sociale* (Lausanne: Payot, 1928).

63 Claude Viel, 'Fourneau (Ernest), Biarritz, 4 Octobre 1872-Ascain, 5 Août 1949', *Revue d'histoire de la pharmacie*, 93 (2005), 646–49.

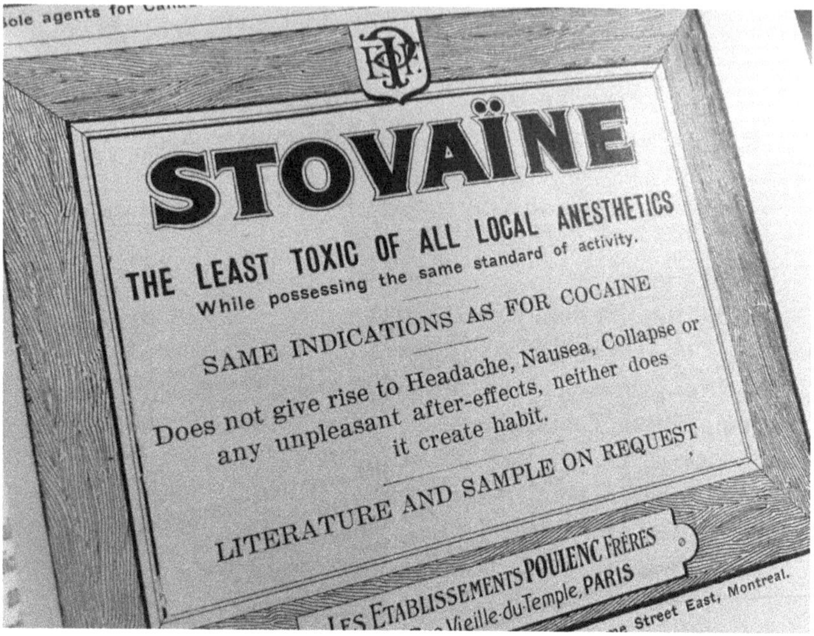

Figure 6.5 Advertisement by Etablissements Poulenc Frères, dating from the early 1920s. In 1928 the company would merge with Société chimique des usines du Rhône and thus change its name. Stovaine continued being used until the 1940s. Courtesy: Université de Montréal, Bibliothèque des livres rares et collections spéciales.

When Fourneau returned to Paris in 1903 as head of research for Poulenc Frères in their Ivry factory, he synthesized a cocaine analogue. To his own surprise, when he sampled it, his tongue became numb. He thought he had invented a local anaesthetic, which predominantly became a panacea for dentists, but it had other applications to stifle pain, such as in epidural injections during delivery of a baby. He named it stovaine, which was used as an analgetic during delivery for about 40 years.[64]

64 Léon Kendirdjy, *L'anesthésie chirurgicale par la stovaïne* (Paris: Masson, 1906); Dan McKenzie, 'The Local Anaesthetic Action of Stovaine', *The British Medical Journal* (May 1906), 1099; Alexander Don, 'Notes on Twenty Cases of Spinal Anæsthesia with Stovaine', *Edinburgh Medical Journal*, 2 (June 1909), 546–50; Lloyd Noland, 'Stovaine Spinal Anesthesia', *Annals of Surgery*, 51 (1910), 449–52; Victor Herbert Veley and Augustus Desire Waller, 'On the Comparative Action of Stovaine and Cocaine as Measured by Their Direct Effect upon the Contractility of Isolated Muscle', *Proceedings of the Royal Society of London. Series B, Containing Papers of a Biological Character*, 82 (1910), 147–51; Sophie Chauveau, 'Les origines de l'industrialisation de la pharmacie avant la Première Guerre Mondiale', *Histoire, économie et société*, 14 (1995), 627–42; Christine Debue-Barazer, 'The Effects of the Success of the Synthesis of Stovaïne in Science and Industry. Ernest Fourneau (1872–1949) and the Transformation of the Field of Medicinal Chemistry in France', *Gesnerus*, 64 (2007), 24–53.

However, the monopoly of stovaine was relatively short-lived because it soon was accompanied by a competitor: novocaine, produced by Hoechst. The stovaine patent (late 1903) brought a windfall to Poulenc Frères. They thus rode the wave of the second industrialization in France and turned from laboratory suppliers into an important pharmaceutical company—the seed for what would become the Rhône-Poulenc company in subsequent decades after World War I.[65]

Symbiosis between Suppliers and Academic Laboratories

To call upon my 30-year experience as a professional chemist: at the time of my training in mid-twentieth century, the chemical laboratory had not changed radically since the period discussed here.[66] I recognize it almost entirely in pictures of, for instance, the Parisian *Laboratoire municipal* in the 1880s. The 1960s were the time when the chemistry laboratory underwent a transformation due to the instrumental revolution.[67] Until then, the stability was remarkable.[68]

Examining in some detail the activity of three *fin-de-siècle* Parisian laboratory suppliers, we have found quasi-symbiotic relationships with the laboratories they supplied. This went beyond good business practice, as there also had to be a great deal of mutual esteem between the manufacturer-merchants and their clients. Such mutual trust endured well into the twentieth century, as witnessed by the former research scientist Alfred Bader

65 Sophie Chauveau, 'Entreprises et marchés du médicament en Europe occidentale des années 1880 à la fin des années 1960', *Histoire, économie et société*, 17 (1998), 49–81.
A word about the term 'stovaïne': the Renard 'fuchsine' had set a precedent. *Fuchs* is the name for a fox in German, the Renard ('fox' in French) brothers used this pun when they devised this synthetic dye. Likewise, the word *fourneau* in French becomes 'stove' in English translation. Fourneau may have wished to distance himself from Willstätter and to make a strong claim for his design of the new drug. Why English? It partook of the ambient Anglophilia that Fourneau shared in; he knew it first-hand from the rich British clientele in the five-star hotel owned by his parents in Biarritz. Jean-Pierre Fourneau, 'Ernest Fourneau, Fondateur de la chimie thérapeutique française: Feuillets d'album', *Revue d'histoire de La Pharmacie*, 75 (1987), 335–55.

66 Jean-Pierre Rioux, 'Devoir de mémoire, devoir d'intelligence', *Vingtième Siècle. Revue d'histoire*, 73 (2002), 157–67; Sacha Tomic, 'Le cadre matériel des cours de chimie dans l'enseignement supérieur à Paris au XIXe siècle', *Histoire de l'éducation*, 130 (2011), 57–83.

67 Peter J.T. Morris, ed., *From Classical to Modern Chemistry: The Instrumental Revolution* (London: Royal Society of Chemistry, 2002).

68 Mary Jo Nye, *Before Big Science: The Pursuit of Modern Chemistry and Physics, 1800–1940* (Cambridge, MA: Harvard University Press, 1999); Peter J.T. Morris, *The Matter Factory: A History of the Chemistry Laboratory* (London: Reaktion Books, 2015).

(1924–2018), who had been part of Robert B. Woodward's group at Harvard and in 1951 founded the Aldrich Chemicals company, which became one of the leading suppliers of laboratory chemicals during the second half of the twentieth century.

Were they manufacturer-merchants? This composite term describes a reality, pointed out above, for the suppliers we focused on. Indeed, designers of laboratory instruments such as Anatole Dupré were not in the business of manufacturing and selling them, not even in small series. Their launching of a start-up company was out of the question; it was definitely not in the spirit of the country, nor of the times. Hence Dupré entrusted his designs for commercialization to Pellin or Alvergniat Frères. For many a Parisian supplier of laboratory chemicals and hardware, the shop was thus the outlet of the factory.

The term 'supplier', while accurate, is incomplete. At least regarding Parisian suppliers of the early 1900s, it does not express their symbiotic interaction with laboratories. Examples will convey this reality. Already some decades earlier, Victor Wiesnegg (1841–1881), who supplied the chemistry laboratory at the *École normale supérieure* on rue d'Ulm from his combined store-workshop nearby on rue Gay-Lussac, at the same time collaborated on various projects with Henry Sainte-Claire Deville and Louis Pasteur.[69] His company specialized in building and supplying laboratory ovens.

In the same Parisian district of the Latin Quarter, Léon Séquard, an 1882 graduate of the *École de physique et chimie industrielle*, directed the factory of Chenal & Douilhet. Thus, Séquard was in a good position to supply equipment to the laboratories of Paul Schützenberger (1829–1897), the first director of PC, and his successors Charles Lauth (1836–1913) and Albin Haller; he was also well-positioned to supply the laboratory of Charles Friedel, a co-founder of PC, at the *École des mines de Paris*.[70]

Turning again to Poulenc Frères: their director of research, Ernest Fourneau, had a most impressive network of academic colleagues. He had become a pharmacist and a chemist on the recommendation of Félix Moureu (1849–1928), a pharmacist in the city of Biarritz. He was a close friend of Charles Moureu (1863–1929), Félix Moureu's brother who was appointed in 1907 professor at the *École de pharmacie* and whose laboratory he would

69 Gilles Pécout, 'L'École normale supérieure au XIXe siècle: réflexions et débats autour d'un 'modèle d'excellence' français', *Annali della Scuola Normale Superiore di Pisa. Classe di Lettere e Filosofia, Serie 5*, 3 (2011), 35–56.

70 Terry Shinn, 'Des sciences industrielles aux sciences fondamentales: La mutation de l'Ecole supérieure de physique et de chimie (1882-1970)', *Revue Française de Sociologie*, 22 (1981), 167–82.

supply. When Fourneau married in 1906, he became brother-in-law of Marc Tiffeneau, chief pharmacist at Hôpital Boucicaut whose laboratory Poulenc Frères would also supply.

What were the problems and the rewards of running such a business during the *Belle Époque*? The challenges were to stock up on many and very diverse items, such as glassware, pottery, sheet metal containers, mechanical and electrical devices, instrumentation of many types, chemicals, and what have you!

Another challenge was posed by nationalistic chauvinism after the French were defeated at Sedan in 1870.[71] Anti-German sentiment ran high, and France was intent upon revenge.[72] It prepared for war, whenever it would come. Hence, purchasing laboratory supplies from Germany was deemed unpatriotic. There was a collective will of independence from German science and technology that nurtured what can only be referred to as 'French insularity'.[73] Wurtz's stunning dictum '*la chimie est une science française*' ('chemistry is a French science') is representative of that mentality. The result was the opening of a vast space for French instrument designers such as Anatole Dupré.[74]

The suppliers not only had to hold their wares but also to catalogue and price all such items. Any look at a suppliers' catalogue from those times makes evident a relatively fast turnover: scientific laboratories are no different from other endeavours and trades in that they are subject to fashion. Equipment popular with one generation can be pushed aside and replaced by a different line. It is the job of a supplier to try to anticipate such changes. This serves to qualify my earlier statement regarding the great stability of the chemistry laboratory, from about 1850 until 1950: some of the equipment varied little, while other pieces of apparatus underwent rapid

71 Alan J. Rocke, *Nationalizing Science: Adolphe Wurtz and the Battle for French Chemistry* (Cambridge, MA: MIT Press, 2001).

72 Bernard Wilkin, 'Organizing for War: France 1870–1914', *French History*, 26 (2012), 130–31.

73 Boris Noguès, 'Elèves ou auditeurs ? Le public des facultés de lettres et de sciences au XIXe siècle (1808-1878)', *Histoire de l'éducation*, 120 (2008), 77–97; Robert Anderson, 'Aristocratic Values and Elite Education in Britain and France', in Didier Lancien and Monique de Saint-Martin, eds., *Anciennes et Nouvelles Aristocraties: De 1880 à Nos Jours* (Paris: Éditions de la Maison des sciences de l'homme, 2018), pp. 261–78.

74 Ad. Wurtz, *Histoire des doctrines chimiques depuis Lavoisier jusqu'à nos jours* (Paris: Hachette, 1869), p. 1. (See also: A. Wurtz, 'Dictionnaire de chimie pure et appliquée', *Le moniteur scientifique: Journal des sciences pures et appliquées*, 11 (1 Feb. 1869), 97–101 (p. 98).) Alan J. Rocke, 'Between Two Stools: Kopp, Kolbe and the History of Chemistry', *Bulletin for the History of Chemistry*, 7 (1992), 19–24; Béatrice Dedinger, 'The Franco-German Trade Puzzle: An Analysis of the Economic Consequences of the Franco-Prussian War', *The Economic History Review*, 65 (2012), 1029–54.

diversification and evolution. The former included, for instance, balances, glassware, polarimeters, and refractometers.[75] The latter obviously included spectroscopes and spectrometers.

As for the rewards, those businesses were undoubtedly profitable. Otherwise, one would not find so many competing at any one time on the Parisian scene. The French State was handsomely funding a renaissance in French science, in chemistry and physics in particular, and merchants came quickly to the feeding trough. The deep influence of the French government and its staunch belief in scientific innovation as the engine for boosting the country to international eminence and power can be illustrated by the aftermath to the explosion that destroyed the Fontaine shop on place de la Sorbonne on 24 March 24 1869: despite a flurry of judicial inquests and suits, no real action was taken to curb the risk posed by these chemical suppliers in the very heart of Paris.[76]

The readers have noted that Parisian suppliers of glassware and chemicals during the *Belle Époque* set shop in the Latin Quarter preferentially. This reflected an ease of convenience for the laboratory supervisors, such as professors at the Sorbonne: they could wander off from their offices and walk to a nearby shop to pick up and carry back almost any desirable item. The Latin Quarter bridges the 5th and the 6th arrondissements of Paris. It included and still includes the Sorbonne, the *École de médecine*, the School of Pharmacy, and the *École normale supérieure* and the *École de physique et chimie industrielles*—to name institutions already mentioned in this account. If the Latin Quarter was the most desirable location for the suppliers' shop, rue de la Sorbonne was the choicest address. To laboratory suppliers, it was High Street, despite its short length of just a few hundred meters. As we saw, both Alvergniat Frères and Chenal & Douilhet had their shops there. Poulenc Frères were a stone's throw away, on rue de Cluny.

Some suppliers originated not only in the provinces but also in rural France: Alvergniat Frères from the Vendômois (near the Loire river), Poulenc Frères from the Aveyron. Their move to Paris at mid-century was typical of the rural exodus propelled by the inheritance tradition (only the eldest son

75 Kijan Malte Espahangizi, 'From Topos to Oikos: The Standardization of Glass Containers as Epistemic Boundaries in Modern Laboratory Research (1850–1900)', *Science in Context*, 28 (2015), 397–425; Catherine M. Jackson, 'Chemical Identity Crisis: Glass and Glassblowing in the Identification of Organic Compounds', *Annals of Science*, 72 (2015), 187–205.

76 Sacha Tomic, 'La gestion du risque chimique en milieu urbain: les conséquences de l'explosion du magasin Fontaine à Paris en 1869', in Thomas Le Roux, ed., *Risques industriels. Savoirs, Régulations, Politiques d'assistance. Fin XVIIe-XXe siècles* (Rennes: Presses universitaires de Rennes, 2016), pp. 279–303.

inherited the family farm), the reduction of deaths in infancy, and railroad availability. They brought with them rural values, with a premium on hard work.[77] That some such as the Alvergniats and the Poulencs achieved upward social mobility was not atypical, as their firms were well-run family businesses.

Another observation is the dominance of pharmacists. The proximity of the two trades, that of druggists and of pharmacists, in Paris during the second half of the nineteenth century is a partial explanation. Another is the rise of health issues in the public consciousness, already mentioned in relation to the establishment of the *Laboratoire municipal.*

Moreover, to be a French pharmacist was synonymous with being a mediator: between the high caste of physicians and their patients, but also between natural products and the public. To mention a characteristic case, persons who found wild mushrooms would bring them for safety inspection to the local pharmacy. Pharmacists as well as laboratory suppliers represented the trading zone, to refer to the useful concept Peter Galison introduced.[78]

Pharmacists' high impact on supplies to chemistry laboratories reflected their overall presence in French chemistry during the entire nineteenth century, and this endured until the mid-twentieth century. Starting with Berthelot and lasting until Alain Horeau (1909–1992), the Collège de France had two chairs in chemistry, with a chemist appointed to one and a pharmacist to the other.

Paris World's Fairs in 1878, 1889 and 1900 were keen to display the entry of France into modernity, due to the second industrialization and to the scientific progress that accompanied it. At the 1889 event—that of the Eiffel Tower—advances in healthcare and hygiene were featured, with proud mention of the *Laboratoire municipal.* The suppliers we have focused on in this chapter participated actively in these exhibitions, most notably in the 1900 Fair.[79]

77 Kiva Silver, 'The Peasants of Paris: Limousin Migrant Masons in the Nineteenth Century', *French History*, 28 (2014), 498–519; Patrice Bourdelais, 'L'industrialisation et ses mobilités (1836-1936)', *Annales. Histoire, Sciences Sociales*, 39 (1984), 1009–19.

78 Peter Galison, *Image and Logic: A Material Culture of Microphysics* (Chicago: University of Chicago Press, 1997).

79 A. Nicot, *La Chimie et la pharmacie à l'exposition universelle de 1889* (Paris: Octave Doin, 1890); Alfred Picard, ed., *Exposition universelle de 1889. Rapport du jury international* (Paris: Imprimerie nationale, 1892); [Ministère du commerce, de l'industrie, des postes et télégraphes], *Rapport du Jury International Exposition Universelle Internationale à Paris de 1900* (Paris: Imprimerie nationale, 1902), vol. 14; Pascal Ory, *L'Expo universelle, 1889* (Paris: Editions Complexe, 1989); Yves Saint-Jours, *Les Expositions universelles à Paris* (Paris: Revue de l'économie sociale, 1990); Charles

While the bulk of their business was in central Paris, those laboratory suppliers also catered to a segment that was far from marginal: the mail orders. Those came from not only the provinces but also from the French colonial empire overseas. French colonization in Africa and in Asia led to imports of valuable raw materials and goods in exchange for export of healthcare, education, and knowledge. The spread of Pasteur Institutes testified to such concerns, however paternalistic. Parisian laboratory suppliers thus answered orders of many kinds of small apparatus from owners of dairies and cheese factories, wineries, oil mills, producers of essential oils and other natural products, and governmental laboratories, located either in France or overseas.[80] In this manner, the Laboratory Revolution extended its reach worldwide from the Latin Quarter: into Algeria and the Maghreb, Madagascar, the Congo, Senegal and Western Africa, India and Indochina, and Latin America.

Those suppliers of glassware and chemicals in turn-of-the-century Paris were activists—that is a key finding from this study. Far from being mere merchants interested primarily in the bottom line, their activities were fostered by the contemporary ideology of progress through science yet reinforced by their participation in the 1889 and 1900 World's Fairs.

As we saw, Chenal & Douilhet advanced not only biochemistry but also inorganic chemistry with their preparation of rare earths. Poulenc Frères helped to develop the art of photography. No wonder that Emile Poulenc, who loved the arts, inspired his son Francis (1899–1963), who became a major composer of the twentieth century.[81]

Another finding of this study is, likewise, the porosity of the interfaces: between the School of Pharmacy and the pharmacists trained in it, on

Guyotjeannin, 'La pharmacie française aux expositions universelles de 1878, 1889 et 1900', *Revue d'histoire de la pharmacie*, 81 (1993), 313–30; Jean Pierre Babelon, Myriam Bacha, and Béatrice de Andia, *Les expositions universelles à Paris de 1855 à 1937* (Paris: Action artistique de la Ville de Paris, 2005). A comprehensive online bibliography can be found at http://www.csufresno.edu/library/subjectresources/ special collections/worldfairs/*bibliographies*.html (accessed on 12 January 2019). See also: Walter Benjamin, 'Paris, capitale du XIXe siècle', in Walter Benjamin, *Das Passagen-Werk* (Frankfurt am Main: Suhrkamp Verlag, 1982), pp. 60–77.

80 Sven Dierig, 'Engines for Experiment: Laboratory Revolution and Industrial Labor in the Nineteenth-Century City', *Osiris*, 18 (2003), 116–34. See also: Maria Kaika, and Erik Swyngedouw, 'Fetishizing the Modern City: The Phantasmagoria of Urban Technological Networks', *International Journal of Urban and Regional Research*, 24 (2000), 120–38; Pierre Laszlo, 'Enquête à partir d'une ancienne carte postale', *L'Actualité Chimique*, 430–1 (2018), 109–14.

81 Elinor Accampo, 'Dawn of the Belle Epoque: The Paris of Monet, Zola, Bernhardt, Eiffel, Debussy, Clemenceau, and Their Friends', *French History*, 27 (2013), 476–77.

one hand, and academic laboratories and their suppliers, on the other hand—Demarçay and Séquard representative of the latter.

In Paris the Laboratory Revolution was part and parcel of a wider, social revolution that the Third Republic embodied: Republican science. Science was the new religion.[82] It had to displace the established Catholic religion, and it is no wonder that the Third Republic enacted the Combes anticlerical laws in 1905. But that is an altogether different story.

About the Author

Pierre Laszlo is emeritus professor of chemistry, University of Liège, Belgium and École polytechnique, France (and also held academic positions at Princeton University and Cornell University). His fields of research are physical organic chemistry and catalysis of the major reactions of organic chemistry by inorganic solids, clays in particular. He recently published *A Life and Career in Chemistry: Autobiography from the 1960s to the 1990s* (Cham: Springer, 2021).

82 Robert Fox, *The Savant and the State: Science and Cultural Politics in Nineteenth-Century France* (Baltimore, MD: Johns Hopkins University Press, 2012).

7 Fighting for Modern Teaching and Research Laboratories in Norway: The Chemistry Laboratory in Political Dispute around 1920

Annette Lykknes

Abstract

Expectations were high when the Norwegian Institute of Technology (NTH) was established in Trondheim in 1910. Large sums were invested in laboratory facilities, which were considered important in training industrial chemists, i.e. chemical engineers. When the laboratory facilities turned out to be too small a decade later and thus inadequate to serve their purpose, their legitimacy became part of a political dispute between representatives from the institute in Trondheim and the Royal Frederick University in Kristiania, which were also in need of new laboratory facilities. At stake were the roles ascribed to the different institutions that educated (industrial) chemists or chemical engineers in the country, and thereby the allocation of resources for modern laboratory facilities.

Keywords: teaching laboratories, higher chemistry education, industry, political dispute, national identity

Introduction

> … a completely modern and well-equipped chemical laboratory so that from now on in Norway, it will be possible to participate in great international work on how to utilize the chemical sciences in the interest of the chemical industry.[1]

1 Anon., 'Den tekniske høiskole', *Tidsskrift for kemi, farmaci og terapi* (*Pharmacia*), 18 (1910), 273–99 (p. 273). All translations from the Norwegian are the author's, if not otherwise indicated.

Berkel, Klaas van, and Ernst Homburg (eds), *The Laboratory Revolution and the Creation of the Modern University, 1830-1940*. Taylor & Francis Group, 2023
DOI: 10.5117/9789463720434_CH07

When the Norwegian Institute of Technology was inaugurated in 1910 as the first technical institute in the country, it was celebrated as an academic institution that trained elite engineers who could establish new industry and thereby build the new, independent Norway. A lot was at stake to make the country self-supporting and economically viable after it gained its independence in 1905. Thus, great hope and optimism for the future surrounded the establishment of the institute. The institute was perceived as the cathedral of technology—symbolized by the building's spires and cathedral-like hall, and its location vis-à-vis the Nidaros Cathedral. The new buildings and laboratories offered the latest technologies, including electricity, which added to the institute being seen as a centre for technology in Norway.

Indeed, laboratories are an important part of chemistry practices and culture. As Peter Morris has argued, laboratories 'are where chemistry is carried out—from the freshman's first inorganic analysis to the most complex of organic syntheses'.[2] The laboratory is the *place* where chemistry makes progress, and at the same time an improved laboratory can *enable* chemistry to move forward.

The establishment of a Norwegian Institute of Technology took place towards the end of a period when laboratories were being built on a large scale at universities all over Europe and when industry started to undertake research—a transformation that has been referred to as the Laboratory Revolution. After 1860 universities in the German states one after another installed laboratories for research in chemistry, in addition to the teaching laboratories that had been created between 1820 and 1860.[3] Some of them have even been referred to as 'chemical palaces' because they were so large and well-equipped. As Peter Morris has described it, they 'contained almost everything that a twentieth-century chemist associated with the term "chemical laboratory"'.[4] At the turn of the twentieth century, proper

2 Peter J.T. Morris, *The Matter Factory: A History of the Chemistry Laboratory* (London: Reaktion Books, 2015), p. 9.

3 Morris, *The Matter Factory*, chs. 5 and 6; Ernst Homburg, 'Two Factions, One Profession: The Chemical Profession in German Society 1780–1870', in David Knight and Helge Kragh, eds., *The Making of the Chemist: The Social History of Chemistry in Europe, 1789–1914* (Cambridge: Cambridge University Press, 1998), pp. 39–76.

4 Morris, *The Matter Factory*, p. 149. In physics, as in mechanical engineering, the implementation of research laboratories was slower; the process of erecting physics laboratories started in the 1870s and continued until the 1880s and 1890s. See: Robert Fox and Anna Guagnini, *Laboratories, Workshops, and Sites: Concepts and Practices of Research in Industrial Europe, 1800–1914* (Berkeley, CA: Office of the History of Science and Technology, University of California, 1999), pp. 42–50. Likewise, mechanical laboratories were demanded from the 1880s on and were established all over Germany from around 1895. See: Wolfgang König, 'Technical Education and Industrial

research facilities were established at the *Technische Hochschulen* ('institutes/universities of technology') all across Germany, creating and strengthening research collaboration between industry and the *Hochschulen*.[5] The 1870s also saw the first purpose-built industrial research laboratory in Europe when the laboratory of the Badische Anilin- und Soda-Fabrik (BASF) in Ludwigshafen, Germany was set up, followed by the establishment of similar facilities in the other main dye companies in Germany.[6] Ernst Homburg has argued that these industrial research laboratories paved the way for a new way of doing research that would be followed by other companies in the subsequent years.[7]

Research was emphasized in technical institutes beyond Germany as well, e.g. at the Royal Institute of Technology (KTH) in Stockholm, Sweden, which in official rhetoric highlighted the importance of science for technology and industry.[8] New, modern laboratories were inaugurated in 1899, although professors had already been provided with research laboratories when new buildings were erected 36 years earlier, in 1863. However, the 1899 facilities included rooms for polarization studies, a 'stink room' and an 'explosion room', which the 'chemical palaces' in the German states had established in the 1860s and 1870s. It has been argued that the new facilities expressed an increased emphasis on science and research at the technical institute.[9] By the time the Scandinavian newcomer—the Norwegian Institute of Technology—was founded some ten years later, the academic-scientific ideal for the Norwegian *Hochschule* was already well established in rhetoric and practice. In fact, it was claimed that the institute in Trondheim had the most modern chemistry laboratory and institute in Scandinavia[10]—sufficiently

Performance in Germany: A Triumph of Hetereogeneity', in Robert Fox and Anna Guagnini, eds., *Education, Technology and Industrial Performance, 1850–1939* (Cambridge: Cambridge University Press, 1993), pp. 65–87 (p. 76).

5 König, 'Technical Education', p. 77.

6 Morris, *The Matter Factory*, p. 234.

7 Homburg, 'Two Factions, One Profession'; Ernst Homburg, 'Chemistry and Industry: A Tale of Two Moving Targets', *Isis*, 109 (2018), 565–76 (p. 574). The first R&D laboratory to be established in the US was set up in 1901 at the General Electric Schenectady plant in New York See: W. Bernard Carlson, 'Innovation and the Modern Corporation: From Heroic Invention to Industrial Science', in John Krige and Dominique Pestre, eds., *Science in the Twentieth Century* (Amsterdam: Harwood, 1997), pp. 203–26.

8 Anders Lundgren, *Kunskap och kemisk industri i 1800-talets Sverige* (Lund: Arkiv förlag, 2017), p. 218.

9 Lundgren, *Kunskap*, pp. 214, 223–24; Morris, *The Matter Factory*, pp. 149–50.

10 Claus N. Riiber, 'Vore kemiingeniørers utdannelse. Foredrag ved immatrikuleringen paa Den tekniske høiskole igaar', *Dagsposten*, 2 September 1916.

equipped and modern to ensure utilization of the chemical sciences in the interest of the budding industry, as stated in the introductory quote.

Large sums had been invested in Trondheim's laboratory facilities, which were considered to be important in the training of industrial chemists, that is, the chemical engineers.[11] When the laboratory facilities turned out to be too small a decade later and thus inadequate to serve their purpose, their legitimacy became part of a political dispute between representatives from the institute in Trondheim and the Royal Frederick University in Kristiania (now the University of Oslo). At stake were the roles ascribed to the different institutions which educated (industrial) chemists or chemical engineers in the country, and thereby the allocation of resources for modern laboratory facilities were of central concern. The debate of the 1920s thus illustrates the importance of an up-to-date laboratory for training chemists or chemical engineers for industry.[12] For what was chemistry, in fact, without the laboratory? And what was an engineering education without proper technical facilities? A lack of up-to-date facilities had by then become unthinkable since laboratories had become a sine qua non at universities. I will discuss how the laboratory became a symbol for chemists and chemical engineers in terms of both personal and national identity building, given that Norway's independence from Sweden had taken place only around fifteen years earlier. I will also consider the importance of geographical location in this debate, which was (and is) particularly relevant in a country as spread out as Norway.

National Pride, Industry, and the Institute of Technology

When the first proposal for a Norwegian polytechnic school was put forward to the Parliament in 1833, the idea was turned down. Very little industry existed in the country; in fact the proposal came ten years before the first industrialization period, during which the Norwegian textile industry became established. The first major breakthrough in Norwegian industrialization, however, took place between 1875 and 1895 and was characterized by growth

11 Annette Lykknes and Joakim Ziegler Gusland, *Akademi og industri: Kjemiutdanning og -forskning ved NTNU gjennom 100 år* (Trondheim: Fagbokforlaget, 2015), pp. 265–77.

12 An early draft of an analysis of this debate was prepared by Annette Lykknes and Ola Nordal for the 6th meeting of STEP (Science and Technology in the European Periphery) in Istanbul, 2008. Nordal is acknowledged for his contribution to that paper. The debate is also discussed in: Thomas Brandt and Ola Nordal, *Turbulens og tankekraft: Historien om NTNU* (Oslo: Pax, 2010), pp. 155–62, and Lykknes and Gusland, *Akademi og industri*, pp. 195–205.

in the textile and metal industries, canned food, and the pulp and paper industry.[13] At the turn of the twentieth century, a professional community of engineers had grown and become institutionalized. Associations such as the Norwegian Polytechnical Association and the Norwegian Engineer and Architect Association (NIAF) had been established to promote scientific thinking, technology, and modern industry in the country and to work for the professions of technicians and engineers in society. NIAF fought long and hard to establish a polytechnic school at the elite level along the lines of the *Technische Hochschulen* established in Germany in the 1870s. Three technical (lower) schools had been established in Norway in the 1870s (and a more specialized one sixteen years earlier), but the top level still did not exist in the country. In 1895 NIAF and the Polytechnical Association joined forces and put forward yet another proposal for such a polytechnic *Hochschule*.[14] Finally, in 1900, the decision was made to establish technical education at the elite level.

The next battle concerned the location of the new institute. Two major action groups were established, one in Kristiania and the other one in Trondheim. Major arguments in favour of Kristiania included the fact that the bulk of Norwegian industry was located around the capital and that the majority of the scientific-academic elite resided there. Other voices claimed that placing it in a smaller town like Trondheim would provide incentive for generating more industry. In the end, however, the decisive factors were of another kind. Norway's larger rural development policy—to strengthen growth in areas outside the major urban centres—was one important reason in Trondheim's favour. Trondheim had recently lost the headquarters of Norges Bank to Kristiania, and it was argued that Kristiania could not be the centre for all cultural activities in the country. Secondly, it was considered important that Trondheim already had a well-renowned lower technical school (Trondhjem tekniske læreanstalt), which was regarded as a *national* school to train (lower) engineers for leading positions all over the country. In that sense, competence was already in place, and the new institute would not need to build its training programme from scratch. And last but not least, the municipality of Trondheim could offer suitable ground for the new school.[15]

13 Brandt and Nordal, *Turbulens og tankekraft*, p. 71; Ragnhild Rein Bore, 'Industrien har begyndt at rotfæste sig i Landet', in Ragnhild Rein Bore and Tor Skoglund, eds., *Fra håndkraft til høyteknologi – norsk industri siden 1829* (Oslo: Statistisk sentralbyår, 2008), pp. 24–44.

14 Brandt and Nordal, *Turbulens og tankekraft*, pp. 86–90.

15 Brandt and Nordal, *Turbulens og tankekraft*, pp. 86–96; Olaf Devik, *N.T.H. femti år: Norges tekniske høiskoles virksomhet 1910–1960* (Oslo: Teknisk ukeblad, 1960), pp. 33–38.

The establishment of the polytechnic (higher) school, which was given the name Norwegian Institute of Technology (NTH), came at a convenient time. Hydroelectric power stations were giving rise to power-intensive and knowledge-based industries, such as The Norwegian Company for the Electrochemical Industry, or Elkem (formerly known as Elektrokemisk), established in 1904, followed by Norsk Hydro one year later. Norway was a young nation; the Swedish–Norwegian union—formed following 400 years of Danish rule in 1814— was dissolved only in 1905. National pride was at its height, and optimism was especially pervasive regarding knowledge, exploration, and industry. Polar researcher Fridtjof Nansen was praised as a national hero after he returned from one of his polar expeditions in 1896.[16] His explorations symbolized what the new country could achieve. The engineers being trained at the new institute ensured that the country would develop industrially as well, especially with the newly available resources supplied by hydroelectric power stations. The inauguration of the institute reflected this optimism and festive mood: the king was present, flags were waving, and schools and shops closed so that the people of Trondheim could take part. For the first time, Norwegians could train as academic engineers in their own country, and expectations were high as to what the institute could offer in terms of candidates who would propel new industry forward.[17] At NTH before World War II, a German engineering ideal emphasizing scientific expertise and broad technical competence prevailed.[18]

One of the symbols of industrial growth in Norway in the early 1900s could be found in the metal alloy producer Elkem, which established research stations from about 1915 with the purpose of experimenting and researching processes of interest for the company. Other firms established

16 Robert Marc Friedman, 'Nansen, National Honour and the Rise of Norwegian Polar Geophysics', in Reinhard Siegmund-Schulze and Henrik Kragh Sørensen, eds., *Perspectives on Scandinavian Science in the Early Twentieth Century* (Oslo: The Norwegian Academy of Science and Letters, 2006), pp. 85–110.

17 In fact, the establishment of the Royal Frederick University in Kristiania also happened in the context of independence, this time from Denmark (the university was founded in 1811). It was considered important to have a university in Norway.

18 An American engineering ideal, which emphasized technical management and unit operations, was debated in the 1920s and 1930s but implemented only after World War II. See (for the case of chemical engineering at NTH): Lykknes and Gusland, *Akademi og industri*, pp. 265–77. See also: Pål Nygaard, *Ingeniørenes gullalder. De norske ingeniørenes historie* (Oslo: Dreyer, 2013), pp. 22–30 (on different engineering ideals and their place in the history of engineers in Norway) and chs. 7 and 8 (on the transformation from a German ideal to an American ideal in the Norwegian engineering profession at large).

research laboratories, too.[19] Professors who left the institute were recruited to conduct research in power-intensive industries and their laboratories. Other professors became consultants and did research for the companies from their laboratories at the institute.[20] Laboratories were thus imperative for developments in the new industries at the time, and many of the chemical engineers graduating from NTH found their place in those laboratories.

According to a survey of practising engineers, the electrochemical industry was the branch of the chemical industry that attracted the most chemical engineers before 1923. Next on the list were the pulp and paper industry and foodstuff industries. For seven of the ten first graduating classes of chemistry students from the institute, more than half of the candidates went into a chemical industry of some sort.[21] The first groups of engineers trained in Norway thus seemed to have successfully established a symbiosis between the technical institute and industry in different parts of the country, in keeping with the expectations for the institute.

The University, its Chemistry Laboratory and the Chemical Community

Before 1910 the Norwegian chemical community was rather small. Training in chemistry was mainly offered at the only university in the country in Kristiania but also at the Norwegian College of Agriculture at Ås just outside Kristiania, which offered courses in agricultural chemistry. As noted, training in technical chemistry at the intermediate level had been offered at a few technical schools around the country since the 1870s. University education in chemistry around the turn of the twentieth century mainly served to support other professions, such as the medical and pharmaceutical professions, or to qualify as science teachers at the upper secondary school level. Educating chemists for the chemical industry was not relevant in

19 Elkem established a pilot plant for research on the pigment titanium white in 1915, and in 1917 Fiskaa Verk, where research and innovation on electrical smelting was conducted. See: Knut Sogner, *Skaperkraft: Elkem gjennom 100 år: 1904–2004* (Oslo: Messel forlag, 2003), pp. 51, 63. Norsk Hydro's central laboratory for research and innovation was in full operation from 1919. See: Ketil Gjølme Andersen and Gunnar Yttri, *Et forsøk verdt: Forskning og utvikling i Norsk Hydro gjennom 90 år* (Oslo: Universitetsforlaget, 1997), p. 66. The pharmaceutical company Nyegaard & Co (Nycomed) started to do research towards the end of the 1920s. Thanks to Knut Sogner for providing this information.

20 Lykknes and Gusland, *Akademi og industri*, ch. 4; Annette Lykknes, 'The Chemistry Professor as Consultant at the Norwegian Institute of Technology, 1910–1930', *Ambix*, 67 (2020), 271–88.

21 Lykknes and Gusland, *Akademi og industri*, p. 141, fig. 4.2.

the beginning since the chemical industry that existed at the time did not require qualified academic chemists.[22]

In 1893 the Norwegian Chemical Society (then named 'Group of Chemists') was created as an extension of the Polytechnical Society, established in 1852. Forty chemists became members of the Chemical Society during the first year, and they included chemists and mineralogists working at the university, at one of the technical schools, or at the Norwegian College of Agriculture, as well as chemists working in breweries, for the city gas works, at the chemical control station, and as officers and pharmacists; a few industry owners were also members.[23] Towns and communities in Norway were spread over large areas, and as in some other countries, the area around the capital Kristiania was considered the nation's hub—the rest of the country was on the periphery. Indeed, all the members of the Chemical Society who registered in 1893 lived in or near Kristiania.

In 1875 the chemistry department at the University had been relocated to new buildings in Frederiksgate near the main university building but also in the vicinity of the Royal Palace, the National Gallery, the National Theatre, and the Parliament, testifying to the importance of the university and its chemical laboratory in the capital of Norway. The move to new premises coincided with the beginning of the breakthrough of industrialization in the country, which was used rhetorically when describing the many opportunities offered by this new facility. The new building was described as 'perfectly adequate to educate chemists and [it] is in every respect on a par with foreign qualifications, in particular German'. As noted, the construction of this laboratory happened at the same time as research laboratories were being established in universities all over Germany.[24] In fact, the rhetoric went even further; it was stated that the scientists graduating from this laboratory would 'find rich applications at the many laboratories that are now arising everywhere in the country'—possibly in reference to the rising pulp and paper industry in the country, where one hoped that chemists

22 An exception might be the pulp and paper industry, which saw explosive growth already in the last quarter of the nineteenth century in Norway. In 1909, 12,000 people were working in this branch of industry. In a survey of Norwegian engineers from 1923, however, only 32 were listed in the pulp and paper industry. It is not clear how many of these were (academic) chemists and when the first chemists were hired. See: Lykknes and Gusland, *Akademi og industri*, p. 142.

23 Bjørn Pedersen, 'Norway: A Group of Chemists in the Polytechnic Society in Christiania. The Norwegian Chemical Society, 1893–1916' in Anita Kildebæk Nielsen and Soňa Štrbáňová, eds., *Creating Networks in Chemistry: The Founding and Early History of Chemical Societies in Europe* (London: The Royal Society of Chemistry), pp. 223–35.

24 Anon., 'Det kemiske laboratorium i Kristiania', *Skillingsmagazinet*, December 1876, quoted in: Ellen Gleditsch, 'Universitetets kjemiske laboratorium i Fredriksgate 32', *Aftenposten*, 17 June 1966.

would be needed, or to research laboratories and test stations for veterinary science, fisheries, agriculture, or hygiene.[25] The pulp and paper industry would experience explosive growth in the last quarter of the nineteenth century, but in 1876 the industry establishments were still only in their initial phase.[26]

The once so successful laboratory at the university soon became crowded, and after 40 years the laboratory was considered truly unsatisfactory. Indeed, the building was crumbling and cramped and suffered from inadequate gas, water, and ventilation systems. These conditions, in the view of the chemists working there, made it impossible to educate chemists and indeed, even to teach the elementary courses.[27] Once again, the education of industrial chemists became part of the rhetoric, only this time as more than merely an ideal. Industry had become more prosperous since the beginning of the century, especially within branches—such as the electrochemical and electrometallurgical industry—that could make use of the newly installed hydroelectric power. Industry was in need of chemists to bring their expertise to the many research laboratories that had been established within the companies. This time more was at stake. As I will demonstrate, it was felt that to participate in the development of the country's chemical industry, the university needed a better laboratory. And candidates from NTH in Trondheim were already filling the positions in industry.

Apart from the electrochemical and electrometallurgical industry, Norway's main chemical industries at the time were (still) the pulp and paper industry (Borregaard was Norway's largest company in 1909, with its industrial plant numbering 2,000 employees) and the foodstuff industry.[28] All in all, chemists were regarded as important for the continuing growth of the Norwegian chemical and electrochemical industries. The good business prospects naturally impacted the chemistry training as well, as laboratory facilities had to meet society's needs. Without proper training in up-do-date

25 Quote from: anon., 'Det kemiske laboratorium i Kristiania'. On the pulp and paper industry, see note 22. On other research laboratories and institutes for veterinary science, fisheries, agriculture, and, in particular, hygiene, see: Tove Elvbakken and Annette Lykknes, 'Relationships between Academia, State and Industry in the Field of Food and Nutrition: The Norwegian Chemist Sigval Schmidt-Nielsen (1877–1956) and His Professional Roles', *Centaurus*, 58 (2016), 257–80 and references therein.

26 Lykknes and Gusland, *Akademi og industri*, p. 142.

27 Annette Lykknes, Lise Kvittingen, and Anne Kristine Børresen, 'Ellen Gleditsch: Duty and Responsibility in a Research and Teaching Career', *Historical Studies in the Physical and Biological Sciences*, 36 (2005), 131–88 (p. 143).

28 Lykknes and Gusland, *Akademi og industri*, pp. 140–44.

Figure 7.1 Professor Ellen Gleditsch with her research assistants Ernst Føyn and Ruth Bakken in the old university laboratory in Frederiksgate in the early 1930s. Courtesy: Norsk Farmasihistorisk Museum/Norsk Folkemuseum.

laboratories at the universities, students could not be expected to perform the kind of laboratory-based research that was required in industry. In other words, without new facilities, university candidates did not stand a chance against the chemists from the NTH.

Two Opposing Views

In a meeting of the Norwegian Chemical Society in 1918, the future of the country's chemical industry was discussed. One expressed claim was that Norway would lag behind internationally if the chemical industries were not modernized.[29] At the same time, there was a real demand for chemists in existing industries, and not enough candidates were trained to fill this

29 I.J. Moltke-Hansen, 'Nydannelser i kemisk industri, Staten og det private intiativ. Kortfattet referat av direktør I.J. Moltke-Hansens foredrag i Norsk Kemisk Selskap, gruppe av P.F. den 12te februar 1918', *Teknisk Ukeblad*, 23 (7 June 1918), 292–95.

need.[30] At the meeting Rector Waldemar Christopher Brøgger, who was professor of geology, took the opportunity to speak up for new chemistry laboratories at the university. Brøgger had been a key actor in the efforts to expand the university campus and buildings around the turn of the century. He was an advocate for independent research, and at the centenary of the university in 1911, one year after the inauguration of NTH, he advocated to appoint more personnel in general.[31]

In the Chemical Society discussions, which were reported in the weekly technical journal *Teknisk Ukeblad* (official journal for NIAF and the Norwegian Polytechnical Association), Brøgger voiced his fear of falling behind in industry and spoke to the need for a closer connection between science and industry in the country. One obvious solution was to equip the university with better laboratories so that they could train chemists for industry and contribute to a desired synergy between university training and industrial research. Brøgger's main reference was the situation in Germany, where as early as the 1870s, most university-trained chemists had found employment in the chemical industry. Few became teachers.[32] This contrasted starkly with the situation in other European countries, where, as was the case in Norway, the universities mainly trained chemists for teaching in secondary schools. This was also the case in the Netherlands until industry started doing research around the turn of the twentieth century. In Sweden, too, the chemical training at the universities was regarded as very different from that offered at the Royal Swedish Institute of Technology, and it was seen as quite out of touch with what was happening in industry.[33]

30 Thv. Lindeman, 'Træk av den kemiske industris utvikling. Foredrag holdt ved Den tekniske Høiskoles immatrikuleringsfest den 2den september 1918', *Tidsskrift for kemi*, 18 (15 September 1918), 277–90.

31 John Peter Collett, *Historien om Universitetet i Oslo* (Oslo: Universitetsforlaget, 1999), pp. 115–17.

32 Jeffrey A. Johnson, 'Academic Self-Regulation and the Chemical Profession in Imperial Germany', *Minerva*, 23 (1985), 241–71. At the turn of the century, too, about 90 per cent of all academically trained chemists had moved to the chemical industry. In all other fields than chemistry, however, universities in Germany educated candidates for teaching more than for industry. See: Peter Lundgreen, 'Engineering Education in Europe and the U.S.A., 1750–1930: The Rise to Dominance of School Culture and the Engineering Professions', *Annals of Science*, 47 (1990), 33–75 (p. 58).

33 For the situation in the Netherlands, see: Geert J. Somsen, 'Selling Science: Dutch Debates on the Industrial Significance of University Chemistry, 1903–1932', in Anthony S. Travis, Harm G. Schröter, Ernst Homburg, and Peter J.T. Morris, eds., *Determinants in the Evolution of the European Chemical Industry, 1900–1939: New Technologies, Political Frameworks, Markets and Companies* (Dordrecht: Kluwer Academic Publishers, 1998), pp. 143–68. For Sweden, see: Anders Lundgren, *Kunskap och kemisk industri i 1800-talets Sverige* (Lund: Arkiv förlag, 2017), e.g. on

One of the chemists who raised their voice in the discussion that continued in *Teknisk Ukeblad* after the meeting was chemistry professor Claus Nissen Riiber at NTH. He was unhappy with Brøgger's view to reserve a minor role for NTH—for who else would be better equipped to train industrial chemists than the institute in Trondheim?[34] After all, the chemists trained there were not only chemists but also chemical engineers, and they would gain the broad technical competence that was needed at the industrial plants. Brøgger, however, had claimed that the university was the best place to specialize in the field of chemistry, arguing that the German chemical industry was founded by university-trained chemists. Although this was true for some of the major companies, he had neglected to mention the fact that the *Technische Hochschulen* were also of considerable importance. Brøgger also ignored the debates in Germany in which representatives from industry had expressed the need for broader chemical and technical training at the universities.[35]

In Riiber's opinion, there were indeed benefits of the training offered in Trondheim. In his response in the technical journal, he reminded Brøgger that the Institute of Technology actually had four chairs for four specialities while the university had only two—and more importantly, most of the university students were not science students; they were studying medicine or pharmacy.[36] Most of the students who did study natural science became teachers after their exam.[37] Riiber questioned how specialized they needed to be, since very few of them would actually become practising chemists.[38]

Brøgger, for his part, responded that he did not accept that the university's chemistry education should be reduced to a place for basic coursework in chemistry; on the contrary, the university should be where real chemists

pp. 225, 230. Lundgren's main argument is in fact that scientific knowledge was of little use overall in chemical industry in nineteenth-century Sweden.

34 Claus N. Riiber, 'Utdannelsen av vor Industris Kemikere. Tilsvar til Prof. Dr. W.C. Brøgger', *Teknisk Ukeblad*, 33 (1918), 422–23.

35 In fact, at the turn of the century the *Technische Hochschulen* were regarded as better suited than the universities to train the candidates modern chemistry needed. See: Johnson, 'Academic Self-Regulation'; Jeffrey A. Johnson, 'Academic Chemistry in Imperial Germany', *Isis*, 76 (1985), 500–24.

36 In 1911, 31 out of 100 students following lectures in chemistry were science students; the rest studied medicine or pharmacy. Of these, only six were advanced students in chemistry, doing a master's equivalent major. For data on the number of students, see: *Universitetet i Oslo: Årsberetninger* [annual report of the University of Kristiania] (1910/11), 125–26.

37 Eivind Myhre, *Universitetet i Oslo 1811–2011: Kunnskapsbærerne 1811–2011* (Oslo: Unipub., 2011), pp. 145–46.

38 Claus N. Riiber, 'Skal en betragtelig Del av de Kemikere vor Industri trænger utdannes ved Universitetet?', *Teknisk Ukeblad*, 26 (28 June 1918).

were hatched, especially now that industry was in need of chemists for research.[39] Riiber considered this to be trespassing on what was regarded as the domain of the Institute of Technology, but to be fair, in his letter to the editor of *Teknisk Ukeblad*, he pleaded for good laboratories both in Trondheim and Kristiania. The discussions stopped short of addressing specific sums and concrete applications for funding, and the discussions in the technical journal thus ended on friendly terms. A few years later, the situation was different, and the discussion now entered the public arena of the daily press, not just the limited readership of the *Teknisk Ukeblad*.

The Chemistry Building at the Norwegian Institute of Technology

In 1910 four buildings had been erected at the technical campus in Trondheim, one of them being the chemistry building. Specially tailored for chemical experiments with laboratories and chimneys, the Institute of Technology was proud to present its chemistry department. It hosted 'the best equipped laboratory of Scandinavia'[40] with electricity, water, and gas supplies, and even three-phase electric power had been installed.[41] The use of mains electricity was quite novel, as the two power stations in Trondheim had been installed in 1901 and 1910, respectively. According to Peter Morris, electricity had become sufficiently established worldwide for use in laboratories only in the 1920s,[42] so 'provincial' Trondheim was ahead of many other respectable laboratories. This is not surprising, however, given the technical profile of the institute and the access to hydroelectric power in Norway. The lecture halls were also well equipped for demonstration experiments. In the main chemistry lecture hall in the chemistry building, 'experiment tables' equipped with gas, water, and electricity were installed, along with a large slide projector from Carl Zeiss and motorized blinds which could close in seven seconds (much faster than today's electric blinds in the Natural Science building at NTNU).[43]

The chemistry department consisted of four sections: organic chemistry, inorganic chemistry, technical-organic chemistry, and technical-inorganic

39 Waldemar Christopher Brøgger, 'Vore Kemikeres Utdannelse. Tilsvar til dr. C.N. Riiber', *Teknisk Ukeblad*, 28 (12 July 1918), 368–69.

40 Claus N. Riiber, 'Vore kemiingeniørers utdannelse. Foredrag ved immatrikuleringen paa Den tekniske høiskole igaar', *Dagsposten*, 2 September 1916.

41 Lykknes and Gusland, *Akademi og industri*, pp. 81–85.

42 Morris, *The Matter Factory*, p. 141.

43 Lykknes and Gusland, *Akademi og industri*, p. 85.

Figure 7.2 One of the student chemistry laboratories at the Norwegian Institute of Technology, undated. All students had to pass the laboratory courses in qualitative and quantitative analysis. The laboratories remained more or less unchanged for 40 years. Courtesy: NTNU University library.

chemistry, each chaired by one professor. Each chair had their own research laboratory, and each section included a student laboratory. The largest one was the laboratory for qualitative and quantitative analyses, coursework that all students at the institute (not just the chemistry students) had to pass at the beginning of their studies.[44]

A decade after its inauguration, the chemistry building at NTH had become cramped. While the institute was built to house 400 students in total (100 per year), the number of admitted students had grown to 700 by 1920—75 per cent more than planned for. This was partly because NTH, as the national Institute of Technology, carried special responsibility for training a sufficient number of engineers and wanted to accommodate as many students as possible. But even with the expanded admissions, only two-thirds of the students who had applied were admitted. In 1917, 90 applicants were rejected, of which 50 wanted to study chemistry.[45] The lack

44 Lykknes and Gusland, *Akademi og industri*, p. 117.

45 Lykknes and Gusland, *Akademie og industri*, pp. 194–95.

of space and the need for more engineers made it crucial to obtain funding to expand the buildings for chemistry. The situation was particularly serious for the advanced laboratory courses in technical chemistry; due to the lack of space, these courses were only offered on an ad hoc basis. Despite the demand for technical training and expertise, equipping a laboratory for technical experiments on a semi-factory scale was thus considered to be out of the question under these circumstances. In the 1920s economy, funding for new building construction was assumed to be almost impossible to obtain. So when the Norwegian Chemical Society published a press release to urge support for the university's need for a new chemistry laboratory in 1922, it was received as a provocation, as we shall see.

The Controversy over New Laboratories (1922)

In 1920 a decision was made to build a new university campus at Blindern on the western side of Kristiania. For the first time, physicists had been promised a building of their own. However, the proposal to build a new *chemistry* building had been turned down. As described above, the chemistry building from 1875 was so unsatisfactory that the situation had become desperate. Professor Eyvind Bødtker therefore turned to the Norwegian Chemical Society for support. At this point, the university chemists had to rent rooms for basic-level laboratory teaching, and associate professor Ellen Gleditsch was not able to accept undergraduate students in radiochemistry in the laboratory.[46] In a meeting of the Norwegian Chemical Society in February 1922, the idea to issue a statement from the society was discussed and a text agreed upon. Among other things, it stated that industry also needed university-trained chemists, referring again to Germany, and that the university's chemical laboratory must have room for technical-chemical research.[47]

Before the statement was released to the press, professor Riiber at NTH, who was not present at the meeting, was given the opportunity to comment on the statement. His reply came to four pages.[48] He was upset that such a statement gave the appearance of coming from the whole chemical

46 Lykknes, Kvittingen and Børresen, 'Ellen Gleditsch', p. 143.

47 Anon., 'Universitetets kemiske institutt', press release discussed at the Norwegian Chemical Society on 6 February 1918, published in: *Tidsskrift for kemi og bergvæsen*, 5 (1922), 77–78.

48 C.N. Riiber to the board of the Norwegian Chemical Society, 14 February 1922, Trondheim, NTNU University library, Special collections, private archive no. TEK 3, Claus Nissen Riiber, Box 13.

community in Norway, since it had only been agreed upon by chemists living in or around Kristiania. He maintained that the university should request better laboratories to continue the kind of training that the university already offered, and that it should not trespass on NTH's purview, that of technical education and research. However, his comments did not change the statement to any degree.

A commentary in the newspaper *Dagbladet* entitled 'Chemistry and the future' appeared shortly after the statement was published.[49] According to the commentary, university-trained chemists could solve nearly any challenges related to the chemical industry. Again, Riiber was provoked and submitted a reply.[50] He was particularly upset that the Norwegian Institute of Technology was not mentioned at all in the commentary, writing, 'It is entirely a complete "quantité négligeable", which is being passed over in silence on the question of industry and its development.'[51] More responses followed, including a commentary in the Trondheim-based newspaper *Trondhjems Adresseavis* entitled 'No assassinations, please!'[52] The editor of the newspaper feared that the Norwegian Institute of Technology would suffer from 'malnourishment' and then 'die'. Students expressed similar thoughts.[53] This dispute echoed a debate that had taken place in the Netherlands some years earlier (in 1904) when the Delft polytechnic emerged as a competitor to the university. One of the strategies used to defend the university's territory was to present chemical science and technical chemistry as an integrated whole that the universities were best suited to take care of.[54] In that case, too, the benefits of the technical (engineering) education at Delft were ignored.

As for the new chemistry laboratories in Kristiania, what, in fact, were the plans? Riiber and others feared that a new technical institute for chemistry was in the making, which Rector Fredrik Stang denied, giving assurances that 'no plans are in the works to expand the chemistry teaching at the university at the expense of the Institute.'[55] However, this is not entirely clear from reading the correspondence between Stang and the former rector in Trondheim Sem Sæland—then a professor of physics at NTH—who

49 Anon., 'Kjemi og framtid', *Dagbladet*, 5 March 1922.

50 Claus Nissen Riiber, 'Industrien og universitetet', *Dagbladet*, 22 March 1922.

51 Riiber, 'Industrien og universitetet'.

52 Anon., 'Snikmord frabedes!', *Trondhjems Adresseavis*, 23 March 1922.

53 Aage W. Owe, 'Kemi og fremtid. Universitet kontra høiskole', *Under Dusken*, no. 6, 25 March 1922.

54 Somsen, 'Selling Science'.

55 Fredrik Stang, 'Kjemisk institutt', *Dagbladet*, no. 71, 24 March 1922; Fredrik Stang, 'Høiskolen og Universitetets nye kemiske institutt', *Trondhjems Adresseavis*, 25 March 1922.

would soon be appointed professor at the university and later become its rector (starting in 1928). There were indeed comprehensive plans for a new technical-chemical training, and Sæland wrote to Stang that he felt the plans had been poorly communicated, and people might interpret them as plans for a new institute.[56] In fact, the plans included a laboratory of 150–200 square meters, a room for smelting and high-voltage experimentation, a room for grinding and beneficiating, and a large room in the basement well suited for technical-chemical investigations. Although Sæland to a certain extent defended Riiber in a letter to Stang, he also signed a letter to the Rector at NTH, stating that Riiber had gone too far in agitating against the university. He probably feared that Riiber's public statements would harm the university's case when it reached the Parliament.[57]

Positioning and Inferiority Complex from the Periphery

Clearly, the press release from the Norwegian Chemical Society and the newspaper commentaries that followed as well as the discussion that took place four years earlier hit a nerve for professor Riiber and others at the Norwegian Institute of Technology in Trondheim. Their sensitivity had many reasons, all of them having to do with the position the institute held compared to the university. Although the inauguration of the institute was highly celebrated in 1910 and hopes and expectations for what it could bring to the new nation-state were high, many people still seemed to consider the classical ideal of *Bildung* superior to the technical-practical ideal. In 1916 NTH Rector Alfred Getz stated in his welcome address to the new students that he had the impression that 'people' generally did not consider studying a technical subject as sophisticated as the subjects offered at university.[58] In their view, knowing about (Henrik) Ibsen—or (Charles) Darwin for that matter—was regarded as more important than understanding the principles of the dynamo or knowing the difference between steel and iron. If Getz saw the need to speak about this in his inaugural address, he must have felt it necessary to assure the students that technical knowledge was equal in status to 'scientific' knowledge.

56 Sem Sæland to Rector Stang, 28 March 1922, Oslo, Public Record Office (Riksarkivet), University of Oslo, University board (Kollegiet), E-0086 Diverse, Planer for nytt kjemisk institutt 1918-1924.

57 Sæland probably felt squeezed between loyalty to the two institutions he had been and would be working at.

58 Anon., 'Immatrikuleringen paa Den tekniske Høiskole', *Tidsskrift for bergvæsen* 9 (1916), 103–05.

A few years later, in 1920—a decade after the opening of NTH—former Rector Sæland reflected on the first ten years of the institute's history. He was particularly worried about the appointment of professors, since in the beginning people *applied* to those positions, but ten years later the institute had to *look for* the candidates.[59] Some professors had also left their job at the institute after a few years for a job at the university or in a centrally located industry. When Riiber later became rector, however, he blamed the vacancies on poor pay, a difficult housing market, the poor climate, and the peripheral location.[60] But given his many comments in the daily press, he probably also suffered from the inferiority complex of comparing the position of the university in society with that of the technical institute.

Another aspect of the positioning battle NTH fought in 1922 is the fact that this was also the year in which the institute gained the right to award the technical PhD degree, which the *Technische Hochschule* in Berlin (as the first one in Germany) had been granted in 1899. The ideal for NTH from the very beginning was to be an academic, elite institution which required the matriculation exam to be admitted and which offered advanced courses in mathematics, physics, mechanics, and chemistry as a basis for specialized study (the German engineering ideal). Indeed, the right to grant PhDs became important for their status as a university-like institution, yet with its own specific mandate like that of the *Technische Hochschulen.*

For most studies offered at the Norwegian Institute of Technology, the courses taught were unique. Nowhere else in the country could one study to become an electro-engineer, mechanical engineer, or metallurgist at the top level. Trondheim and NTH were the place to go, and there was no doubt that the country needed this expertise. However, in the case of chemistry, there was competition, as the university had trained chemists for more than 100 years. Clearly, most of them became science teachers, but it was also true that it was only in the twentieth century that industry progress became so promising that it was actually attractive to train chemists for this line of work. These developments were happening at the same time as NTH became established as an institution to train engineers for exactly that, thus competing with the university. As a result of the renewed interest in industry at the university, NTH and its professors had to legitimize their engineering courses and make a case for why the institute in some cases was better suited than the university to educate industrial chemists.

59 Sem Sæland, 'Omkring Høiskolens 10-aarsdag', *Teknisk Ukeblad*, 35 (1920), 438–42.

60 Anon., 'Norges tekniske høiskole. Er dens anseelse i avtagende? Et interview med skolens rektor, professor C.N. Riiber', *Trondhjems Adresseavis*, 13 November 1923.

How could the representatives from NTH convey the benefits and distinctive qualities of the training provided at the institute? Distinguishing between 'pure' and 'applied' chemistry was a strategy which was used in the discussions both in 1918 and in 1922 as one way of emphasizing what was special about the training offered at NTH—or at the university for that matter. The label 'applied science', which was widely used in the US from the 1870s on, has been used both to refer to applications of 'pure science' (pertaining to the pure science ideal) and to the application of scientific methods to produce technical relevant knowledge (in keeping with the ideal of a research engineer). Ronald Kline has shown that engineers alternated between keeping the distinction between pure and applied science and blurring it, depending on what would favour their profession and help them raise money or gain status.[61] As noted earlier, this was the case for the Dutch debate as well. Likewise, for the chemists at Kristiania, blurring the distinction was one way of demonstrating that the university could train good industrial chemists, even though they did not offer engineering education as such. In Germany, however, the industry itself demanded a change, and it was in fact the universities and not the *Technische Hochschulen* that had to adapt and include more technical courses.[62]

For Riiber and NTH, the chosen strategy in the debate over the need for laboratories was to emphasize the special (technical) curriculum and branches of industrial chemistry offered at the institute so as to claim autonomy. As for his own identity as a professor at the Norwegian Institute of Technology, Riiber embraced both the academic tradition and the technical-industrial. He became the most publishing chemistry professor before 1925 and frequently received research funds, trained as he was in the university tradition.[63] He was also one of two (out of the four) chemistry professors who were members of the Royal Norwegian Society of Sciences and Letters (DKNVS), testifying to his standing as a scientist. At the same time, he took the expectations of gaining technical expertise seriously, filed several patents, and even established his own alkali chloride business.[64] He was known to be an agitator, and it seems that in this case his motive was first and foremost to support his institute and not necessarily to protect his own interest in the chemistry landscape, although his profile would have

61 Ronald Kline, 'Construing "Technology" as "Applied Science": Public Rhetoric of Scientists and Engineers in the United States, 1880–1945', *Isis*, 86 (1995), 194–221.

62 Johnson, 'Academic Self-Regulation', p. 262.

63 Lykknes and Gusland, *Akademi og industri*, p. 147.

64 Lykknes and Gusland, *Akademi og industri*, p. 152.

fit both at the university and at the technical institute. The laboratories, however, were crucial for both institutions.

The Values of a Laboratory

The laboratories were the spaces where technical-scientific training and research were conducted, even more so after their place in universities had become more prominent. Well equipped, modern laboratories indicated a wealthy society with hopes for the future. Indeed, the chemistry department at the Norwegian Institute of Technology in Trondheim was recognized as having the most modern chemistry laboratories in Scandinavia at its inauguration in 1910. In that sense, the laboratory became a symbol for what the institute could achieve, just as the once new and modern laboratory at the university had in 1876. Similarly, the lack of a laboratory symbolized poverty and an inability to compete when it came to training candidates for the emerging industry.

So, what could one do when this important symbol and necessary space had become inadequate for chemical investigations and could no longer fulfil its function? Quite often, institutions found a pragmatic, temporary solution, for example teaching in borrowed and less satisfactory rooms. As the priority debates raged in 1918 and the 1920s, such temporary solutions had already been in place for a long time, both in Kristiania and Trondheim. The only way to really solve the problem was to obtain funding for new laboratories. Although rivalry would probably always exist between the competing institutions, the competition became far worse in hard financial times such as the post-World War I years. On top of differences in training and qualities of education, the geographical differences in Norway added to this polarization, with questions related to centre and periphery sharpening the dispute.

Both the university's chemists and Riiber at NTH found the new important issue to be the need for more chemists in the emerging chemical industry. According to professor of technical-inorganic chemistry Thorvald Lindeman, the number of chemists required in industry was more than the Institute in Trondheim could supply.[65] Both NTH and the university tried to argue why their laboratory was more important to fill the mission of increasing and developing new industry in the country; their arguments were based on what their education could offer—and on how desperate their situation

65 Lindeman, 'Træk av den kemiske industris utvikling'.

was. The engineers trained at NTH would come away with broad technical competence and the opportunity to specialize in a technical branch of chemistry. The university candidates, it was argued, profited from more extensive thesis work.[66] As in the Netherlands, dichotomies characterized the debate.

In Norway the university drew the longer straw in the fight for new laboratories. In May 1922 the sum of one million Norwegian kroner from the lottery was allocated to a new chemistry building. However, it took more than a decade before the combined physics and chemistry building was inaugurated in 1935 at the new campus at Blindern, and in the end it was largely externally funded. Although the new laboratories at the university were well equipped, they were not radically different from their precursors, and they did not become the mecca for technology that Riiber had feared. Raising money for new laboratories at the Institute of Technology took much longer than anticipated and was even further delayed because of World War II. In fact, the first stage of the building process was only financed in 1946–47, and the five new building blocks were only erected between 1951 and 1968.[67] These laboratories, even more than their 1910 precursors and the laboratory erected at the university in 1935, represented something new, with advanced instruments that had become more common in chemical laboratories in the post-war years (referred to as the 'instrumental revolution').[68] This modern laboratory came as a result of the long delay in establishing new laboratories at NTH and was not part of the debate in the 1920s.

The debate came to a close in November 1923 when the University and NTH issued a joint statement. It specified that the university and the institute should both be entitled to train full-fledged chemists for industry, with the institute having special responsibility for the technical-chemical exams. As for research, it was agreed that the two institutions should accept competition. NTH would be given special responsibility for technical-chemical research.[69] So although the university was prioritized in the fight

66 These arguments were provided on more than one occasion. See for example (in favour of the university education): Brøgger, 'Vore Kemikeres Utdannelse' and L.J. Dorenfeldt, 'Kemien og høiskolerne', *Tidens Tegn*, 31 March 1922, and (for NTH): Riiber, 'Utdannelsen av vor Industris Kemikere'.

67 Lykknes and Gusland, *Akademi og industri*, pp. 277–85.

68 Peter J.T. Morris, ed., *From Classical to Modern Chemistry: The Instrumental Revolution* (Cambridge/London/Philadelphia: Royal Society of Chemistry, Science Museum and Chemical Heritage Foundation, 2002).

69 Statement from the committee scrutinizing the plans for the new chemistry building to the Ministry of Church and Education, 1 November 1923, Trondheim, NTNU University library, Special collections, private archive no. TEK 3, Claus Nissen Riiber, Box 13.

Figure 7.3 Laboratory course in chemistry at the University of Oslo in autumn 1952, in the new physics and chemistry building inaugurated in 1935 (officially inaugurated the year after). Note the fume hoods to the right. Courtesy: Museum of University History.

for new laboratories, the Institute of Technology was acknowledged for its trademark as a specialized chemistry (engineering) training tailored to industrial applications. But was this sufficient to reassure the professors at the institute?

Claiming that the laboratory is an important part of a chemist's identity is hardly controversial; very few people would think it possible to graduate in chemistry without having the necessary laboratory training. In the specific situation around 1920 and in the public debates that arose, I argue that in the eyes of the actors, the laboratory even became a symbol for the nation. They argued that without a proper chemical laboratory, it was not possible to train industrial chemists—which meant that industry and the whole country would suffer. An up-to-date laboratory, especially for technical chemistry, would attract students, chemists, and funding to NTH in peripheral Trondheim and help define its unique role among institutions for higher education. Far away from the main industry in Norway and a long journey from Kristiania, the material culture that the laboratory represented was important in terms of offering a proper education, its role as a symbol of the institute's excellence, and its utility for the nation-state. NTH chose to adhere to its ideals of technical know-how *and* scientific knowledge, despite being regarded as inferior by some and suffering to a certain extent

Figure 7.4 Crowded laboratory course at the Norwegian Institute of Technology in 1958, in one of the newly erected chemistry buildings. Every laboratory bench was equipped with small fume cupboards and sinks. Courtesy: Thorleif Anthonsen / NTNU.

from brain drain. But the praise and expectations for the institute and for the chemistry programme, which was also partly a university field, were not enough. For all the rhetoric about the importance of technical-practical expertise, in practice the ideal of the 'pure' scientists had won, at least in the minds of decision-makers and the public. The laboratory itself, however, did not take a stand. It was equally important to both the central Kristiania and peripheral Trondheim institutions. Indeed, the laboratory served the whole country.

Acknowledgements

I am grateful to Ernst Homburg for his helpful comments and careful editorial work on this chapter, to Klaas van Berkel for the final edits, and to Mentz Indergaard for providing many of the images.

About the Author

Annette Lykknes is a historian of science and professor of chemistry education at NTNU-Norwegian University of Science and Technology in

Trondheim. Her main research interests include the history of women in chemistry, the periodic system, and 20th century chemistry and chemistry education in Norway. Recently she co-edited a special issue of *Ambix* on chemistry, women and gender in the Enlightenment and the era of professional science, where she published an article about the first generations of women chemical engineers at the Norwegian Institute of Technology.

8 Religion and the Laboratory Revolution: Towards a Physiological Laboratory at a Calvinist University in the Netherlands, 1880–1924

Ab Flipse

Abstract

Originally orthodox Christians were ambivalent about the modern research laboratory, which many of them dismissed as a symbol of 'materialism' and disbelief. It was only in 1918 that the Calvinist Vrije Universiteit in Amsterdam established its first laboratory, for physiology, and F.J.J. Buytendijk became the first professor of physiology. Although it was precisely in the chosen field of animal psychology that some distinctive, Christian emphasis could be placed, the most important consequence of this step was that the university was more than before adapting to what was already customary elsewhere. It turned out that the foundation of the laboratory instigated the Vrije Universiteit's own 'laboratory revolution'.

Keywords: Laboratory, religion, Neo-Calvinism, animal psychology, F.J.J. Buytendijk

Introduction

Over the course of the nineteenth century, the laboratory came to be seen as a symbol of scientific progress and innovation. Initially its role was restricted to the fields of chemistry and pharmacy, but in the second half of the century, laboratory work acquired significant authority in medicine as well. It was in the laboratory, proponents believed, that medicine finally seemed to find a scientific basis. The experimental method promised to

Berkel, Klaas van, and Ernst Homburg (eds), *The Laboratory Revolution and the Creation of the Modern University, 1830-1940.* Taylor & Francis Group, 2023

DOI: 10.5117/9789463720434_CH08

open the way to a golden future for medicine, and the laboratory was the right place for it.[1]

However, the laboratory also carried a more negative image. Especially in religious circles, it was often seen as a symbol of 'materialism' and disbelief. After all, medicine was practised in a fundamentally reductionist way in the laboratory. This was especially true of new forms of experimental physiology, which had captured a place in Germany around the middle of the century (championed by Carl Ludwig, Emil du Bois-Reymond, and others) and for which specialized laboratories were built. A characteristic of this work was the disappearance of a fundamental distinction between life and non-life. In the reductionist approach, there was no longer any room for teleological and vitalist principles or something like a separate life force. Everything, including features of living entities, had to be explained in terms of physico-chemical processes. Pathological research, based on cell theory, also found its way into the laboratory, where it was mainly based on intensive microscopic research, as advocated by Rudolf Virchow. This new anatomy therefore also possessed a reductionist character.[2]

All this was accompanied by a great increase in vivisection, animal experiments. This caused a backlash from both anti-vivisectionists and some religious groups. The 'vivisector in the lab' was sometimes compared to a new priest of science, the laboratory the temple, and animal experimentation a 'sacrificial rite demanded by the false god of a pseudo-religion'.[3] And nineteenth-century life sciences and medicine generally seemed to blur the lines between humans and animals. This primarily followed from the theory of evolution and eventually led to extensive research into animal behaviour and the animal psyche. Around 1900 comparative psychological research with laboratory animals took off in full force.[4]

1 On the nineteenth-century laboratory revolution in medicine: Andrew Cunningham and Perry Williams, eds., *The Laboratory Revolution in Medicine* (Cambridge: Cambridge University Press, 1992); William F. Bynum, *Science and the Practice of Medicine in the Nineteenth Century* (Cambridge: Cambridge University Press, 1994), pp. 92–117 and Frans van Lunteren's chapter in this volume.

2 Lynn K. Nyhart, *Biology Takes Form: Animal Morphology and the German Universities, 1800–1900* (Chicago/London: University of Chicago Press, 1995), pp. 67–90; Richard L. Kremer, 'Physiology' and Russell C. Maulitz, 'Pathology', in Peter J. Bowler and John. V. Pickstone, eds., *The Modern Biological and Earth Sciences. The Cambridge History of Science, vol. 6* (Cambridge: Cambridge University Press, 2009), pp. 342–66, 376–81; cf. Frederick Gregory, *Scientific Materialism in Nineteenth Century Germany* (Dordrecht: Reidel, 1977), pp. 164–88.

3 Chien-hui Li, 'Mobilizing Christianity in the Antivivisection Movement in Victorian Britain', *Journal of Animal Ethics*, 2 (2012), 141–61 (p. 151).

4 Judy Johns Schloegel and Henning Schmidgen, 'General Physiology, Experimental Psychology, and Evolutionism: Unicellular Organisms as Objects of Psychophysiological Research, 1877–1918',

Starting in Germany, the new laboratory medicine was also able to conquer its place in the Netherlands, starting with the appointment of a new generation of professors of physiology and the construction of associated laboratories at the universities. This started in Groningen, with the appointment in 1851 of Izaac van Deen. Elsewhere—Amsterdam, Utrecht, Leiden—physiology also conquered its independence, with its own laboratories in which research was conducted on a reductionist basis.[5]

When the Vrije Universiteit ('Free University'; VU), a university with a distinctive Calvinist identity, opened its doors in Amsterdam in 1880, it was faced with the question of what to do with this 'materialistic medicine'. A decision did not have to be made immediately, because the VU initially included only three faculties: theology, the arts, and law. Discussions, deliberations, and pondering would take place over several decades, but in 1907 the first medical professor was appointed, and a clinic was opened three years later. In 1918 the first—physiological—laboratory followed.

This chapter is a case study of how orthodox Christians related to the laboratory revolution in medicine. Which considerations guided discussions within the VU, and what did this Christian university have to offer against the 'materialism' of laboratory science? Answering these questions may provide insight into the role of the laboratory in the science–religion debate of the early twentieth century.

Plans for a Medical Faculty at the Vrije Universiteit

Although the VU had started with three humanities faculties, the absence of two complementary faculties, those for medicine and the natural sciences,

Isis, 93 (2002), 614–45; Robert A. Boakes, *From Darwin to Behaviourism: Psychology and the Minds of Animals* (Cambridge: Cambridge University Press, 1984); Michael Pettit, 'The Problem of Raccoon Intelligence in Behaviourist America', *The British Journal for the History of Science*, 43 (2010), 391–421.

5 H. Beukers, 'Een nieuwe werkplaats in de geneeskunde: de opkomst van laboratoria in de geneeskundige faculteiten', *GEWINA / TGGNWT*, 9 (1986), 266–77; C.A. Pekelharing, 'De physiologie in Nederland in de laatste halve eeuw', *Nederlandsch Tijdschrift voor Geneeskunde*, 50 (1907), 8–19; Stefan van der Poel, *Tussen zieken, boeken en kikkers. De fysiologie van een leven: Izaac van Deen (1804-1869)* (Groningen: Barkhuis, 2012); Stefan van der Poel, '"Doch deze onderscheidt zich op eene eervolle wijs." Izaac van Deen (1804-1869): de eerste Joodse hoogleraar in Nederland', in L.J. Dorsman and P.J. Knegtmans, eds., *De menselijke maat in de wetenschap: De geleerden(auto) biografie als bron voor de wetenschaps- en universiteitsgeschiedenis* (Hilversum: Verloren, 2013), pp. 74–97; Laurens de Rooy, *Snijburcht: Lodewijk Bolk en de bloei van de Nederlandse anatomie* (Amsterdam: Amsterdam University Press, 2011), pp. 30–43.

both necessary to form a 'full' university, was discussed since its foundation. Its most prominent founder and first rector, Abraham Kuyper (1837–1920), delivered an inaugural speech on 20 October 1880, in which he outlined the ideal of a complete university with five faculties. Kuyper wanted 'a separate scientific development' not only for theology but for all academic fields. In Kuyper's view, one's religious starting point makes a difference in all fields of science and scholarship. In medicine, Kuyper stated, Christian researchers believe that 'it is not a sick mammal that medical science would help, but a person created in the image of God'.[6] The aim of the VU was therefore to develop a Christian science and scholarship, founded on the so-called Calvinist principles. Medical research would be shaped differently than elsewhere. For the faculty of medicine, this implied that it could not restrict itself to clinical work and training Christian doctors, however important these tasks might have been. The basic scientific research topics also had to be given a place, precisely because of the perceived virulent materialism at the other universities that manifested in these fields. This stance is characteristic of Kuyper's so-called neo-Calvinist view of science. This philosophy of science had a radical character and differed from, for example, the approach to science among kindred Calvinists in the Anglo-Saxon world, who advocated a shared common-sense epistemology between Christians and non-Christians.[7]

The 'Society for Higher Education on the Basis of Reformed Principles' (*Vereeniging voor Hooger Onderwijs op Gereformeerden Grondslag*) was responsible for and governed the VU through a board of directors (elected in the general meeting of the Society). A board of curators (appointed by the directors) supervised the work at the university. The Society was also the channel through which the relation between the VU and its Calvinist constituency was maintained. This was important because the university could not survive without its supporters. In this period it was not funded by the government and remained completely dependent on donations.

6 Abraham Kuyper, *Souvereiniteit in eigen kring. Rede ter inwijding van de Vrije Universiteit* (Amsterdam 1880), p. 33. Translated as: Abraham Kuyper, 'Sphere Sovereignty', in James Bratt, ed., *Abraham Kuyper: A Centennial Reader* (Grand Rapids, MI: Eerdmans, 1998), pp. 463–90 (p. 487).

7 The literature on Kuyperian, neo-Calvinist philosophy of science include: Jacob Klapwijk, 'Abraham Kuyper on Science, Theology and University', *Philosophia Reformata*, 78 (2013), 18–46; Del Ratzsch, 'Abraham Kuyper's Philosophy of Science', in Jitse M. van der Meer, ed., *Facets of Faith and Science*, 4 vols. (Lanham, MD: University Press of America, 1996), vol. 2: *The Role of Beliefs in Mathematics and the Natural Sciences: An Augustinian Perspective*, pp. 1–32; Ab Flipse, *Christelijke wetenschap. Nederlandse rooms-katholieken en gereformeerden over de natuurwetenschap, 1880-1940* (Hilversum: Verloren, 2004), pp. 52–9.

The neo-Calvinist scientific ideal was guarded by the university's administrators—the directors and curators—while professors were expected to integrate these norms and values in their own work.[8]

The curators noted as early as 1881 that the appointment of a medical professor was a very pressing need, especially for teaching anthropology, psychiatry, and physiology, because of 'the unlimited influence of materialism in those subjects at the state universities'.[9] Two years later, they informed the directors that the establishment of a medical faculty was urgently needed.[10] However, two problems arose: firstly, there was a lack of suitable candidates, and secondly, the expense of such a faculty, with all its facilities, instruments, and laboratories, would far exceed the costs of the three existing faculties.

In 1888 the curators met specifically to discuss the minimum requirements for founding a medical faculty. Starting with a single professor was seen as pointless. After all, several subjects were considered as important: anatomy, physiology, anatomical pathology, therapy, and psychiatry. The battle against materialism needed to be waged across the full range. Moreover, it was clear that no suitable candidates would be available. As the curators put it: 'Whence do we acquire a Virchow?', meaning, of course, a Christian version.[11]

These problems were also underlined by Abraham Kuyper himself in a speech to the general meeting of the Society in 1891. It was not merely a matter of appointing a professor who was only good at microscopy or anatomy, Kuyper stated. After all, a different kind of science had to be built up, precisely in those areas that were 'in the hands of materialists'. Nor should the university start—as had been suggested—with a psychiatrist. Ideally one would first deal with physiology and pathology, 'which explains

8 Arie van Deursen, *The Distinctive Character of the Free University Amsterdam, 1880–2005: A Commemorative History* (Grand Rapids, MI: Eerdmans, 2008), pp. 9–14.

9 Minutes meeting curators, 7 January 1881, Amsterdam, VU, Curators' Archives: 'de onbegrensde invloed van het materialisme in die vakken aan de staatsuniversiteiten'. In the near future, the VU Archives will be moved to the Amsterdam City Archives, where they will be available for consultation. What follows in this paragraph is partly based on: Leo van Bergen, *Van genezen in geloof tot geloof in genezen. De medische faculteit van de Vrije Universiteit, 1880-2000* (Diemen: Veen Magazines, 2005), pp. 59–92; Mart J. van Lieburg, *Barmhartigheid en wetenschap. De onvoltooid verleden tijd van de Faculteit Geneeskunde VU* (Amsterdam: Vrije Universiteit, 1990); J.C. Rullmann, *De Vrije Universiteit. Haar ontstaan en haar bestaan 1880-1930* (Amsterdam: De Standaard, 1930), pp. 159–74.

10 Minutes meeting curators, 15 January 1883, Amsterdam, VU, Curators' Archives.

11 Minutes meeting curators, January 1888, Amsterdam, VU, Curators' Archives: 'van waar bekomen wij een Virchow?'

the origin of disease, because it is precisely in this area that the materialists are diametrically opposed to the Holy Scriptures'. The next step could then be the appointment of a psychiatry professor because of this field's focus on the relationship between body and soul.[12]

The Rotterdam doctor Th.G. den Houter emphasized this strategy in a brochure from 1895 on 'Medical science and the Vrije Universiteit'. He too emphasized 'the materialist nature of medicine' and the fact 'that man was regarded as a mammal' at other universities.[13] Like Kuyper, Den Houter was also critical of what he considered to be unnecessary, unbridled vivisection. After all, such experiments, in which animals were tormented and tortured, gave only limited insight into the sick patient and, even worse, trained students to be cruel instead of caring, an attitude which would push them towards atheism.[14] He pointed to the fundamental science subjects as the place where 'the evil of materialism' was already being sown. Den Houter pragmatically suggested starting with a chair for history, philosophy, and encyclopaedia of medicine, precisely because the foundation of laboratories seemed unattainable for the time being.[15] Despite hesitation, financial hurdles, and different opinions about the proper first steps, in the first decades following the VU's foundation, the shared conviction of all those involved was that one day, the university should comprise a complete medical faculty with its own laboratories.

In this endeavour the VU also followed the pattern of other confessional, mainly Catholic, universities outside the Netherlands. When the Catholic university of Freiburg (Fribourg), Switzerland began establishing a medical faculty in 1893, Kuyper set this as an example for the VU's supporters: 'What is now going on in Freiburg once more calls us to action.'[16] Kuyper might also have mentioned the Catholic University of Leuven in Belgium,

12 'Is de aanstelling van een hoogleeraar in de Psychiatrie geraden, zoolang de Medische faculteit als zoodanig niet is ingesteld?' *Twaalfde Jaarverslag van de Vereeniging voor Hooger Onderwijs op Gereformeerde Grondslag* (Amsterdam: 1892), pp. XXXVII–XLII: 'die de oorsprong der ziekte verklaart wijl juist daarin de materialisten lijnrecht ingaan tegen de Heilige Schrift.'

13 Th.G. den Houter, *De medische wetenschap en de Vrije Universiteit* (Leiden: Donner, 1895), pp. 12–13.

14 Den Houter, *De medische wetenschap*; A. Kuyper, 'Uit de pers', *De Heraut*, 2 November 1890; cf. Amanda Kluveld, *Reis door de hel der onschuldigen. De expressieve politiek van de Nederlandse anti-vivisectionisten, 1890-1940* (Amsterdam: Amsterdam University Press, 2000), pp. 168–72.

15 Den Houter, *De medische wetenschap*, pp. 26–8. Cf. M.J. van Lieburg, 'Reformatorische traditie, geneeskunde en geneeskunst. Enkele historische kanttekeningen', *Beweging*, 48 (1984), 67–70 (p. 69).

16 Abraham Kuyper, 'Medische faculteit', *De Heraut*, 3 September 1893: 'Wat thans te Freiburg gaande is, brengt ons weer een roepstem, die tot handelen prikkelt.'

where all sorts of laboratories were being built during these decades.[17] In the Netherlands a Catholic university (in Nijmegen) was not established until 1923, with a medical and science faculty being added only after World War II. However, the tensions at these Catholic universities were not as great as they were at the VU, because the ideal of a 'Catholic science' was not considered as the main duty of the faculties, as historian Geert Vanpaemel writes in his book about the Leuven science faculty.[18] The neo-Calvinist ideal of a Christian science was more radical than this Catholic implementation. And it remained vital during the foundation and expansion of the VU.

During the first 25 years of its existence, the VU did not take concrete steps towards founding a medical faculty. It was only through collaboration with the Association for Christian Care for the Insane and Neurotics (*Vereeniging tot Christelijke Verzorging van Krankzinnigen en Zenuwlijders*) that the first professor of medicine would be appointed in 1907. The objective of this Association, which already administered several clinics in the Netherlands, was slightly different from that of the VU. The training of Christian psychiatrists was their primary concern. The Association had repeatedly urged the VU to create a chair of psychiatry. In 1907 the collaboration between the VU and this Association led to the appointment of Leendert Bouman (1869–1936) in a joint position as professor at the VU and as medical director of a newly built clinic. Whereas the three existing faculties of the VU were housed in a converted ancient canal house in the city centre, this psychiatric-neurological clinic was built in the new neighbourhood Amsterdam-Zuid at the Valeriusplein. The clinic was officially opened on 3 November 1910; it was later called the Valerius Clinic (Valeriuskliniek).[19] Although the Association provided for the costs of the clinic, the expenses for the VU also rose sharply. However, it was precisely during this period that the VU came into possession of wealthy pastor C.L.D. van Coeverden Adriani's (1843–1911) estate. A foundation in his name provided the funds

17 Geert Vanpaemel, *Wetenschap als roeping. Een geschiedenis van de Leuvense faculteit voor wetenschappen* (Leuven: Lipsius, 2017), p. 95.

18 Vanpaemel, *Wetenschap als roeping*, p. 91. See also: Flipse, *Christelijke wetenschap*, pp. 126–8.

19 On the history of the Association for the Christian Care for the Insane, and of the Valerius Clinic, see: G.A. Lindeboom and M.J. van Lieburg, *Gedenkboek van de Vereniging tot Christelijke Verzorging van Geestes- en Zenuwzieken 1884-1984* (Kampen: Kok, 1984); W.J. Wieringa, 'Lotgevallen van de Valeriuskliniek', in W.J. Wieringa, ed., *Een halve eeuw arbeid op psychiatrisch-neurologisch terrein, 1910-1960. Gedenkboek uitgegeven ter herdenking van het vijftigjarig bestaan van de Valeriuskliniek te Amsterdam uitgaande van de Vereeniging tot christelijke verzorging van geestes- en zenuwzieken in Nederland* (Wageningen: Zomer en Keuning, [1960]), pp. 11–87.

Figure 8.1 On the left: the Psychiatric-Neurological Clinic (*Valeriuskliniek*) in Valeriusplein in Amsterdam-Zuid, which opened in 1910. On the corner: the Physiological Laboratory, opened in 1918. Picture from 1925. Courtesy: VU, Collection HDC | Protestant Heritage.

for adding new faculties to the university.[20] This way, the medical faculty was allowed to grow in the coming years.

The question remained how the new faculty, with Bouman at its head, would give substance to Christian medicine and whether it would undergo its own laboratory revolution. For Bouman, it was clear that a clinic was not enough, that the execution of research was essential, and that the aim should be a medical faculty with laboratories. Inspired by what Kuyper and others had previously argued, yet with a slight shift in emphasis, he developed his vision of psychiatry and medicine in general over the following years.

Development of the Medical Faculty of the Vrije Universiteit

Leendert Bouman had an outspoken conception of psychiatry.[21] On the basis of his strong Christian convictions, he was clearly opposed to a strictly

20 A.H. Bornebroek, *Als een goed rentmeester. Een schets van de Van Coeverden Adriani Stichting en haar oprichter* (Amsterdam: HDC, 1991), p. 45.

21 On Bouman and his view on psychiatry: Timo Bolt, 'De pendel, de kloof en de kliniek: Leendert Bouman (1869-1936) en de "psychologische wending" in de Nederlandse psychiatrie',

biological, brain-anatomical psychiatry, the approach that had flourished in the mid-nineteenth century and was still influential around 1900. Thus, he refused to reduce the 'soul' to physical-chemical processes in the brain. Although Bouman was not the only one who advocated this approach—he himself gratefully noted that there was a wider movement at the time that paid more attention to the special character of the human *psyche*—for him, as a devout Christian, it was of course especially important.[22] In his inaugural address in 1907, entitled *De wetenschappelijke beoefening der psychiatrie* ('the scientific approach to psychiatry'), he sketched his ideas, which in his view were more 'scientific' than those of the 'materialists', who strove for a 'Psychologie ohne Seele' ('psychology without a soul'). A truly scientific psychiatry acknowledged the existence of 'the soul'. Consequently, psychiatry had to be based primarily on clinical research, according to Bouman, although he also acknowledged the importance of laboratory research for the field, within certain limitations. According to him: 'Anatomy, chemistry, and bacteriology, and also the endogenous factors that are assumed in heredity, can only provide explanations in the domains of anatomy, physiology, or bacteriology, but they are powerless in providing scientific understanding of aberrations in the psychological domain.' And therefore: 'Neuropathological, psychological-chemical, bacteriological, and histopathological investigations are very important for the clinician, but they function only as auxiliary sciences.'[23] In line with this view, and much like his later colleague Frederik Jacobus Johannes Buytendijk, Bouman did not principally object to research on animals and vivisection, although he admitted that there were also some drawbacks to the experiments, as he himself had gradually become more opposed to causing pain to animals.[24]

Nevertheless, as 'auxiliary sciences', the basic sciences and laboratory experiments (on animals) were important for psychiatry, and for medicine

Studium, 3 (2010), 82–99; J.A. van Belzen, *Psychopathologie en religie. Ideeën, behandeling en verzorging in de gereformeerde psychiatrie, 1880-1940* (Kampen: Kok, 1989), pp. 37–47.

22 Cf. Hans de Waardt, *Mending Minds. A Cultural History of Dutch Academic Psychiatry* (Rotterdam: Erasmus Publishing, 2005), pp. 89, 97–103.

23 L. Bouman, *De wetenschappelijke beoefening der psychiatrie. Rede bij de aanvaarding van het hoogleraarsambt aan de Vrije Universiteit te Amsterdam, den 27en september 1907 uitgesproken* (Kampen: Kok, 1907), p. 18: 'Anatomie, chemie en bacteriologie, ook de endogene factoren, die men in de herediteit wil aannemen, kunnen echter alleen verklaringen geven op anatomisch, physiologisch of bacteriologisch gebied, maar zijn machteloos voor het geven van een wetenschappelijk inzicht in de afwijkingen op psychisch gebied. Neuropathologische, psychologisch-chemische, bacteriologische en histopathologische onderzoekingen zijn voor den klinicus van veel gewicht, maar ze fungeeren alleen als hulpwetenschappen.'

24 Kluveld, *Reis door de hel*, p. 184.

in general, but always as part of the broader picture. Thus, Bouman certainly did not underestimate or neglect this kind of research. Within the clinic he had already furnished rooms as 'laboratories': there was an anatomical laboratory, a chemical laboratory, and a biological laboratory. For the latter two laboratories, dr. J.A. van Hasselt and the aforementioned F.J.J. Buytendijk were appointed as his assistants in 1913.[25]

Inside the laboratories Bouman gave his assistants a lot of freedom to do research. Especially when Frits Buytendijk (1887–1974), at the time a 26-year-old promising physician, extended his research in the following years, it quickly became clear that a larger laboratory was needed, as the room in the clinic soon became inadequate.[26] In 1914 Buytendijk was appointed lecturer (*lector*) in biology, and shortly afterwards, he presented specific and detailed plans and calculations for a new 'biological' (or physiological) laboratory to the directors. This proposal was part of a wider plan for the expansion of the medical faculty, which Buytendijk and Bouman presented to the directors, including ideas about laboratories and professorships and the costs these would involve.[27]

In these plans, an expanded medical faculty would initially train students for a bachelor's degree (*kandidaatsexamen*) only. At an earlier stage, there had been plans to first develop a programme for the master's degree (*doctoraal-examen*) and the medical finals. This would have been more expensive, and it would require a hospital, but it would also have appealed more to the imagination of the Calvinist supporters of the university, who, after all, were more interested in (Christian) patient care than in fundamental research. Buytendijk's plans, however, were not cheap either. They required several laboratories: at least a physiological, an anatomical, and a pathological lab. Because the medical study also involved many subjects in the natural sciences, laboratories for physics, chemistry, botany, and zoology would be needed as well in the long run. Buytendijk explained to the directors that the possibilities in the psychiatric clinic (the 'laboratories') were already too limited for current research. There was not enough space, and the rooms

25 VU, Curators' Archives 1913/1914, inv.nr. 22. See also Wieringa, 'Lotgevallen', p. 34.

26 On Buytendijk and his view of science: W.J.M. Dekkers, *Het bezielde lichaam. Het ontwerp van een antropologische fysiologie en geneeskunde volgens F.J.J. Buytendijk* (Zeist: Kerckebosch, 1985); Flipse, *Christelijke wetenschap*, pp. 196–209; Ruud Abma, 'Frederik Buytendijk (1887-1974)', in V. Busato, M. van Essen, and W. Koops, eds., *Van fenomenologie naar empirisch-analytische psychologie. Pioniers van de Nederlandse gedragswetenschappen*, vol. 2 (Amsterdam: Bert Bakker, 2014), pp. 27–101; Sebastiaan Broere, 'Synthesis and Race: Barge, Buytendijk, and the "rassenvraagstuk" of the 1930s', *Studium*, 9 (2017), 185–201 (pp. 188–93).

27 Meeting documents, d.d. 1 April 1915, Amsterdam, VU, Directors' Archives, inv.nr. 22; meeting documents, d.d. 8 November 1915, Amsterdam, VU, Directors' Archives, inv.nr. 58.

were unsuitable for animal experiments. As Buytendijk convincingly argued when he was invited to a meeting with the directors, much more specific equipment was needed in the future.[28]

Design and Equipment of the Physiological Laboratory

Laboratories come in various shapes and sizes, depending on the character of the research pursued. It is therefore interesting to see what kind of laboratory Buytendijk had in mind. At the request of the directors, Buytendijk expounded his plans in some detail in a 22-page notebook with the title 'Explanatory annex to the plans for the construction of a Physiological Laboratory', dated 12 June 1916. To find inspiration for his plans, he had visited other physiological laboratories in the Netherlands (in Utrecht and Groningen) and he had spoken with professors there: Hendrik Zwaardemaker and Hartog Jacob Hamburger.[29] Because of the costs, he was naturally supposed to aim for a laboratory that was as small and cheap as possible, but at the same time it had to be a 'sufficiently big laboratory', suitable for modern, contemporary research and teaching as well as for scientific meetings, summer courses, and other similar purposes.

In the planned laboratory building, there would be a lecture room for about 50 students and several laboratory rooms: a physical laboratory, a chemical laboratory, and an 'animal and psychology laboratory'. In addition, the plans outlined a dark room, an instruments workshop, a room for tissue research, a library, a room for microchemistry, and accommodation for animals. There had been communication with an architect, and a provisional floor plan was added. Because the possibility had arisen of taking a long lease for the piece of land next to the Valerius Clinic, a speedy decision was needed. The directors decided that the laboratory was going to be built, in particular because they were convinced that the research that Buytendijk envisaged was important for the university.[30]

28 Minutes, 30 March 1915, Amsterdam, VU, Directors' Archives.

29 Meeting documents, 1916, Amsterdam, VU, Directors' Archives, inv.nr. 35: 'Memorie van toelichting bij de plannen voor den bouw van een Physiologisch Laboratorium', and minutes directors, 20 June 1916. On Hamburger: Klaas van Berkel, *Universiteit van het Noorden. Vier eeuwen academisch leven in Groningen*, 4 vols. (Hilversum: Verloren, 2014–22), vol. 2: *De klassieke universiteit, 1876-1945*, pp. 259–63. On Zwaardemaker: G. Grijns, 'In memoriam: prof.dr. H. Zwaardemaker Czn', *Nederlandsch Tijdschrift voor Geneeskunde*, 74 (1930), 4752–5.

30 Minutes meetings directors, 18 December 1915, 2 May 1916, 20 June 1916, Amsterdam, VU, Directors' Archives.

On 29 November 1916 the first stone was laid in the presence of the architects Th. Groenendijk and Th.J. Lammers, the VU directors and professors, and the directors of the Van Coeverden Adriani Foundation, the most important financial backer.[31] More than a year later, on 22 January 1918, the Physiological Laboratory was officially inaugurated, on which occasion Buytendijk gave an address on 'instinct and life'. Just as Bouman had done earlier, Buytendijk gratefully noted that 'the science of our time is no longer characterized by materialism'. It was generally recognized 'that life is more than force and matter, more than a collection of physico-chemical processes'. But this insight had thus far hardly resulted in a different type of research. This is precisely where the new laboratory could take initiative.[32]

The same year, Buytendijk obtained his PhD in Utrecht with Zwaardemaker with a dissertation on *Proeven over gewoontevorming bij dieren* ('Experiments into habit formation in animals'), largely based on research he had actually done at the VU.[33] Now the road was also open to an appointment as professor at the VU.

Research Programme and Teaching

The university wanted to provide Christian higher education, but it also wanted to do research as part of developing a 'Christian science' based on Calvinist principles. Buytendijk therefore had to convince the directors that his research was going to be relevant in that context as well. Shortly before his appointment as professor, he produced a 'Sketch of a method for creating a Christian biology'. Here he explained that there are different 'schools' in biology, which often arrive at different results in the domain of natural philosophy and scientific theories, but they also differed in the choice of experiments. 'Materialist', 'psycho-monistic', 'energetic-monistic', and, on the other hand, 'theistic' convictions led to different forms of science at every level. Physiologists with those beliefs, but also someone like Ivan Pavlov with his reflex response research on behaviour, opted for strictly mechanistic explanations, which fitted in a materialist worldview. According to Buytendijk, theists had a wider perspective and looked for different

31 Bornebroek, *Als een goed rentmeester*, p. 57.

32 F.J.J. Buytendijk, 'Instinkt en leven', *Orgaan van de Christelijke Vereeniging van Natuur- en Geneeskundigen in Nederland*, (1918), 1–18 (p. 2): 'Het heet, dat de wetenschap van onzen tijd niet langer in het teeken van het materialisme staat', 'Dat het leven meer is dan kracht en stof, meer dan eene verzameling physico-chemische processen'.

33 F.J.J. Buytendijk, *Proeven over gewoontevorming bij dieren* (PhD diss., Utrecht, 1918).

types of explanations, which involved immaterial factors as well, such as teleological (vitalist or holist) explanations of life or of the psychological domain. His own research, Buytendijk claimed, 'converged towards these [theist, AF] natural-philosophical principles and thus towards Scripture itself'.[34] In this way, Buytendijk managed to convince the directors of the importance of the new laboratory and, incidentally, of his own reliability as well. These points had involved a certain amount of debate. The directors wondered: was his quite technical work sufficiently inspired by Calvinist principles, and was it sufficiently in line with the character of the VU?[35]

On 9 May 1919, on the occasion of his inauguration as a professor at the VU, Buytendijk gave an address entitled *Oude problemen in de modern biologie* ('Old problems in modern biology'). He stated again that 'theist philosophy' had its own approach concerning 'the theory of life'. He discussed extensively what modern biology had to say about 'the old problem of the inner life of animals'. These days, he pointed out, 'the existence of an animal psyche' was once again recognized. This meant that 'the machine explanation of life' had been abandoned. In addition, research seemed to show that there was a fundamental difference between the human and the animal psyche.[36] It was this research into the animal psyche that Buytendijk wanted to replicate and further develop in his own laboratory.

With Buytendijk's appointment, the medical faculty in the making counted two professors. The core of the faculty remained Bouman's psychiatric work in the clinic and, in principle, the work in the laboratory was conducted in the service of the work in the clinic. Buytendijk therefore gave various lectures in physiology that were deemed important for prospective psychiatrists. In the academic years 1916–17 and following, a variety of lectures given by Buytendijk were listed in the *Series Lectionum*, the university roster, including 'General biology' (metabolism, nervous system, vitalism), 'Physiology' (e.g. respiration, energy exchange, heat regulation, eye and ear), 'Animal psychology' (e.g. demonstrations about observation and action, differences in psychological functions between monkeys and children), and 'Philosophy of nature' (e.g. evolution, individuality, essential differences between humans and animals). These lectures were attended

34 Meeting documents, 1919, Amsterdam, VU, Directors' Archives, inv.nr. 2: 'Schets eener methode voor het tot stand brengen eener Christelijke Biologie', quotation on p. II: 'convergeeren naar deze natuurphilosophische principia en dus naar de Schrift zelf'.

35 Bornebroek, *Als een goed rentmeester*, pp. 56–7; Van Bergen, *Van genezen in geloof*, p. 210.

36 F.J.J. Buytendijk, *Oude problemen in de modern biologie. Rede bij de aanvaarding van het hoogleraarsambt aan de Vrije universiteit te Amsterdam, den 9en mei 1919 uitgesproken* (Haarlem: Bohn, 1919), pp. 7, 18–19.

Figure 8.2 F.J.J. Buytendijk in the Physiological Laboratory, standing next to (probably) a micro respirometer, 1919. (Thanks to Mart van Lieburg for his suggestion regarding the apparatus.) Courtesy: VU, Collection HDC | Protestant Heritage.

by 10 to 40 students; some of them were students from other faculties, while others were medical students who combined their studies at the VU with their programme in medicine at the municipal University of Amsterdam.[37]

The choice of research topics was partly determined by the fact that the lab was an offshoot of the psychiatric clinic. The clinic viewed physiology as an auxiliary science and as such important for Christian physicians. For psychiatrists in training, research into animal psychology seemed most relevant. Buytendijk capitalized on this point with his demonstrations during his lectures. Despite these teaching links, the laboratory was quite independent of the clinic. The lab was also formally part of the university, as it was part of the nascent medical faculty. Within the university this faculty occupied a special place, as it was very different from the other faculties, if only because much more money was involved. Most of this money still had to be raised by the Calvinist rank-and-file, although in 1922 Bouman and Buytendijk were the first to get a government research grant.[38] The way the work was organized was also very different from the

37 *Jaarverslag van de Vereeniging voor Hooger Onderwijs op Gereformeerden Grondslag* (1918–24).

38 J. Roelink, *Een blinkend spoor. Beeld van een eeuw geschiedenis der Vereniging voor wetenschappelijk onderwijs op gereformeerde grondslag, 1879-1979* (Kampen: Kok, 1979), pp. 163–4.

humanities faculties. Instead of professors doing their work individually, here there was a complete research group, with assistants and laboratory staff, working under the guidance of a professor. This had implications for the role of Calvinist principles of the university, because only the professor was vetted to see if he subscribed to these principles. A number of young scientists and doctors with a variety of philosophical views gathered around Bouman and Buytendijk during these years, and several of these individuals would later become professors at other universities.[39]

The Research Carried out in the Laboratory

The plans Buytendijk had presented were well received, but what sort of research was actually carried out in the lab? In a lecture given a few years after the opening of the lab, Buytendijk expanded on this point and explained why he had made those particular choices. He stated: 'We have, both in the design and in the equipment of the laboratory, made a definite choice about the direction that our research was going to take.'[40] The research had developed in two directions: metabolism and animal psychology. Both choices were inspired by the Christian character of the VU, Buytendijk claimed, as both had to be seen as so-called *Ganzheitserscheinungen* ('holistic phenomena'), and the research was partly influenced by this starting point.

Nevertheless, the actual research on metabolism appears to have been barely influenced by religious considerations. It is true that the physiological research that Buytendijk carried out was unique, as it was not yet conducted at other universities in the Netherlands, but the choice had been made mainly for pragmatic reasons; it was a niche, which was also relatively cheap. The research focused, among other things, on the influence of sports, in particular rowing, on metabolism in humans. For this, a respiration calorimeter was built, modelled on that of the American chemist W.O. Atwater and with several improvements. It was the first respiration calorimeter of its kind in Europe. Many other kinds of physiological measurements were also taken in animals and humans, including measurements on the heart

39 H.E.S. Woldring, *Een handvol filosofen. Geschiedenis van de filosofiebeoefening aan de Vrije Universiteit in Amsterdam van 1880 tot 2012* (Hilversum: Verloren, 2013), p. 87; De Waardt, *Mending Minds*, pp. 99–101; Van Deursen, *Distinctive Character*, p. 110.

40 F.J.J. Buytendijk, 'Iets over den arbeid in het physiologisch laboratorium der Vrije Universiteit', *Orgaan van de Christelijke Vereeniging van Natuur- en Geneeskundigen in Nederland* (1922), 7–13 (p. 7): 'Wij hebben dan ook èn bij den bouw èn bij de inrichting van dit laboratorium een zeer besliste keuze gedaan in de richting waarin onze onderzoekingen zich zouden bezighouden.'

and lungs. Despite its hands-on character, this kind of research, according to Buytendijk, nevertheless touched upon fundamental questions, because metabolism involved 'coherence and cooperation of everything present in the organism' as it aimed at a purpose. In addition, this research could also lead to 'recognition of God's greatness in the works of his hands'.[41]

The research into animal psychology was more distinctive; it mainly focused on 'habit formation' (*gewoontevorming*), the term Buytendijk used to denote the learning abilities of animals.[42] In the late nineteenth century, a great deal of comparative psychological research was done in this area, particularly in America. It was a field of research derived from the work of Charles Darwin and George Romanes, as evolutionary theory made it interesting to investigate 'animal intelligence'. Buytendijk wanted to do similar research but starting from different principles, and he had already referred to Robert Yerkes's (1876–1956) laboratory for animal psychology at Harvard in his proposals to the directors of the VU.[43] What Buytendijk had in mind was training animals in multiple-choice arrangements, or in mazes and puzzle boxes. Buytendijk was the first to introduce this type of research in the Netherlands.[44]

Buytendijk did research with a wide variety of animals: water fleas, snails, dogs, monkeys, birds, fishes, amphibians. Some of the research concerned the senses, such as smell and scent, sight, and hearing. But much of the research was genuine animal psychology. He investigated whether birds could learn behind which hatch there was food and for how long they could remember this. Or how quickly a snail emerged from a test tube. And whether monkeys could make links between colours and shapes and whether they could discover patterns in the locations where they could find food.[45]

41 Buytendijk, 'Iets over den arbeid', pp. 7–11 (p. 7): 'samenhang en samenwerking van alles wat in het organisme aanwezig is'; quotation on p. 11: 'erkenning van Gods grootheid in de werken van zijn handen.'

42 Buytendijk, 'Iets over den arbeid', pp. 8–9; Buytendijk, 'Instinct en leven', pp. 20–2; D.R. Röell, 'F.J.J. Buytendijks (1887-1974) ontwerp van een christelijke diepsychologie', *Gewina*, 15 (1992), 34–50.

43 Cf. *Zevenendertigste Jaarverslag van de Vereeniging voor Hooger Onderwijs op Gereformeerden Grondslag* (Amsterdam: 1917), p. XXVII; Buytendijk, 'Instinct en leven', p. 11. On Yerkes: Boakes, *From Darwin to Behaviourism*, pp. 148–58.

44 D.R. Röell, *The World of Instinct: Niko Tinbergen and the Rise of Ethology in the Netherlands (1920-1950)* (Assen: Van Gorcum, 2000), pp. 147–52.

45 For a list of Buytendijk's publications and a detailed description of the research, see: J.A. Bierens de Haan, 'Sieben Jahre tierpsychologische Arbeit in Amsterdam', *Zeitschrift für angewandte Psychologie*, 27 (1926), 236–67. Thanks to W.J. van der Schoor for letting me use his MSc thesis: *Bezield gedrag. Theorie en experimentele praktijk in de dierpsychologie van F.J.J. Buytendijk (1887-1974)* (Unpublished thesis, Leiden, 1984).

In his book *Psychologie der dieren* ('The psychology of animals'), published in 1920 and partly based on his own research, Buytendijk explained that the psyche directs life. In sensory perception and behaviour, some psychological element must be active. Mechanist theories are therefore incorrect: behaviour cannot be explained purely mechanically. The second point that he emphasized was that the greatest achievements of animals, however surprising, are still incomparable to those of human beings. In addition, it was not the case that animals that are supposedly—evolutionarily—closer to humans are always better able to learn. In this context he criticized Yerkes's interpretation of certain experiments with orang-utans.[46]

According to Buytendijk, his research produced two main results. On the one hand, the results challenged the mechanist explanations of life, and on the other hand, the experiments demonstrated the special status of humans.[47] The religious and philosophical convictions that inspired Buytendijk where in harmony with his research. They had influenced the choice of experiments but to a certain extent also their interpretation.

Although Buytendijk's experiments were similar to those of Yerkes and other early animal psychologists, there were also important differences, not only regarding the (religious) interpretation but also regarding methodology. Buytendijk often refrained from analysing his results quantitatively with the use of statistics and learning curves, as was done in Harvard and other laboratories in the US. His claims were mainly based on qualitative descriptions of behaviour. In the end Buytendijk's approach to animal behaviour was what might be called a 'Verstehen' of the phenomena. Buytendijk was convinced that in the life sciences, in addition to the method of 'Erklären', a 'verstehende' approach was needed in order to read 'the book of nature'. 'Verstehen' was for Buytendijk a synthetic approach in which the totality was disturbed as little as possible. The analytical method failed because nature was, according to Buytendijk, 'a book with many letters and not with a number of ink blotches'.[48]

In 1924 Buytendijk moved to Groningen, where he was appointed professor of general physiology as the successor of Hamburger. In Groningen he initially

46 F.J.J. Buytendijk, *Psychologie der dieren* (Haarlem: Bohn, 1920), passim, esp. pp. 195–222.

47 See also: F.J.J. Buytendijk, *Bijdrage tot een onderzoek naar het wezensverschil van mensch en dier. Referaat gehouden op de Wetenschappelijke samenkomst [van de Vrije Universiteit] op 12 Juli 1922* (Amsterdam: Kirchner, 1922).

48 F.J.J. Buytendijk, *Over het verstaan der levensverschijnselen. Rede uitgesproken bij de aanvaarding van het ambt van hoogleeraar in de physiologie aan de Rijksuniversiteit Groningen op 17 januari 1925* (Groningen: Wolters, 1925), p. 11: 'een boek met vele letters en niet met een hoeveelheid inktvlekken'. See also: Röell, 'Buytendijks ontwerp', p. 44; Dekkers, *Het bezielde lichaam*, p. 66.

continued his research on animal psychology; he later focused on human psychology. He also developed in the realm of religion, and he would later convert to Catholicism. There were various reasons for his departure from the VU. Buytendijk gradually felt less at home with Calvinism, especially because the Calvinist theologians in the 1920s took a more fundamentalist stance than before.[49] Moreover, there was the fact that there seemed to be little prospect of further growth for the medical faculty at the VU. After Buytendijk's departure the Physiological Laboratory was used for a variety of other functions for several decades. Physiological research took place only on a very small scale. The Valerius Clinic continued to function, although Bouman departed as well, but to Utrecht. (Bouman continued to give some lectures at the VU as an extraordinary professor.)[50] Buytendijk's work in animal psychology incidentally had some followers outside the VU. J.A. Bierens de Haan, who had worked in Buytendijk's lab, adopted his approach, and he continued his research in his own laboratory in the Artis Zoo in Amsterdam for some time.[51]

In the 1930s 'ethology' emerged as an independent discipline that had its own approach to animal behaviour. The most important representatives of this new discipline were Niko Tinbergen and Konrad Lorenz.[52] Remarkably, 'ethology' and 'animal psychology' did not merge, but 'animal psychology' also did not survive on its own. In his study *The World of Instinct*, on the origin of ethology in the Netherlands, René Röell suggests several explanations for this course of events. Many who were critical of the work of the pioneers of 'animal psychology' in the Netherlands took issue with the anti-mechanistic (vitalistic or holistic) and teleological framework in which study of the animal soul was placed by the field's most important representatives.[53] In the case of Buytendijk, this was even an explicitly Christian framework, even though one could say that did not seem to have influenced the practice of the experiments very much. Nevertheless, it may have put off other physiologists and biologists.

49 Flipse, *Christelijke wetenschap*, p. 216. See also: Stuart Mathieson and Abraham C. Flipse, 'Religious Controversy in Comparative Context: Ulster, the Netherlands and South Africa in the 1920s', *History*, 106 (2021), 429–55.

50 Peter Hellema, *Spreek en houd het niet voor je. De loopbaan in de psychiatrie van Lammert van der Horst (1893-1978)* (Amsterdam: Vesuvius, [2019]), pp. 78–9; Wieringa, 'Lotgevallen van de Valeriuskliniek', pp. 52–60; Connie Pieksma, *Het fysiologisch laboratorium VU/VUmc. Feiten en gebeurtenissen* (Amsterdam: VU, 2007).

51 Röell, *The World of Instinct*, pp. 152–3.

52 Richard W. Burckhard Jr., *Patterns of Behavior: Konrad Lorenz, Niko Tinbergen, and the Founding of Ethology* (Chicago: University of Chicago Press, 2005), pp. 34–86; Röell, *The World of Instinct*, pp. 34–86.

53 Röell, *The World of Instinct*, pp. 172–3.

When the VU expanded again in 1930, it chose to establish a science faculty. A condition in the Law of Higher Education of the Netherlands of 1905 turned out to be the deciding factor for action. This law had recognized the validity of the VU degrees (the so-called *effectus civilis*). One important condition was that the VU had to expand gradually to a 'complete' university, with five faculties: a fourth faculty was to be established not later than 1930, and a fifth not later than 1955 (both comprising at least three chairs).[54] Therefore, by 1930, not founding a fourth faculty was no longer a serious option. Now that the medical faculty had almost disappeared, the directors opted for starting a science faculty as the fourth faculty. This was not easy either, but it seemed slightly more realistic than a complete medical faculty with all the different specializations. For the science faculty, new laboratory buildings were erected for chemistry and physics, next to the Physiological Laboratory. Professors of mathematics, physics, and chemistry were appointed, and the first students arrived in 1930.[55]

Conclusion: The Distinctive Character of the Laboratory and its Lasting Heritage

Although a laboratory and working in one were novel phenomena within the Vrije Universiteit in 1918, it seems that the introduction of the lab, notwithstanding earlier hesitation and decades of debate, proceeded relatively smoothly. It was of course a big step which required a lot of money. Moreover, a discipline like theology, but also a hospital or a clinic, appealed more to the imagination of the Calvinist constituency of the university than physiological research in a laboratory. Above all, in the discussions at the time, a certain suspicion is evident against this new form of experimental science, which was associated with 'materialism'. On the other hand, it was ultimately realized that a physiological laboratory was part and parcel of a modern medical faculty. However, the neo-Calvinists also believed that it was possible to create a radically different research strategy grounded in Calvinist principles in their own laboratory. This confidence was reinforced by the

54 A.C. Flipse, 'Against the Science-Religion Conflict. The Genesis of a Calvinist Science Faculty in the Netherlands', *Annals of Science*, 65 (2008), 363–91 (pp. 371–2). The law was passed by Parliament during the period of a coalition cabinet of which Kuyper himself was Prime Minister.

55 Ab Flipse, *'Hier leert de natuur ons zelf den weg'. Een geschiedenis van Natuurkunde en Sterrenkunde aan de VU* (Zoetermeer: Meinema, 2005), pp. 48–74.

fact that the intended research in animal psychology seemed especially relevant and could be used to take a distinctive course.

It turned out that the foundation of this first laboratory was in a sense revolutionary, and it instigated the Vrije Universiteit's own 'laboratory revolution'. For the first time, the university encountered an academic discipline in the sciences. Although the chosen field of animal psychology was precisely an area in which a distinctive emphasis could be imposed, this step was crucial because it demonstrated that the university was increasingly adapting to what was already customary elsewhere. Therefore, even though the physiological lab was used only briefly, it paved the way for the future science faculty, in 1930, and for the further growth into a university that comprised all faculties, later in the century.

About the Author

Ab (Abraham) Flipse (PhD, Vrije Universiteit Amsterdam) is a historian of science at, and the university historian of, the Vrije Universiteit (VU) Amsterdam. His research focuses on university history (especially of the VU) and the historical relationship between science and religion. He recently published, with Stuart Mathieson, 'Religious Controversy in Comparative Context: Ulster, the Netherlands and South Africa in the 1920s', *History*, 106 (2021), 429–55.

Part III

Laboratory Values

9 Aspects of the Social Organization of the Chemical Laboratory in Heidelberg and Imperial College, London

Peter J.T. Morris

Abstract
This chapter deals with the social organization of chemical laboratories, in particular the hierarchy of laboratory workers and the relationship between professors and technicians. This study focuses on two major chemical departments of the late nineteenth and early twentieth centuries, namely Imperial College London (and its predecessors) and the University of Heidelberg. This pairing allows us to explore the differences (and similarities) between laboratories in England and Germany. Over time, the hierarchy of chemical laboratories has become flatter as relationships between the teaching staff, students, and technicians have become more socially equal.

Keywords: hierarchy, technicians, stratification, Heidelberg University, Imperial College

Introduction

When I wrote my book *The Matter Factory* about the physical aspects of the chemical laboratory, I relied very largely on pictorial evidence for two reasons.[1] Archival evidence is limited and there is hardly any laboratory of the nineteenth century (or earlier) which remains in the same state as it was when it was built. Unfortunately, this approach does not work for the

1 Peter J.T. Morris, *The Matter Factory: A History of the Chemistry Laboratory* (London: Reaktion, 2015).

Berkel, Klaas van, and Ernst Homburg (eds), *The Laboratory Revolution and the Creation of the Modern University, 1830-1940*. Taylor & Francis Group, 2023
DOI: 10.5117/9789463720434_CH09

social organization of laboratories. While there are group photographs of the staff in university chemical laboratories, they are almost invariably limited to faculty and postgraduate students. A few may include the head technician and/or the superintendent, but almost none will compass all the non-academic staff. This is also largely true of the staff lists found in university yearbooks, which also tend not to list temporary or non-salaried staff. Some information can be gleaned from individual histories of the laboratories and from autobiographies, but it is usually scanty.[2]

The Sociologist's View

The social structure of the chemical laboratory has been studied, not surprisingly, by sociologists. In *Laboratory Life*, Bruno Latour and his co-author Steve Woolgar acknowledge the complex group dynamics in the laboratory, but their main concern is with scientific creditability (as measured by citations), and hence their focus is on the dynamics between different scientists who are seen as entrepreneurs dealing in credit rather than capital.[3] They note that technicians intensely seek credit but usually cannot get it owing to the lack of academic qualifications, and they are poorly paid; hence their main concern is their salary, and there is a high turnover in technicians. However, they can 'trade' in replaceability, and a technician with highly desired skills will be placed higher in the hierarchy ('supertechnicians') to retain them. Conversely, a departing scientist will seek to take good technicians with them, thus undermining their original laboratory. Latour and Woolgar admit that technicians' 'importance in the production of facts is usually underestimated' but refuse to consider the issue any further.[4] In his later *Science in Action*, Latour is concerned with the questioning and validity of science. There is a whole section called 'Laboratories' but the only important person in it is the 'Professor', although clearly other people

2 For a general discussion of the role of the 'invisible workers' in science, see: Klaus Hentschel, 'Wie kann Wissenschafts- und Technikgeschichte die vielen "unsichtbaren Hände" der Forschungspraxis sichtbar machen?', in Klaus Hentschel, *Unsichtbare Hände* (Diepholz: GNT-Verlag, 2008), pp. 11–24. For two studies of the role of technicians in scientific research, albeit mostly medical research, see: N.C. Russell, E.M. Tansey, and P.V. Lear, 'Missing Links in the History and Practice of Science: Teams, Technicians and Technical Work', *History of Science*, 38 (2000), 237–41; E.M. Tansey, 'Keeping the Culture Alive: The Laboratory Technician in Mid-Twentieth-Century British Medical Research', *Notes and Records of the Royal Society*, 62 (2008), 77–95.

3 Bruno Latour and Steve Woolgar, *Laboratory Life. The Construction of Scientific Facts* (Princeton, NJ: Princeton University Press, 1986; originally publ. 1979), ch. 5.

4 Latour and Woolgar, *Laboratory Life*, p. 323.

are vaguely present and at one point a technician is held responsible for a mistake.[5] The current chapter will show that this is not a complete picture of the social organization of the laboratory, which is complex and changes over time. It also varies from university to university, and in particular from country to country.

The Social Stratification of the Laboratory

One of the first things we can say about the social organization of the laboratory is that it is marked by a set of social demarcations. In the eighteenth century and even earlier, there was simply a single professor, and his laboratory servant would clean the furnaces and apparatus and also set up apparatus for lectures. Hence no particular technical skills were necessary for the assistant. But these functions should not be overlooked; cleaning used apparatus and setting up lecture demonstrations for lecturers remain important responsibilities of technicians even today. From the time of Lavoisier onwards, however, these assistants began to possess specific technical skills to assist the researchers or teachers. There were two major skills that quickly came to the forefront. One was dealing with chemicals, including the ordering, making, and storing of the chemicals. Sometimes the technician would even be able to sell chemicals to the students or even the public. Bunsen's technician (called a *Diener*, see below) at Marburg, Johann Bretthauser, sold coal cylinders for Bunsen batteries in the 1840s.[6] The other important skill was glassblowing. Until the mid-twentieth century much standard glassware was made in the laboratory, and for specialized glassware such as tubing to handle vacua, a glassblower (external or internal) was essential. Furthermore, the glassblower (or even just an ordinary laboratory technician) would repair glassware or adapt it for specific uses and experiments.[7]

Obviously there has always been a divide between the teachers and the taught, which was also often a commercial relationship, as students paid the lecturer, not the university, for their education. By the mid-nineteenth century, the long-established teacher–client relationship between academic

5 Bruno Latour, *Science in Action. How to Follow Scientists and Engineers through Society* (Cambridge, MA: Harvard University Press, 1987), p. 66.

6 Personal email from Christoph Meinel dated 27 September 2017.

7 Examples of this repaired or modified glassware can be found in the Science Museum's collections.

chemists and their pupils began to give way to the teacher–student relationship we know today. At the same time, a new class of students—research students—became increasingly important. Their relationship with the teaching staff was more like that of a master and their apprentices, with a period of training sometimes followed by a period of post-doctoral collaboration as an assistant or teaching fellow.[8] At Owens College, Manchester, during Henry Enfield Roscoe's time as professor of chemistry, there was a definite pattern of the best students undertaking research in an extra year or so as an undergraduate, followed by postgraduate work at a German university and then a period of two or three years as an assistant lecturer at Owens before finding a more permanent position.[9]

The Laboratory Leadership

From the 1860s onward the chemical laboratory grew in size and in status, first in Germany and then in other countries that were mainly seeking to copy the German model.[10] As these new 'chemical palaces' had specialized rooms, for combustion analysis, chemical stores, balance rooms and dark rooms for spectroscopy, so specialized tasks were created. Laboratory assistants would undertake the combustion analyses, as they were too hot and time-consuming for the senior academics to do, and they might also carry out routine techniques such as spectroscopy or polarimetry. These buildings also had libraries and even chemical museums, hence creating other new functions: librarians and curators. Rather than employ a full-time member of staff, these roles were often given to junior academic staff, thus slightly blurring the demarcation between the academic and the service staff. There were also amanuenses and artists, who drew laboratories, apparatus, and experiments.[11] As chemical laboratory buildings grew even larger, new service roles were created for housekeepers, porters, and cleaners. As

8 For a detailed account of the role of the assistant in German academia generally and the evolution of this role, see: Klaus Dieter Bock, *Strukturgeschichte der Assistentur* (Düsseldorf: Bertelsmann Universitätverlag, 1972).

9 Henry Enfield Roscoe, *Record of Work Done in the Chemical Department of the Owens College, 1857–1887* (London: Macmillan, 1887).

10 Morris, *The Matter Factory*, chs. 6–7.

11 For amanuenses see: Bock, *Strukturgeschichte*, pp. 147–9 and Hentschel, 'Wie kann Wissenschaft- und Technikgeschichte'. For an example of an artist (at Heidelberg), see: Christine Nawa and Christoph Meinel, eds., *Von der Forschung gezeichnet: Instrumente und Apparaturen in Heidelberger Laboratorien, skizziert von Friedrich Veith (1817–1907)* (Heidelberg: heiBooks, Universitätsbibliothek Heidelberg, 2020), available online at https://doi.org/10.11588/heibooks.793.

the administrative burden grew, chemical laboratories might even acquire typists or secretaries. (See the chapter in this volume by Bas Nugteren.)

With this increasing size came a greater sense of hierarchy and perhaps a greater need for a chain of command. We have to digress a little here to outline the different academic structures in Germany and Britain.[12] It must be stressed that in both countries, academia was very small even at the beginning of the twentieth century, and academic chemists had even smaller numbers. There were perhaps (then) two thousand academics in Britain and 3,800 in Germany.[13] The German universities were funded by the individual states even after 1871 (Baden in the case of Heidelberg).[14] There was a divide between the academic faculty, with its ordinary professors who were academic equals within a broader faculty (initially philosophy in the case of Heidelberg and then mathematics and natural sciences), and the chemical institute, where the laboratories were located. Under the professors there were extraordinary professors who had no part in the governance of the faculty and were either paid by the state or had to earn their income from student fees—or both. The next rank down were the *Privatdozenten* (sometimes translated as instructors in English, but essentially untranslatable), who aspired to have an academic career. They were allowed to give lectures on specific topics as a way of gaining teaching experience. They were wholly dependent on student fees. In Heidelberg they were usually not allowed to work in the university laboratory. As their courses were in their own research specialism and usually not well attended as the introductory courses taught by the professors, their financial position was usually dire, and they had to make ends meet by becoming a teaching

12 For an examination of German education in its European context and a comparison of the German and British situations, see: Fritz K. Ringer, *Education and Society in Modern Europe* (Bloomington, IN: Indiana University Press, 1979). For a comparison of German and French academia, see: Alan J. Rocke, *The Quiet Revolution: Hermann Kolbe and the Science of Organic Chemistry* (Berkeley, CA: University of California Press, 1993) and especially Alan J. Rocke, *Nationalizing Science: Adolphe Wurtz and the Battle for French Chemistry* (Cambridge, MA: MIT Press, 2001).

13 For Britain, see: A.H. Halsey and M.A. Trow, *The British Academics* (Cambridge, MA: Harvard University Press, 1971), p. 139, and for Germany in 1910, see: Alexander Busch, *Die Geschichte des Privatdozenten* (Stuttgart: Ferdinand Enke, 1959), p. 76, table 4.

14 For the political, social, and economic context of German academia in this period, see: Charles E. McClelland, *State, Society and University in Germany, 1700–1914* (Cambridge: Cambridge University Press, 1980). For the specific case of Baden, in which Heidelberg was located, see: Peter Borscheid, *Naturwissenschaften, Staat und Industrie in Baden (1848-1914)* (Stuttgart: Ernst Klett, 1976). For an earlier perspective, see: Johannes Conrad, *The German Universities for the Last Fifty Years* (Glasgow: David Bryce, 1888) and Friedrich Paulsen, *German Education, Past and Present* (London: T. Fisher Unwin, 1908), both of which are available on the Internet Archive.

assistant or a section head (or even finding a job outside the university). As promotion to a professorship, even an extraordinary professorship, was far from certain, the position of the *Privatdozenten* by end of the nineteenth century was often pitiful.[15]

By contrast in Britain, Imperial College (and its predecessor The Royal College of Science) was one of the few fully state-funded universities in the nineteenth century.[16] There was no distinction between the teaching side and the laboratory side; they were both part of the chemistry department.[17] At the top of the department were the professors who were few in number until after the Second World War. If there was more than one professor, one would be the head of department, initially a permanent post, but in some universities the post eventually rotated between the professors. Under the professors were the lecturers. Lecturers were sometimes promoted to the rank of senior lecturer on the basis of seniority. This carried a small uplift in salary but no meaningful increase in status. Lecturers were also (rarely) raised to the rank of reader if they gained academic fame, usually through their research, but the post was effectively an honorific one and is of relatively recent origin, first appearing in the 1920s.[18] At the bottom of the teaching hierarchy were the demonstrators (sometimes called assistant lecturers), who supervised the undergraduate's laboratory work and did the basic teaching.

Hence the single professor in charge of the laboratory was gradually replaced by a director of the chemical institute in Germanic countries and by a head of department in English-speaking countries.[19] The German model was clearly more hierarchical and very much tied to the person of the director, meaning that they would live in or near the laboratory and would remain in charge for a long period.[20] This was the case at Heidelberg where

15 For a detailed study of the *Privatdozent* in German academia, see: Busch, *Geschichte des Privatdozenten.*

16 For a general history of Imperial College and its predecessors, see: Hannah Gay, *The History of Imperial College London, 1907–2007* (London: Imperial College Press, 2007).

17 In contrast to Germany, there is very little on the history of British academia, but see Halsey and Trow, *British Academics*. For the rejection of the German model of the all-powerful professor by British universities, see: Joseph Ben-David, *Centers of Learning. Britain, France, Germany, United States* (New Brunswick/London: Transaction Publishers, 1992; originally publ. 1977), p. 105.

18 Halsey and Trow, *British Academics*, table 7.4, p. 151.

19 Jeffrey A. Johnson, 'Academic Chemistry in Imperial Germany', *Isis*, 76 (1985), 500–24 (p. 508).

20 General late nineteenth-century trends promoting directorial power in the German institutional hierarchy are summarized in: Jeffrey A. Johnson, 'Hierarchy and Creativity in Chemistry, 1871–1914', in Kathryn M. Olesko, ed., *Osiris. Science in Germany: The Intersection of Institutional and Intellectual Issues*, 5(1) (1989), 214–40, esp. pp. 222–24.

Bunsen was in charge for 36 years from 1852.[21] The director was directly appointed by the state but on the basis that he was also a full professor. He was in the powerful position of being the only representative of chemistry in the university's higher governing body, the Senate. By contrast, the British head of department was more collegial, although perhaps less so in the case of a forceful professor such as Roscoe, who modelled himself on Bunsen. As the work of the director or head of department became more onerous, they would appoint a deputy to help with the administration and perhaps a research assistant to do some of their research. Heidelberg had a deputy director by 1920—namely Johannes Rissom, who had previously been listed as an assistant—but the post did not exist at Imperial until after the Second World War. The research assistant post had been created at Imperial by 1908.[22] In Heidelberg (and in German academia generally), the situation was more complex. From February 1859 Bunsen had a third assistant, who was not on the official staff list, who was employed to analyse Baden spa waters, in which Bunsen and the physicist Gustav Kirchhoff discovered caesium in 1860 and rubidium in 1861.[23] In general, the professor would give their doctoral students topics which were of interest to themselves, so the students were effectively doing research on the professor's behalf.[24] This was not unusual and was later commonplace, for instance, in the United States. To give a famous example, Robert Burns Woodward was hopeless at the bench but carried out his brilliant multistage syntheses through his PhD students and his postdocs. In Britain PhD students were rare until the 1920s, hence there was a need for a paid research assistant. In Germany the best doctoral students would in time become teaching assistants, which could limit their ability to do their own research. However, in Victor Meyer's

21 Theodor Curtius and Johannes Rissom, *Aus der Geschichte des Chemischen Universitätslaboratoriums zu Heidelberg seit der Gründung durch Bunsen* (Heidelberg: Verlag Rochow, 1908) provides a good history of the chemistry department up to 1908, but it is unfortunately very rare; there are no copies in the UK for example. For a more recent history of the chemistry department at Heidelberg in Bunsen's time, see: Christine Nawa, 'A Refuge for Inorganic Chemistry: Bunsen's Heidelberg Laboratory', *Ambix*, 61 (2014), 115–40. Also see the excellent booklet produced for the dedication of Bunsen's laboratory as a Chemical Landmark: Christine Nawa, *Robert Wilhelm Bunsen und sein Heidelberger Laboratorium, Heidelberg, 12. Oktober 2011* (Frankfurt a.M.: Gesellschaft Deutscher Chemiker, 2011).

22 The London-based institution which began as the Royal College of Chemistry and via the Royal College of Science became the Imperial College chemistry department. For simplicity I will usually refer to it as Imperial.

23 Curtius and Rissom, *Geschichte des Chemischen Universitätslaboratoriums zu Heidelberg*, p. 15. For the role of the private assistant in German laboratories, see: Bock, *Strukturgeschichte*, pp. 153–58.

24 Bock, *Strukturgeschichte*, p. 37; Johnson, 'Hierarchy and Creativity', p. 225.

time at Heidelberg, the teaching assistants paid by the state were provided with the facilities to do their own research, which one would suspect was probably closely aligned to that of the professor. When Theodor Curtius moved from Bonn to Heidelberg in 1898, he brought his research assistants with him, and they were put on the Heidelberg University payroll, although it is not clear whether they were research assistants or teaching assistants. So clearly research assistants (sometimes paid out of the professor's pocket) existed at other German universities (Wilhelm Hofmann had them in Berlin),[25] but there is no firm evidence for them in Heidelberg before the arrival of Curtius.[26] It appears that Meyer had research assistants at ETH Zurich but not in Heidelberg.[27]

Nonetheless, the number of senior academics was very low until the early twentieth century when the number started to rise. There was only one professor in the main chemistry institute (i.e. the laboratory) at Heidelberg even in 1885, namely Bunsen, and just one professor and an assistant professor at Imperial in 1891.[28] The assistant professor is not a standard British academic post. By 1905 there were six heads of departments under the director (Theodor Curtius) in Heidelberg, five of whom were extraordinary professors (three with a state salary and two without) and one remarkably a *Privatdozent*, Ernst Mohr, who shortly thereafter became an extraordinary professor. By contrast, in 1908 there were three assistant professors and a lecturer at Imperial.[29] Jeffrey Johnson has indicated that the university chemists were the most hierarchical of all university fields, with an actual decrease in the number of senior (Ordinarius) professors

25 For Hofmann's use of assistants in Berlin, see: Charles Loring Jackson, 'August Wilhelm Hofmann', *Proceedings of the American Academy of Arts and Sciences*, 28 (May 1892–May 1893), 411–18, esp. pp. 413, 415.

26 Curtius and Rissom, *Geschichte des Chemischen Universitätslaboratoriums zu Heidelberg*, p. 52. See also: Jackson, 'August Wilhelm Hofmann', pp. 413, 415.

27 Richard Meyer, *Victor Meyer: Leben und Wirken eines deutschen Chemikers und Naturforschers*, in Wilhelm Ostwald, ed., *Grosse Männer: Studien zur Biologie des Genies* (Leipzig: Akademische Verlagsgesellschaft, 1917), vol. 5 has references to Victor Meyer's research assistants (*Privatassistenten*) in Zurich (pp. 76 and 84) but none to any research assistants in Heidelberg.

28 To be clear, the famous historian of chemistry Hermann Kopp was a Ordinarius professor of chemistry at Heidelberg between 1864 and 1892, but he was not in the Chemistry Institute, and Wilhelm Delffs was director of another Chemistry Institute between 1851 and 1889, but this was for medical chemistry and not chemistry.

29 All the figures for Heidelberg in this paper are from: *Adreßbücher der Universität Heidelberg* 1818–1922 , <https://www.ub.uni-heidelberg.de/helios/digi/unihdadressbuch.html> (accessed 5 February 2019), sampled at 1845, 1860, 1864, 1885, 1905, 1920. Likewise, all the figures for Imperial College are from the college calendars in the Imperial College archives, Sherfield Building, Imperial College, and sampled for 1881, 1891, 1920 and 1938.

Figure 9.1 The 70th birthday celebration held by the Chemical Institute for Paul Jannasch (middle of the front row) in October 1911. He had come to Heidelberg with Victor Meyer in 1898. The director Theodor Curtius is standing at the far left of the second row. The heads of departments (professors) were Jannasch; Robert Stollé sitting between Curtius and Jannasch; Georg Bredig (the third man on the right standing); Emil Knoevenagel (second man on the right, behind Bredig). The assistants were Johannes Rissom (on the far right standing); August Darapksy (man with a beard behind Curtius); Otto von Mayer (two rows behind Jannasch with a walrus moustache); and Ernst Muckermann, who was also a *Privatdozent* (on the far left behind Curtius). The women were presumably their wives, pictured together with the younger members of Jannasch's family. Courtesy: Universitätsbibliothek Heidelberg.

between 1864 and 1890, while the numbers of lower-ranking chemists nearly tripled from 27 to 72.[30] By 1910, despite a modest increase in the numbers of Ordinarius professors, the system became even more hierarchical as the lower-ranking academic chemists more than doubled their numbers, from 72 to 147. Contrast to the non-laboratory sciences, in which senior positions increased faster than those in lower ranks during the same period. Notably, in the *Technische Hochschulen*, which had been very un-hierarchical in 1890, the numbers of lower-ranking academic chemists nearly quadrupled between 1892 and 1910. Johnson has also argued that the increasingly hierarchical nature of the German academic chemistry laboratory is related to the increasing proportion of collaborative research after the 1870s, as directors more and more integrated students and junior

30 Johnson, 'Academic Chemistry', p. 507, table 2.

staff into their research—indicated by higher numbers and proportion of multi-authored publications, while directors published fewer individually authored papers after the 1890s.[31]

In 1920 Heidelberg had five heads of departments, one less than in 1905, three of whom were extraordinary professors with state salaries and two were extraordinary professors without a salary. While numbers then remained static at Heidelberg, they continued to grow at Imperial with two full professors each with their own laboratory, three assistant professors who now also had the senior academic title of Reader, and eight lecturers by 1938. This also shows that Imperial was moving away from its distinctive quasi-German system to the standard British academic structure, probably under pressure from the University of London, which Imperial had joined in 1929. Like Heidelberg, the number of professors remained relatively static at Utrecht (for the figures for Utrecht, see the chapter by Bas Nugteren in this volume). After the Second World War, there were also laboratory superintendents who ran the laboratories on a day-to-day basis at Imperial. By contrast, Heidelberg had a housekeeper (*Hausmeister*) from 1908 who was responsible for the upkeep and running of the building as a whole rather than the laboratory.

Junior Academic Staff

There was a demarcation between the permanent tenured staff and the temporary teaching staff who would hold junior academic positions for a few years at most before moving to another university or a non-academic position. As laboratories grew in size and careers in chemistry became more complex, there was a growing number of these temporary staff in laboratories. In German universities they were usually called assistants. They acted as lecture assistants and as teaching staff (for example marking papers), and they also carried out their own research insofar as their other duties permitted. Ostwald was notable for guaranteeing his assistants 50 per cent free time to do their own research so that they could develop as independent scholars and win professorships[32] One of the problematic developments noted at the time was a tendency for assistantships to become permanent, as junior faculty found it increasingly difficult to find independent professorships at other universities. In the largest institutes

31 Johnson, 'Hierarchy and Creativity', pp. 226–27.
32 Johnson, 'Hierarchy and Creativity', p. 225.

Figure 9.2 The organic chemistry staff and researchers at Imperial College, 1921–2. The professor Jocelyn Field Thorpe (middle of the front row) was a leading organic chemist. Christopher Kelk Ingold (on Thorpe's right), a pioneer of organic reaction mechanisms, worked with Thorpe at Imperial College between 1913 and 1924. He married Hilda Usherwood (second from the right, front row) in 1923. John W. Baker (third left, second row) discovered the Baker–Nathan effect with Wilfred Nathan in 1935. The laboratory assistant Ernest Coppen (called Crippen by the students because of his resemblance to the infamous murderer) is at the back at the far right. He was later dismissed for attempting to form a trade union for technicians. Courtesy: Archives, Imperial College London.

(especially Munich and Berlin, but also elsewhere), the professors divided the institute into two or three specialized sections for teaching purposes, usually presided over by *Extraordinarien* (Associate Professors), formally employed as section heads (*Abteilungsleiter*) with salaries larger than the typical assistantship and paid as part of the regular institute budget. By 1910 there were about twenty of these positions in German universities. For many years Heidelberg only had two assistants. Bunsen had an assistant for lectures and another for the practical teaching. Victor Meyer brought a group of assistants with him from Göttingen in 1889, and there was a shift to the more labour-intensive organic chemistry and a new, larger laboratory.[33] Then under his successor Curtius, a new medical chemistry

33 For the assistants under Victor Meyer, see: Curtius and Rissom, *Geschichte des Chemischen Universitätslaboratoriums zu Heidelberg*, p. 35 and for the new laboratory building see pp. 27–34.

building was erected.[34] Hence the number of assistants rose and there were eight assistants in 1905, three of whom were *Privatdozenten*. There were nine assistants in 1920, of which only one was a *Privatdozent* and four of them were doctoral students.

Imperial was originally unusual in having the position of assistant rather than the more usual British post of demonstrators who assisted the professor during lectures (hence the title) and supervised the students' laboratory work. The German director of the Royal College of Chemistry, Wilhelm Hofmann, initially used his assistants (and sub-assistants) according to the German model, and they even acted as his part-time laboratory servants.[35] He also paid for private assistants out of his own pocket; they were usually German. In 1891 Imperial had two demonstrators, three assistants, and two teaching scholars. By 1908 there were no less than twelve assistant demonstrators (who presumably replaced the earlier assistants) alongside the two demonstrators. However, the post had disappeared (or they were not listed) in 1920 when there were just two demonstrators, in contrast to nine assistants at Heidelberg.

The Technicians

There was also the time-honoured distinction between the teacher and researcher, on one hand, and the technical staff, on the other. In the largest laboratories, the technical staff had their own hierarchy from the superintendent, who was in charge of the day-to-day operation of the laboratory building, and the chief technician to the junior technicians (junior in status not necessarily in age), who would do menial tasks such as clearing away and washing apparatus.[36] Such a simple task as washing dirty apparatus is actually complex in social terms, since junior researchers (at least) and students would be expected to clean their own equipment and put it away in their allocated cupboards. As Johannes Thiele once remarked to Richard Willstätter, students were accorded more respect in the laboratory than the

34 For the Mediziner-Bau see: Curtius and Rissom, *Geschichte des Chemischen Universitätslaboratoriums zu Heidelberg*, p. 42.

35 Hannah Gay and William P. Griffith, *The Chemistry Department at Imperial College, London: A History, 1845–2000* (London: World Scientific, 2017), p. 16. Also see: Hannah Gay, '"Pillars of the College": Assistants at The Royal College of Chemistry, 1846–1871', *Ambix*, 47 (2000), 135–69.

36 Personal discussions in the 1990s with Alec Campbell (Newcastle University), who was a technician for many years before being belatedly appointed to the academic staff by Norman Greenwood.

Privatdozenten, presumably because they were usually paying customers, although formally they were at the bottom of the academic hierarchy.[37]

In addition to their traditional roles of room cleaning, apparatus washing, housekeeping duties such as turning the heating on and off and making tea or coffee, maintaining equipment (both scientific and domestic), setting up lectures, and handling chemicals, the technicians also began to work up reaction mixtures which had proved intractable—either to obtain the desired product or to render it safe—and this was perhaps one of their most important functions by the early twentieth century. As machinery for ventilation, fume cupboards, and boiling and refrigeration became more prominent in the laboratory, the mechanic or machinist became an important member of the technical staff.[38] At Heidelberg the *Mechanikus* Peter Desaga was keeper of the physics cabinet while also having his own instrument-making firm, which flourished after Bunsen gave him the rights to market the Bunsen burner and later the Bunsen thermostat. There was a machinist in the chemistry department in 1905, but oddly not in 1920.

Perhaps surprisingly, technicians were not expected to be qualified for this multi-skilled role. They would typically join the laboratory as a laboratory boy at a young age and work their way up the technical hierarchy while learning skills in the job or perhaps in evening classes. Hence a technician was an essential member of staff but one with low status and pay. In Heidelberg, as in many German universities, the role of laboratory servant (or *Diener*) survived well into the twentieth century when the laboratory-based aspect of this role began to change into the position of the modern laboratory technician (or *Laborant*).[39] Willstätter related how the laboratory stewards in Munich would only obey Adolf von Baeyer and the inspector (i.e. the superintendent) Fehl but ignored the other senior professors and maltreated the *Privatdozenten*.[40] By the 1900s, if not earlier,

37 Richard Willstätter, *From My Life: The Memoirs of Richard Willstätter* (New York: W.A. Benjamin, 1965), p. 64.

38 For an example of a *Mechanikus* in a physics department, see: Klaus Hentschel, 'The "Invisible Hand" of Carl Friedrich Gauß – Retracing the Life of Moritz Meyerstein, a 19th Century Instrument Maker and Universitäts-Mechanicus', in Christian Forstner and Mark Walker, eds., *Biographies in the History of Physics: Actors, Objects, Institutions* (Cham: Springer, 2020), pp. 13–36. I wish to thank Klaus Hentschel for a copy of this paper.

39 For the duties of the *Diener* in Marburg during the nineteenth century, see: Christoph Meinel, *Die Chemie an der Universität Marburg seit Beginn des 19. Jahrhunderts: Ein Beitrag zu ihrer Entwicklung als Hochschulfach* (Marburg: Elwert, 1978), pp. 112–15, 212–13.

40 Willstätter, *From My Life*, p. 144. For a history of the chemistry laboratory at the University of Munich (and concomitantly the Bavarian Academy of Sciences), see: Wilhelm Prandtl, *Die Geschichte des chemischen Laboratoriums der Bayerischen Akademie der Wissenschaften in*

there was also the *Hilfsdiener*, an assistant to the *Diener*, who is explicitly called the *Laborant* by 1920. At Utrecht the number of technicians and analysts rose sharply in the interwar period, but the role of technician has been less visible at Imperial. At the Royal College of Chemistry, there was a porter who served much the same role as the *Diener*, and in the early days at least, he sold chemicals to the students.[41] The porter was replaced by a laboratory technician around 1900.[42] Although it is difficult to tell if all the technicians are recorded, it is surprising how few there were at either Heidelberg or Imperial until the interwar period.

Clearly this situation could give rise to tension within the laboratory especially if the technician believed the researcher was being given credit for something done largely by the technician. One aspect of this which has been considered by historians is the invention of apparatus. Doubtlessly researchers would say 'I need something which can do this, and here is a rough diagram or crude example of what I am looking for.' The technician would go away and convert the idea into a working instrument, but needless to say, the academic chemist would get the credit. The example often given is the invention of the Bunsen burner by Desaga on the basis of instructions from Bunsen.[43] Ironically, Desaga did not work directly under Bunsen but was the *Mechanikus* for the *Modell Kabinet* directed by Kirchhoff. There are several aspects of this question which should make us cautious. Above all the naming of apparatus rarely stemmed from the original inventor but rather is the result of chemical apparatus suppliers needing to distinguish one piece of apparatus from the other.[44] Furthermore, technicians were sometimes given the right to make and sell apparatus developed in their laboratory as a profitable side-line, most notably Heinrich Aubel, Justus Liebig's laboratory *factotum* (and amanuensis) at Giessen, who made a small fortune from selling chemicals and apparatus.[45] In Heidelberg it was the other way round: Desaga had already set up his instrument business before becoming keeper of the Physical Cabinet in 1840, but he did not work in the chemical laboratory.

München (Weinheim: Verlag Chemie, 1952), but this is a high-level account which has nothing to say about the junior staff and technicians who worked there.

41 Gay and Griffith, *The Chemistry Department at Imperial College, London*, p. 17.

42 Gay and Griffith, *The Chemistry Department at Imperial College, London*, p. 189, n. 21.

43 Moritz Kohn, 'Remarks on the History of Laboratory Burners', *Journal of Chemical Education*, 27 (1950), 514–16; Georg Lockemann, 'The Centenary of the Bunsen Burner', *Journal of Chemical Education*, 33 (1956), 20–2; and the crucial evidence presented in: L.M. Dennis, 'The Origin of the Bunsen Burner', *Industrial & Engineering Chemistry*, 17 (1925), 651–52.

44 Morris, *The Matter Factory*, pp. 142–44.

45 Personal communication from Christoph Meinel dated 1 December 2020.

Conclusions

My description of the laboratory is somewhat paradoxical. All the people working in the laboratory are demarcated in various ways: their status in the department, their function in the laboratory, their academic qualifications, whether they are permanent or temporary, and so forth. In that sense the laboratory can be seen as divided into different groups. Yet at the same time, the laboratory can only function properly if they all collaborate. Latour's archetypical professor can only produce his scientific results with the help of his research assistants and technicians. But there is hardly anything unusual about this hierarchical collaboration that also exists in the factory and in the hospital ward, which has many structural similarities to the laboratory.

This picture can only seem strange if we have an image of an individual researcher or professor in mind, which has not existed for the most part since the mid-nineteenth century. Even nineteenth-century amateur scientists usually had an assistant and/or a technician. However, the laboratory does differ from the factory or the hospital in one important respect. Whereas the senior managers of the factory or the consultants in the hospital only form a small fraction of the total workforce, in the laboratory the professors were comparable in number to the assistants or the technicians, if not the students, well into the twentieth century and probably even today in many academic chemistry laboratories. The chemical laboratory is a productive and dynamic social community, but it is not a beehive!

Acknowledgements

I would like to thank William H. Brock, William Griffith, Ernst Homburg, Jeffrey Johnson, Christoph Meinel, and Christine Nawa for their help with this chapter, which has greatly improved it. Any remaining errors or misjudgements are entirely mine.

About the Author

Peter J.T. Morris holds honorary appointments at the Science Museum (London) and University College London, having been in charge of the chemistry collections at the Science Museum for 24 years. He is interested in the history of the chemical industry (mostly twentieth century) and chemical instrumentation. He published *The Matter Factory: A History of the Chemical Laboratory* in 2015.

10 Of Growing Significance: The Support Staff in the Laboratories and Institutes of Utrecht University during the Interwar Period

Bas Nugteren

Abstract
This chapter focuses on the development of support staff at Utrecht University. It was a period of rapid growth for the university and profound modernization of scientific practice. This was also evident in the development of support activities and support staff. Their size grew in absolute numbers and in proportion to scientific personnel. New positions emerged, such as analyst or glassblower, and existing positions became more specialized, such as technician and laboratory clerk. This development also contributed to new forms of organization: the 'family-like' unit of the nineteenth century faded and was replaced by a business-like organization. Technicians, analysts, and clerks were not part of the *civitas academica*, but they became more recognized for their contributions to research and education.

Keywords: Utrecht University, support staff, specialization, analyst, technician, Interbellum

Introduction

In 1922 professor of botany and director of the Botanical Laboratory at Utrecht University Friedrich Went wrote a letter to the university's Board of Trustees (*College van Curatoren*) saying:

Berkel, Klaas van, and Ernst Homburg (eds), *The Laboratory Revolution and the Creation of the Modern University, 1830-1940*. Taylor & Francis Group, 2023
DOI: 10.5117/9789463720434_CH10

> Mr. De Bouter is a very capable craftsman and a first-rate instrument maker, for whom the Laboratory has much to be grateful for.... His work has contributed to giving the Botanical Laboratory in Utrecht an excellent reputation in scientific circles in the Netherlands and abroad.[1]

He wrote these words of praise to justify his request to promote P.A. de Bouter, the laboratory amanuensis, to the post of chief instrument maker. Such a promotion would give him recognition for his work—and a higher salary. Went's statement can be seen as a small sign of the changes taking place among the support staff, especially among the technical personnel, and of the increased appreciation for their work.[2]

The interwar period was a time of significant change in the scientific practice as well as in the development of the university. The focus of this chapter will be the effects of these processes of modernization on the support staff, including the social effects. This chapter will deal with the university support staff, specifically that of Utrecht University, during that period.[3] It aims to determine which changes the support staff experienced during the interwar period, especially regarding their numbers, job descriptions, positions, qualifications, and organization. It will also examine whether, as the petition by professor Went cited above suggests, the support staff actually enjoyed increased recognition and appreciation as partners in the processes of research and education. It deals with support personnel in the laboratories, instrument makers in the workshops, clerks (*bedienden*) at the library, gardeners in the Hortus, and other support staff employed at the university, such as stokers, amanuenses, and typists.[4]

1 Letter from Went to Trustees, 22 March 1922, Utrecht, HUA 59/1465. HUA = Het Utrechts Archief (The Utrecht Archives); RUU = Rijks-Universiteit te Utrecht (now: Utrecht University).

2 Support staff is that part of the university staff that has a non-scientific position.

3 This chapter is based on research conducted by the author as part of his PhD studies into the history of academic support at Utrecht University between 1876 and 2011, and the professionalization of that support. The main sources for this essay where the university's archives, including the Budgets and Annual Reports and the local newspaper, *Utrechts Nieuwsblad*. Both can be found at Het Utrechts Archief (HUA, 59/...); the *Utrechts Nieuwsblad* can also be accessed on the website of HUA (hetutrechtsarchief.nl).

4 The scope of the essay does not include research assistants (*assistenten*), as they performed scientific and academic work and were usually either students (4th and 5th year) or PhD candidates. The personnel in the medical clinics are not included, as the vast majority of them were not employed by the university.

The Interwar Period: Growth and Modernization

The photograph below from 1913 (fig. 10.1) shows three employees of Went's Botanical Laboratory. Two of them were support personnel: the clerk D.P. Löbel, who was hired in 1900, and Mrs. Pietje Dietz, who had been cleaning the laboratory since 1882.[5] The third was a research assistant. In addition to these three, the laboratory employed an amanuensis, who would be succeeded by De Bouter in 1918. Another four gardeners and a *hortulanus* were employed in the Hortus Botanicus, which also belonged to Went's institute. In retrospect, the moment of the photograph could be considered the end of a period of decades of slow growth at the university in Utrecht (and the Netherlands as a whole).[6]

In a short film from 1932, we encounter Went again and see him at work in 'his' laboratory.[7] The approximately twenty-minute documentary depicts him arriving at the laboratory, then follows him through all of his day's activities: giving lectures, supervising laboratory work, administering exams, discussing with the technician, listening to presentations and discussing them afterwards, etc. Through it all we catch occasional glimpses of a clock, giving the viewer a good impression of the busy life of a professor and the hectic comings and goings at the institute. The film is a precious gem, as there are no other 'live' depictions of laboratories and the work in the greenhouses from that time. The documentary also zooms in on several of the laboratories' instruments and research installations, such as the clinostat constructed by De Bouter and several electrical switchboards.[8] It is clear that the laboratory had introduced the latest scientific technologies and methods.

When we compare the film from 1932 to the photo from 1913, we can certainly see the growth and division of the laboratory during the period. It reflects a larger process of growth and a mechanism of scientific specialization.

5 RUU Budget 1913, nominative attachment, HUA 59/2240.

6 Between 1876 and 1915 (a period of 40 years), the university grew from 500 students to 1,000 and from 34 fulltime (ordinary) professors to 44, while the number of support personnel grew to 70: G. Jensma and H. de Vries, *Veranderingen in het hoger onderwijs in Nederland tussen 1815 en 1940* (Hilversum; Verloren, 1997), pp. 190–210; RUU Annual Reports (*Staat der RijksUniversiteit*) 1876–7, 1915–16; RUU Budget 1915, HUA 59/2242.

7 'Prof. Went en het botanisch laboratorium' (1932), eyefilm.nl, via YouTube [video], <https://www.youtube.com/watch?v=uIlZoSNgj2A> (accessed 7 November 2022).

8 A clinostat allows researchers to study the influence of gravity on plant growth. In 1922 Went published an article about the clinostat built by De Bouter in the *Verslagen van de Koninklijke Akademie van Wetenschappen* (Reports of the Royal Academy of Arts and Sciences), titled 'Over een nieuwen klinostaat volgens het stelsel De Bouter'. The film also shows the Warburg apparatus and Fernandes installation.

Figure 10.1 The Botanical Laboratory in 1913. From left to right: Löbel, clerk, Van Raalte, assistant, and Dietz, charwoman. Courtesy: HUA 86334, GUA photographic service.

In 1913 the support staff consisted only of a clerk, an amanuensis, and a charwoman in the laboratory (*werkvrouw*) and five men in the hortus. By 1932 there were two clerks, Went's letter had resulted in the promotion of De Bouter to the rank of technician, and a draughtsman had been added to the staff along with a woman who kept the library and administration. This expansion largely took place in 1921.[9] Two charwomen were now needed to keep the laboratory clean. The adjacent Hortus Botanicus still employed four gardeners in addition to the *hortulanus*. Another hortus had also been donated to the university: the Cantonspark at Baarn, with eight gardeners on the staff. The latter hortus was led by Professor August Pulle, who was also director of the Botanical Museum and Herbarium, which had been separated from the Botanical Laboratory and housed elsewhere in the building on the Lange Nieuwstraat in Utrecht. In addition to Pulle, the Botanical Museum and Herbarium employed a curator, a technician, an amanuensis, and one or two charwomen.[10]

The developments within the Botanical Laboratory reflect a larger process of growth and scientific specialization. During World War I, in which the Netherlands remained neutral, the gradual pace of development began to

9 RUU Budget 1920, 'nieuwe aanvragen personeel', HUA 59/2247.
10 RUU Budget 1930, HUA 59/2257; RUU Annual Reports 1931–2, *Lotgevallen*.

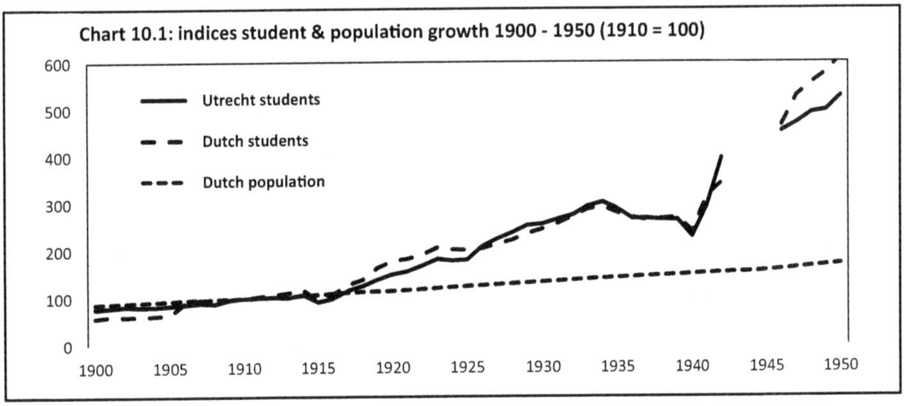

Chart 10.1 Indices of student and population growth, 1875–1950 (1910 = 100). Source: Jensma and De Vries, *Veranderingen in het hoger onderwijs in Nederland*, pp. 190–210; CBS, Statistics Netherlands.

accelerate, as is shown in chart 10.1. Starting in 1916 the number of students began to grow rapidly in Utrecht (and the rest of the country), nearly tripling to 2,900 by 1935.

This rate of growth in the number of students was much higher than that of the population at large and was facilitated by a growing economy that also demanded a more educated workforce. New legislation contributed as well by reducing restrictions on student intake. After 1917 Greek and Latin were no longer required for medicine and natural science studies.[11] The economic growth in the 1920s was not a phenomenon limited to the Netherlands, nor was the development of technology that contributed to further changes in scientific practice. To give an impression of the latter: in his farewell address in 1939, *Vijftig jaren revolutie* ('Fifty years of revolution'), the Utrecht chemist Ernst Cohen recalled the enormous technical progress that had been achieved in his field over the previous decades, which

> of course displays a profound correlation with the revolution in science.... It suffices to list the new tools available: high and low temperatures, high and low pressure, liquefied gases, electrical energy, X-ray technology, ultra-microscopes, electron microscopes, micro- and semi-micro analysis, polarography, photography, cinematography, chromatography, *e tutti quanti*.[12]

11 This was a consequence of a 1917 amendment to the Higher Education Act of 1905, initiated by the liberal member of Parliament, J. Limburg. It is therefore sometimes called 'De wet Limburg'.

12 E.J. Cohen, *Vijftig jaren revolutie. Overgedrukt uit het Chemisch Weekblad van 27 Mei en 22 Juli, 1939* (Amsterdam: D.B. Centen, 1939), p. 14.

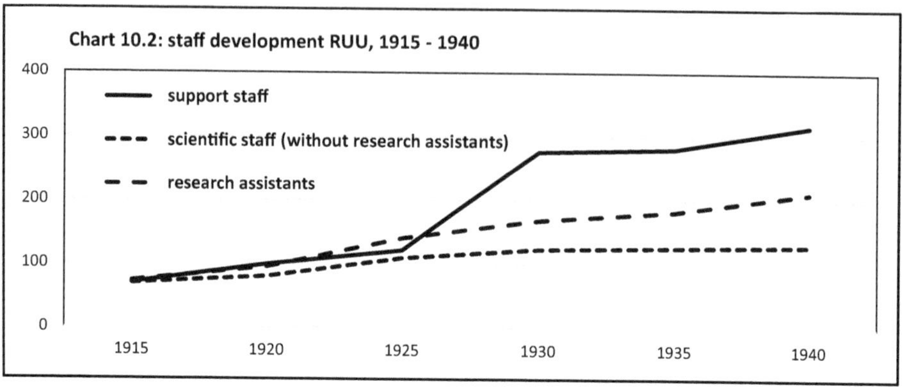

Chart 10.2 Staff development at Utrecht University, 1915–40. Source: RUU Budgets.

The increased specialization was also reflected in the growth in the number of ordinary professors from 44 in 1915 to 64 by 1940, the growth in professors by special appointment from 14 to 49, and the increase of private lecturers (*privaatdocenten*) from 20 to 56![13] Each of them had their own specialization.[14] But it was also manifest in the backbone of the university organization: the institutes and laboratories. Their numbers more than doubled.[15] The total staff (scientific and support) grew from just over 200 in 1915 to 650 in 1940, and there was a considerable change in its composition: in 1915 scientific staff, research assistants, and support staff each accounted for one third. In 1940 the division was roughly one-fifth, one-third, and a half.[16]

The extraordinary growth of support staff (chart 10.2) after 1926 from 120 (1925) to 275 (1930) was partly the effect of the start of the Faculty of Veterinary Medicine, which was incorporated into the University in 1926. The support staff of this new faculty counted around 60, mostly stablemen

13 RUU Annual Report 1939–40, *Staat der Rijks-Universiteit.*

14 Specialized teaching assignments were instituted with the Higher Education Act of 1876, and private lecturers were appointed by the Minister with their own teaching assignment (Jensma and De Vries, *Veranderingen in het hoger onderwijs in Nederland*, p. 272). The significance of these developments was also recognized at the time (*Jaarboek RUU 1921–1922*, 117).

15 In 1915 there were ten laboratories in the natural sciences, five pre-clinical laboratories at the Faculty of Medicine, and two institutes created at the Faculty of Arts. By 1935 the number of science laboratories had doubled to twenty, while there were then nine medical laboratories and seven other institutes. These 36 institutes and laboratories were accompanied by another twelve at the Faculty of Veterinary Medicine, which was incorporated into the university in 1925.

16 The scientific staff consisted mostly of professors and lecturers; research assistants were mostly students (4th, 5th year) and PhD candidates. They not only did research but also assisted with practicums and courses.

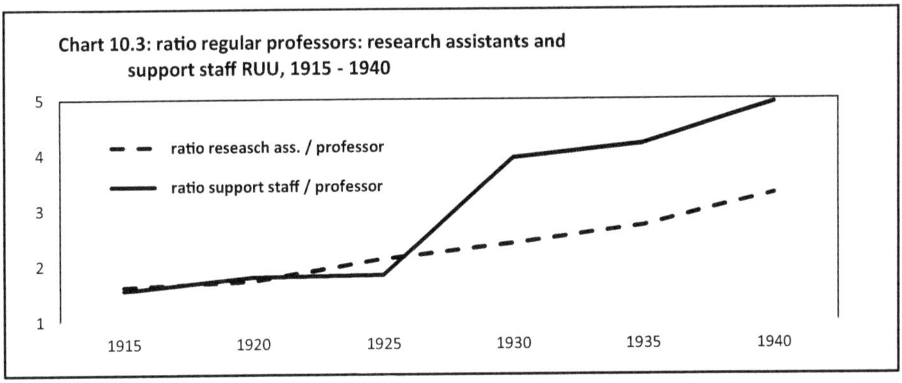

Chart 10.3 Ratio of (ord.) professors and research assistants and support staff at Utrecht University, 1915–40. Source: RUU Budgets.

and laboratory clerks (*laboratorium bedienden*). Nevertheless, we can see a significant increase of research assistants and support staff in comparison to that of the scientific staff (chart 10.3). The absolute and relative growth of the number of research assistants can be explained by a growing number of students, who all had to participate in practicals, and by an increase in research activities. The growth of the support staff seems to have been particularly spurred by growing laboratory and workshop activities. In the next section we will see that the growth was mainly caused by new positions and the specialization of existing ones.

New Positions and Qualifications

In 1936 the former 'President-Curator' (president of the Board of Trustees) of the university, Joachim Fockema Andreae, described the period following World War I as a 'period of robust expansion and radiant growth'.[17] That also applied to the support staff, both in regard to the growth in numbers, to around 300 in 1940 (not counting cleaning staff and apprentice clerks), and to the development of their ranks.[18] New positions were added, such as analysts, and existing positions were upgraded, such as that of stoker, originally a position that required no training and which was usually combined with other duties such as those of a clerk. Starting in 1920 the position of stoker

17 G.W. Kernkamp, *De Utrechtsche Academie 1636-1936*, 2 vols. (Utrecht: Oosthoek, 1936), vol. 2, p. 144.
18 RUU Budget 1940, HUA 59/2267; there were 66 charwomen and 42 apprentice clerks (*leerling bedienden*) employed at the time.

occasionally appeared as an independent position, one which also required the ability to conduct minor technical maintenance. That was the case for the stoker at the Botanical Laboratory and greenhouses, who was hired in 1919.[19]

With the arrival of well-trained analysts and technicians, staff competency was developed and standardized. In general, more attention was being paid to competency and standardization of positions and job requirements. In 1921 a national commission submitted a proposal for educational prerequisites for civil servants, which at the time included university employees.[20] As a result, some minimum standards were introduced on a national scale. The chart below (10.4) illustrates the innovation in support positions at Utrecht University. More than two-thirds of the traditional positions were clerks and amanuenses. The growth, however, mainly came in the form of 'new' positions, such as technicians (like instrument makers and glassblowers), analysts, laboratory clerks, and administrative and secretarial positions. The occasional draughtsmen can also be included in this list. To be clear, 'new' does not imply that the work had never been done before. All these positions, with the possible exception of the analysts, had predecessors at the university, such as the glassblower employed as an amanuensis. What was new was that the positions were now assigned to a single or main task (fulltime), with its own job classification. Chart 10.4 shows the shift from traditional positions to new ones.[21] The other positions that appear starting from 1930 include stablemen and apprentice clerks, not including the approximately 70 charwomen employed by the university.[22]

The following chart (10.5) shows the rise of the analyst in the laboratory. It also shows that the existing position of 'ordinary' clerk was 'replaced' by that of specific laboratory clerks (*laboratorium bediende*), to be distinguished from the position of analyst.

The post of analyst was a new position and is evidence of the growing importance of chemistry in business and research. 'The sizeable expansion of laboratory work in countless departments ... gives Analysts a vital role to play,' wrote Professor Wilhelm Ringer, director of the Laboratory for

19 Due to his experience as a stoker, later operator, machinist, and electrician. See letter from Went to Board of Trustees, 3 September 1919, HUA 59/1464.

20 *Rapport der subcommissie ingesteld door de plenaire vergadering van de Salariscommissie voor Burgerlijke Rijksambtenaren op 30 Januari 1920*, p. 6, The Hague, National Archives 2.14.54-1.

21 The traditional positions included amanuensis, clerk, librarian, and gardener. The new positions included administrative and secretarial staff, analysts, laboratory assistants, technicians, and others such as draughtsmen.

22 The former Veterinary School was incorporated as a university faculty in 1925. Apprentice clerks were not included in the personnel files before 1930, because the position was unpaid.

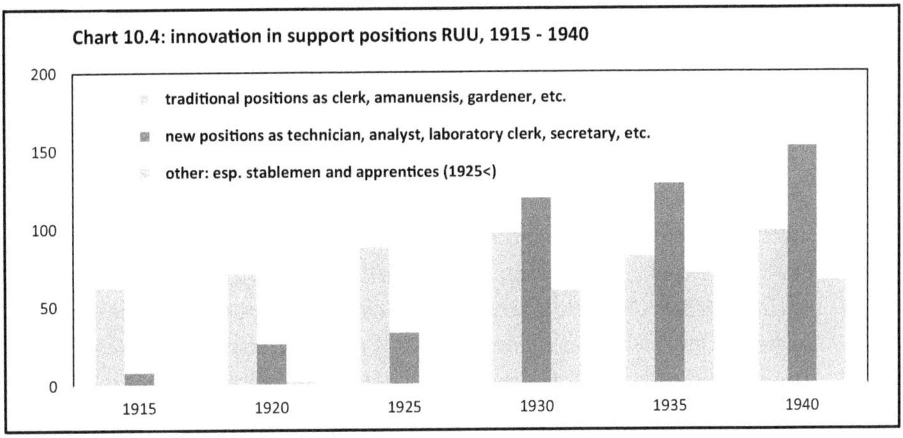

Chart 10.4 Innovations in support positions at Utrecht University, 1915–40. Source: RUU Budgets.

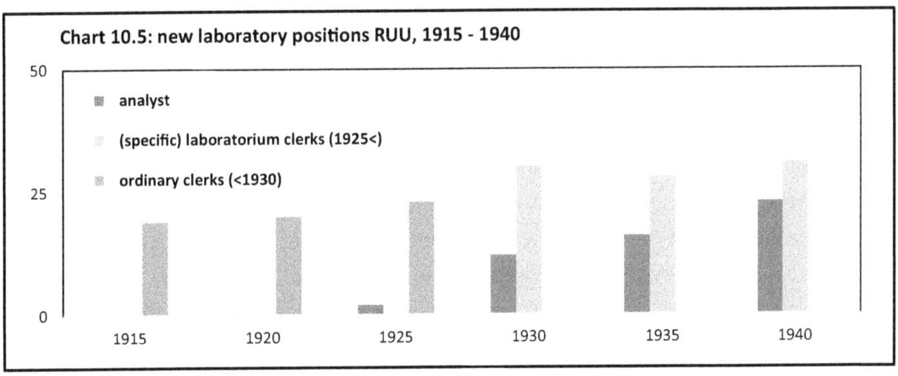

Chart 10.5 New laboratory positions at Utrecht University, 1915–40. Source: RUU Budgets.

Physiological Chemistry in December 1925.[23] In 1917 the *Nederlandsche Chemische Vereniging* ('Netherlands Chemical Society') introduced an exam for analysts. Passing this exam was a requirement for employment by Utrecht University and contributed to raising the level of expertise among support staff.[24] Probably the first analyst in Utrecht was Miss B. Grutterink, who joined the Physiological Laboratory in 1921.[25] Much was expected of the analysts due to the rapid pace of development, especially in Ringer's field of study. They 'must constantly test an overwhelming number of new

23 Letter from Ringer to Board of Trustees, 18 October 1925, HUA 59/1446.

24 Letter from Faculty of Mathematics and Physics to Board of Trustees, 24 November 1925, HUA 59/1446.

25 RUU Budget 1921, explanation of the budget, p.4., HUA 59/2248; RUU Budget 1922, exploitation no. 21, HUA 59/2249.

research methods for their value and reliability'. Their own contributions to the research process were also noted because, as Ringer wrote, assistants were in short supply and 'the Leader of the Clinic not only lacks the time, but undoubtedly in most cases lacks the vital laboratory experience as well.'[26]

Sometime between 1925 and 1930, the laboratory clerk made his entrance and took over the laboratory work from the 'ordinary' clerk. Detailed information is scarce, but a job description from 1932 implies that a laboratory clerk was expected to deal specifically with chemicals and glassware: 'issue of chemicals & glassware, lab. supplies, cleaning glassware. Stoking central heating. Courier duties, doorman, [and] *maintenance of the lecture hall*'.[27] The latter duties are not specifically related to a laboratory but the former are, and it is understandable that some knowledge and skill in dealing with chemicals was expected. The sources do not indicate that there was any difference in payment.

The growth and modernization of laboratories was also indicated by the demand for more glassblowers. Glassblowers were grouped together with technicians and instrument makers. The demand for their skills is illustrated by the fact that in 1926, one of the labour unions set up a course with the university's support to train (or re-train) amanuenses, technicians, and possibly even clerks to work as glassblowers. The course proved to be extremely popular; in the first year, ten university employees enrolled, along with several from outside the university. It remained popular well into the 1930s, which was only to be expected according to the organizers: 'when one realizes that most laboratories are increasingly demanding some knowledge of glassblowing as a requirement for employment.'[28]

Around 1915–20 the amanuensis was, aside from clerk, the most common support position. Amanuenses were not limited to the workshops either; it was also a rank. In 1920 the University Library counted six amanuenses among its staff, and the amanuensis at the observatory was expected to have 'a fairly in-depth knowledge of mathematics'.[29] The amanuensis in the workshop eventually became the instrument maker or technician, a step that we saw De Bouter took as well. This was not only a change in title but also in the mastery of one's trade. In chart 10.6 one sees the 'rise and fall' of the technician and amanuensis in the workshop. Just as analysts were expected

26 Letter from W. Ringer to Board of Trustees, 18 October 1925, HUA 59/1446.

27 Notes on request for dismissal of Verheul, laboratory clerk, as of 1 January 1933, HUA 59/1453. Emphasis in the original.

28 'Glasblaascursus', *Utrechtsch Nieuwsblad*, 1 October 1927.

29 RUU Budget 1926, introduction, HUA 59/2253.

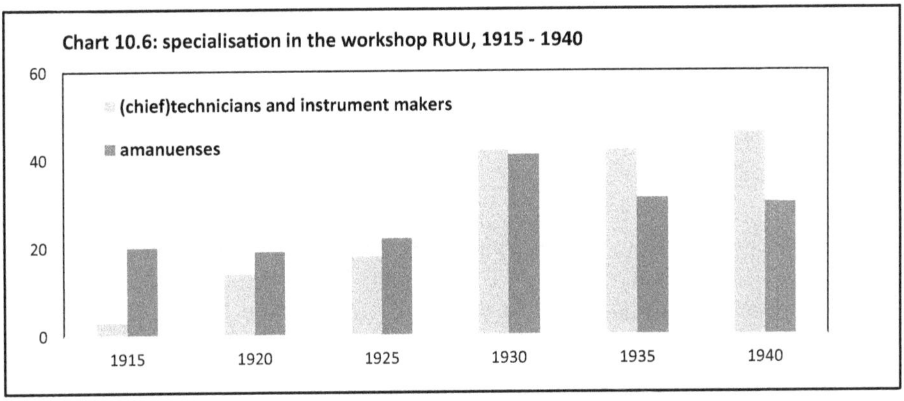

Chart 10.6 Specialization in the workshop at Utrecht University, 1915–40. Source: RUU Budgets.

to pass the national analyst exam, technicians had to be in possession of the so-called 'Leiden Certificate' or meet the equivalent requirements.[30] The 'Leiden Certificate' was the diploma issued upon completion of the training course for instrument maker, glassblower, or electrical technician at the Leiden Instrument Maker's School founded by Heike Kamerlingh Onnes in 1902.[31] The faculty in Utrecht also utilized the course, possibly because there was not enough demand in Utrecht itself to sustain a training programme similar to the one in Leiden. This would change after World War II, however. The aforementioned glassblowing course was in a certain sense an exception: the course was offered in a building at Utrecht University, but participants still had to travel to Leiden for the exam.

The administrative and secretarial sector also grew during the interwar period. This growth reflected the larger number of institutes, on one hand, and the need for more administration and correspondence arising from the increase in scale and demands, on the other. For example, in 1929 the Rector drew attention to 'the increasingly demanding administration of the institutes.... The directors are overwhelmed with guidelines that force them to devote ever-more time to administration.'[32] Concerns about unbridled regulations are apparently not unique to our age. During World War I the

30 Letter from Faculty of Mathematics and Physics to Board of Trustees, 24 November 1925, HUA 59/1446.

31 D. van Delft, 'The Blue-Collar Boys: The School of Instrument-Makers at the Leiden Physics Laboratory of Heike Kamerlingh Onnes', in Bart Grob and Hans Hooijmaijers, eds., *Who Needs Scientific Instruments* (Leiden: Museum Boerhaave, 2006), pp. 151–6.

32 RUU Annual Report 1929–30, p. 187. This coincided with the introduction of the Compatibility Act of 1927, which required the institutes and laboratories to hand over responsibility for their finances to the Board of Trustees.

university administration and financial management were still conducted by the Secretary of the Board of Trustees from his private home. A dedicated office was only furnished in 1920. After 1910 the Secretary could rely on the assistance of a single clerk. By 1940, however, the university bureau had grown to employ a total of five administrators and three clerks. In that year most institutes also had a writing clerk or secretary, many of them working part-time, with a total of 35 working at the university; in 1920 there had been only a handful.[33]

The position of writing clerk was one of the few in which women were employed at that time. In 1915 the number of women working in support positions was limited to less than 10 per cent of the support staff, not counting the charwomen. These included two amanuenses at the University Library and a calculator at the Institute of Astronomy. By 1940 their number had grown to around 25, but they still represented only 10 per cent of the total staff. Most of the female employees were either writing clerks (ten) or analysts (eight).[34]

Rationalization

The university organizational structure did not change during the interwar period. Institutes and laboratories remained the foundation and support was organized within their often narrow walls. These institutes and laboratories were expected to be as self-sufficient as possible and were therefore thriftily organized based on their own research and education activities, with the professor at the helm. In a way, the university resembled an archipelago of separate individual islands.[35] For example, the Botanical Laboratory, Botanical Museum, and Herbarium all had their own charwomen, although they were housed in the same building. When in 1932 the Trustees reduced the cleaning budget for one a bit more than for the other, the decision resulted in protest but not in collaboration.[36]

We know few details about the day-to-day affairs of the institutes and laboratories, with the exception of two of the largest: the Physics Laboratory and the Botanical Laboratory. To mark the re-opening of the renovated

33 RUU Budget 1940, HUA 59/2267.

34 RUU Budgets for 1915, HUA 59/2242 and 1940, HUA 59/2267, nominative attachments. In 1940 there were approximately 70 cleaners.

35 On this development, see the chapter by Van Berkel in this volume (Ch. 5).

36 Letter from A. Pulle to Board of Trustees, 2 December 1932, HUA 59/2058.

Physics Laboratory in 1926, a booklet was commissioned about the new building, its furnishings, and its organization. The building's functionalist approach is of special interest. It was designed entirely around the needs of research and education and possible future use, as all the 'rooms are connected at the corners by porcelain bushings, to avoid the need to break through walls at a later date.'[37] The laboratory featured several workshops: the general workshop, a joinery shop, a forge, a woodworking shop, a precision machine shop, and a workshop for glassblowers. These were accompanied by storage rooms, including rooms intended specifically for chemicals and photography. Two rooms were designated for administrative purposes to support these workshops.

The large room was furnished with the cabinets necessary for storing correspondence and forms, the card catalogue for the instrument cabinet, and there was also a collapsible typewriter desk and a spare typewriter especially for mathematical work. The administration room housed the books in which orders were recorded as well as the requested books for the production and repair of instruments.[38]

The organizational structure was also characterized by this sense of rationalization, with specialization and detailed instructions for the division of tasks. This was facilitated by the laboratory's growth; in 1915 the Physics Laboratory employed an instrument maker, an amanuensis, a clerk (who also acted as a stoker), and a charwoman.[39] By 1926 the support organization had been divided into two departments: the workshop under the leadership of the chief instrument maker, with two instrument makers, a joiner, a glassblower, and two apprentice technicians, and the household service under the leadership of an amanuensis/custodian (*beheerder*). This department employed another amanuensis charged with the storage rooms, chemicals, and photography rooms; a mechanic/switchboard operator; three charwomen; and a stoker. A stenographer rounded out the institute's support staff.

The institute built as many of its own instruments as it could. If the institute could not build the instrument itself, then one would be purchased and customized as needed.

If a scientific worker needed an instrument, he could consult a catalogue to see if it was already in stock, and if so, where. If in-house construction

37 A.M. van Dyck-Huffnagel, *Het natuurkundig laboratorium der Rijks-universiteit te Utrecht; uitgegeven ter gelegenheid van de heropening van het laboratorium op 9 februari 1926* (Utrecht: [s.n.], 1926), p. 41.

38 Van Dyck-Huffnagel, *Het natuurkundig laboratorium*, p. 27.

39 RUU Budget 1915, HUA 59/2242.

was desired, then the researcher could note the instrument to be built in a book designated for that purpose. The professor-director would then issue personal written orders to the workshop based on these requests, stating which craftsman should build the instrument and with whom he should consult in addition to the chief instrument maker.[40]

Social Standing and Appreciation

We have noted how the support staff grew in numbers and composition and the (average) level of competency increased, in part as the result of the introduction of exams and specialization and sometimes as the result of new positions being created. Their expertise and skills became increasingly important, and especially those of technicians and analysts became essential. And with the growth in numbers and the modernization of the scientific practice, there was need for a more rational organization of labour. We saw the examples of the Physics Laboratory in 1926 and the Botanical Laboratory in 1932. This raises the question of whether these developments had an influence on the recognition or prestige of the support staff within the university. By 'prestige', we mean the degree to which support staff were acknowledged and appreciated. Did the increase in competency, specialization, and new positions with more stringent qualifications contribute to increased respect and appreciation of the support staff? Or did they remain 'subordinate personnel', as they were described in the introduction to the 1920 budget?[41]

The 'growth' here is in comparison to the period before World War I. The recognition of the support staff had always been determined mainly by the relative isolation of the small-scale institutes and laboratories. In the words of University of Groningen historian Klaas van Berkel, they were 'small communities' with 'the professor as pater familias'.[42] This seems to be an apt description of the relationships between staff members, at least up until World War I. The institute could be considered as something like a (bourgeois) household with staff, just as the professor had his own domestic staff at home. The professors generally came from the affluent upper and upper-middle classes (*standen*) and certainly became members

40 Van Dyck-Huffnagel, *Het natuurkundig laboratorium*, p. 24.
41 RUU Budget 1920, Introduction p. 4, HUA 59/2247.
42 Klaas van Berkel, *Universiteit van het Noorden. Vier eeuwen academisch leven in Groningen*, 3 vols. (Hilversum: Verloren, 2014-22), vol. 2: *De klassieke universiteit 1876-1945*, p. 344.

of these classes once they were appointed as professors. The notion 'class' is to be understand as Max Weber's notion of 'Stand' and not as economic classes. Around 1900 it was customary for the upper-middle class to employ domestic servants.[43] We know, for example, that the Utrecht scientist, professor, and progeny of a prominent family of clergymen Christophorus Buys Ballot (1819–1890) would lead 'a religious service' (*godsdienstoefening*), which was also attended by the staff before breakfast in the morning.[44]

Work in the institute may not have begun with prayer, but the recognition or respect of the stoker, gardener, clerk, and even the amanuensis would not have been much different than if they worked in the professor's private household. They had neither special training nor a say in the institute's administration. In the background we can see the outlines of the class-based society (*standensamenleving*) that still characterized the Netherlands at the time. The social distance between the professor and his personnel was also reinforced by the fact that at the university level, the *civitas academica* shaped the organizational culture. This *civitas* was the self-governing society of professors, and although in principle it also encompassed the students and alumni to form the community of teachers and students, that was more an ideal than a reality. In any case, the support staff were not considered a part of it.

The question of whether support staff enjoyed growing recognition and appreciation is difficult to answer, as primary sources are lacking. There was no university newspaper, there are no autobiographical works or correspondences written by amanuenses or stokers, there are no 'vignettes' of the time, and there is very little previous research on the topic. Nevertheless, there are indications that their prestige was changing. First, there were the statements by professors such as Went and the observations of Professor Ringer, and there were more. There may be several explanations for why the archives contain letters of recommendation from the interwar period but not from the previous period. There are also other indications of an increase in appreciation. If awarding a royal distinction (*koninklijke onderscheiding*) can be considered as a reflection of appreciation for an individual's work, then the pattern of awards in the interwar period does indeed suggest that more attention was paid to support staff at the university level.

Professors had long been nominated for a royal distinction because of their scientific achievements and importance, but it was exceptional for support staff to be nominated (and honoured) before the interwar period.

43 I. Montijn, *Leven op stand: 1890-1940* (Amsterdam: Thomas Rap, 1998), pp. 220–1.
44 E. van Everdingen, *C.H.D. Buys Ballot 1817-1890* (Den Haag: Daamen, 1953), p. 131.

(We know of nominations in 1913, 1914, and 1922.) However, starting from the mid-1920s, one or more support employees were nominated (almost) annually for a such a honour.[45] This rise in nominations fitted in a national trend towards a growing number of honours awarded to people in the lower classes.[46] A nomination was not honoured automatically, as that had to be decided by the ministry, but a nomination itself can be seen as a sign of appreciation. (As a rule, the nominee was not informed about the nomination.) Although it concerned a lower rank in the rather complicated Dutch honours system and the professors got a much higher-ranked honour, it was very much appreciated by the *decorandi* knowing that the institute director (the professor) and university had nominated him or her.[47] Without exception, the nominations were justified by the nominee's many years of loyal service, combined with the impeccable performance of his or her duties. The nomination was often also accompanied by a glowing recommendation, such as that offered by Professor Leonard Ornstein, director of the Physics Laboratory in 1930: 'I believe that it is not only a richly deserved crown placed on her own career, but that my entire staff will rejoice at this fine industrious woman being honoured in such a manner.' This 'fine industrious woman' was Miss W. de Ruyter, the laboratory's charwoman for the past forty years.

A similar pattern of increased recognition is, with a bit of good will, observable in the obituaries for deceased employees published in the annual university reports in the 1920s (*Verslag van de Lotgevallen der Rijksuniversiteit*, 'Report on events at the university'). That the institutes and laboratories made mention of the passing of a clerk or other employee is not unusual for the time, but at the university level, such mentions only began in 1919 and remained relatively rare. Students and professors, including former professors, had always been memorialized during the inauguration of a new Rector. On the passing of a sitting professor, the Senate even called a special memorial session. Over the course of the 1930s, mentioning the deaths of support personnel was discontinued, perhaps due to the sheer growth in numbers. It is noteworthy that most such mentions referred to technicians, which can perhaps be explained by the fact that technicians' contributions to research were more tangible. For example, in 1927 the Rector

45 'Koninklijke Onderscheidingen', 1911–1940, HUA 59/547.

46 Kees Bruin, *Kroon op het werk. Onderscheiden in het Koninkrijk der Nederlanden* (Amsterdam: Boom, 1989), pp. 86–91.

47 The professors mostly got a knighthood in de 'Orde van de Nederlandse Leeuw'; this was appreciated higher than the 'Orde van Oranje Nassau'. Both orders knew several distinctions and the 'Orde van Oranje Nassau' also included several Medals of Honour, especially introduced for the lower classes in 1892. Support personnel were mostly nominated for the silver Medal.

took a moment to reflect on the passing 'of the capable chief technician of the Physics Laboratory, Mr. G. Koolschijn.... He was an artist in his trade, producing instruments that drew attention even from abroad.'[48]

A Company Union of Higher Education Personnel

But there was yet another reason that the prestige of the support staff seemed to be growing. The end of World War I and the years after was characterized by significant social unrest and the rise of political and trade organizations, both in the Netherlands and abroad. In general, there were two types of trade organizations. The first were national labour unions founded on ideological or confessional grounds, with local chapters such as the chapter in Utrecht. The second focused on their own specific company or organization and its employees, often referred to as a 'company union'. Such unions were common among large companies, but there was also one founded by personnel of the universities. The grounds for organizing into a company union were primarily based on the employees' position in their 'own' company and not on their views on their position in society or on an ideal of social emancipation.

In 1917 a 'Salary Committee' convened by the Minister found that 'the higher education personnel also organized themselves locally in Delft, in Groningen and in Utrecht; in Leiden the Academic Civil Servants at the National Museums also began organizing themselves.'[49]

The local organization referred to in Utrecht was the 'Association of Higher Education Personnel in Utrecht: Strength through Unity' (*Vereeniging Hooger Onderwijs Personeel te Utrecht: Door Eendracht Sterk*), founded in the early spring of 1917. Their colleagues in support positions at other universities also began to organize themselves into unions and associations. In Groningen an 'Association of Civil Servants' (*Vereeniging van Beambten*) had already been founded in 1908, and Delft had the 'Association of Civil Servants and Clerks at the Technical University of Delft' (*Vereeniging van beambten en bedienden aan de Technische Hoogeschool te Delft*). Two years later, the members of the association in Utrecht decided to dissolve the association in favour of the possibility of joining the national civil servants union affiliated with the social-democratic 'Netherlands Association of Trade Unions' (*Nederlandse*

48 RUU Annual Report 1926–7, p. 277.

49 *Rapport der commissie ingesteld bij beschikking van Zijne Exc. den Minister van Staat, Minister van Binnenlandse Zaken van 26 september 1916*, Utrecht/Amsterdam/'s-Gravenhage, 1917, p. 41.

Verbond van Vakverenigingen). But that was a step too far for some, and at the insistence of several civil servants working at Utrecht University who did not wish to join the Central Netherlands Civil Servants Union, the staff decided to found 'a Neutral Association of Higher Education Personnel in this city' (*Neutrale Vereniging voor Hoger Onderwijs Personeel*).[50]

Soon after the foundation of the new association, on 14 June 1919, the associations in Utrecht, Groningen, and Delft joined to establish the national 'Union for Higher Education Personnel H.O.P.' (*Neutrale Bond voor Hoger Onderwijs Personeel; H.O.P.*), intended specifically for the technical and administrative personnel at the universities and the Technical University of Delft.[51] In 1920 they were joined by similar associations from Leiden and Wageningen. Membership was limited to support personnel working in higher education. The foundation of what would later be abbreviated as the H.O.P. was certainly an expression of growing self-confidence among the support staff, who had climbed over the figurative walls between the separate institutes and laboratories to unite together, regardless of whether one was a clerk or an amanuensis. Or, as the later chairman of the union recalled during the tenth anniversary in 1929: 'Various categories are represented here, and persons from every field are assembled in the organization of non-teaching H.O.P, as the struggle for development can only be conducted through solidarity.'[52] It is possible to give only a limited account of the number of H.O.P. members.[53] In 1927 the H.O.P. (all universities) numbered around 250 members. The Utrecht chapter had 79 members, representing all the job categories, as of 1 January 1931.[54] The Utrecht chapter of the social-democratic Civil Servants Union (*Centrale Nederlandse Ambtenaarsbond*) had fewer members in this period (60 as of 31 March 1931).[55] The H.O.P. was more rooted in the university system than any of the other unions.

The H.O.P. only had a very marginal influence in shaping national policy regarding labour relations at the universities. At the local level, however, they

50 'Neutrale Vereen. Hoger Onderwijspersoneel' and 'Neutrale Bond van Hooger Onderwijspersoneel', *Utrechts Nieuwsblad*, 11 April 1919 and 23 February 1928. 'Neutral' meant that the union was not based on political or religious ideologies.

51 'Hooger Onderwijspersoneel', *Utrechts Nieuwsblad*, 17 June 1919.

52 'Neutrale Bond van hooger onderwijspersoneel', *Utrechts Nieuwsblad*, 15 June 1929.

53 Accounts in documents pertaining to the trade organizations, HUA 59/1364-1365; available data are incomplete.

54 Most members were either clerks (28), technicians (19), or amanuenses (17).

55 These included almost no technicians (1), a handful of amanuenses (6), around a dozen clerks and other positions such as gardener, stoker, charwomen (9), servants, etc., some of whom were employed at the academic clinics.

played a vital role in gaining awareness for support staff within the university and earning recognition for the group of employees they represented. The same seems to have been the case elsewhere as well. In 1928, at the annual national conference of the H.O.P. chapters at the University of Groningen, the Rector of that university noted that:

> In the university organism, the corps of civil servants plays a vital role. What would the professors in the laboratories do without their instrument makers, and what would a Rector Magnificus be without the beadles![56]

As a company union, the H.O.P. focused emphatically on its own local work environment and activities. We have already discussed the glassblower course set up by the H.O.P. in collaboration with the university. But the union also paid attention to the members' social life, which is noteworthy, as social activities can help build social cohesion and enhance an organization's performance.

The national board and the chapter board at Utrecht University made every effort to add new members and to keep them on board in the 1920s and 1930s. They organized cultural evenings, excursions, lectures, and courses to promote solidarity among the members (wrote the later chairman of the H.O.P., Ger Ruigrok).[57] One of these social events was the celebration of the union's tenth anniversary on 14 June 1929 in Utrecht. Along with representatives of several student associations, the chairman and secretary of the university's Board of Trustees were in attendance together with their counterparts from Delft, who offered their best wishes to the union's members.[58] Speeches recalled the union's growth and successes, and of course the necessity of self-organization. The Rector and the secretary of the Board of Trustees also attended the next celebration ten years later in 1939, where the former expressed his appreciation for the union's work 'on behalf of the Academic Senate'. He recalled the 'fruitful collaboration that exists between the leaders of the laboratories, on the one hand, and the personnel, on the other. The speaker reminisced on the excellent relationship that the parties enjoy.' In his response, the chairman of the H.O.P. gave 'the assurance that every member of the union feels bound to the field of Higher Education'.[59]

56 'School- en Kerknieuws', *Utrechtsch Nieuwsblad*, 10 April 1928.
57 'Uit de archiefdoos van de afdeling Utrecht', *AC-HOP bulletin*, 1 (2010).
58 'Neutrale Bond van hooger onderwijspersoneel', *Utrechtsch Nieuwsblad*, 15 June 1929.
59 'Bond van Hooger Onderwijs Personeel', *Utrechtsch Nieuwsblad*, 14 April 1939.

Sol Iustitiae

Although the sources are silent on the matter, it is highly likely that the increased appreciation and recognition for support staff was closely linked to the fact that the staff gained a 'face' and grew in importance. This growing respect should not be confused with a paean to the support staff, however. The relationships between academics and staff were still largely distant, or as F.J. van Burken would later relate, until the 1950s the 'class differences ... did not permit academic personnel and non-academic staff to treat each other informally.'[60] There was still a strict social order. When on the occasion of the university's tricentennial in 1936 a total of fourteen (different) distinctions were awarded, the distribution reflected the social order of the time and the hierarchy within the university: strictly top-down.[61]

During the interwar period, the support staff were not considered to be part of the *civitas academica*. When the university attempted to revive the *civitas* spirit in 1946, the chairman of the Utrecht H.O.P. wondered when 'people will realize that the *civitas academica* isn't only made up of lecturers and students'.[62] That was his reaction to the first edition of the university magazine *Sol Iustitiae*, a reference to the university's motto: *Sol Iustitiae, Illustra Nos*. The magazine was subtitled 'Organ of the Utrecht University Community' and was an attempt by the university board to facilitate internal communication and social cohesion. But the spirit of the times was clearly expressed by the fact that it was only distributed to 'all professors, lecturers, and students'.

Conclusion

In the development of science and universities, the interwar period was an era of a profound transition in general as well as for Utrecht University

60 F.J. van Burken, *Honderd jaar organische chemie in Utrecht, 1876-1976; een periode van groei, bloei en stagnatie* (Utrecht: 1983), opposite of p. 48.

61 The President-Curator was promoted to Grand Officer in the Order of the Netherlands Lion, the Rector was named Commander of the same order, and the Secretary of the Board of Trustees was awarded the honour of Knight. Another six professors were also knighted. The librarian was named a Knight in the Order of Orange-Nassau; Evert Homoet, chief administrator, received a gold honorary medal in the Order of Orange-Nassau; the technician C. Kerssen and amanuensis J.C. Kreugel were each awarded a silver honorary medal; and H.J. van Aanholt, clerk at the University Library, received the bronze honorary medal. (RUU Annual Report 1935–6, *Lotgevallen*.)

62 *Sol Iustitiae*, 18 April 1946.

specifically. Developments that originated in the nineteenth century intensified in this period. The number of students and staff tripled, and the modernization of science practice broadened and deepened. The continuing modernization of science was also expressed in the development of support activities and support staff. Their population grew in absolute numbers as well as in proportion to the scientific personnel. New positions were introduced, such as those of analysts and glassblowers, and existing positions were transformed into more specialized jobs, such as technician and laboratory clerk. Traditional positions such as clerks and amanuenses remained, but they were outnumbered by the new positions which were more qualified; specialized preliminary training and diplomas became standard.

More and more, the expertise of the support staff, at least that of technicians and analysts, was needed for the scientific practice. Their knowledge and skills became more valueable and sometimes even irreplaceable. Their role became more important; as Professor Ringer described in 1925, 'in most cases' the analyst possessed 'indisputably the necessary laboratory experience', while said experience was (sometimes) lacking among the scientific staff and assistants.[63]

The growth in numbers and developments in scientific practice, including specialization and division of labour among the support staff, stimulated a more rationalized organization of the laboratories and workplaces. At least that was the case at the Physics Laboratory and the Botanical Laboratory, as we observed. The 'family-like' unit faded over time and was replaced by a business-like organization. This inevitably commercialized the (social) relations.

All this also contributed to the support staff's social standing. But this was also influenced by the social changes society underwent. The massacres of World War I contributed to the erosion of the class society. This and economic developments stimulated the emancipation of large groups in society, especially the labour force. By extension, many of the university's support staff did organize themselves in unions, the largest being a company union. Clearly this contributed to the emancipation of the support staff too.

In response, the university began showing some modest appreciation of the efforts and contributions of the support staff, especially for technicians, but they remained 'support staff' in the sense of a class society and were not considered as colleagues or members of the *civitas academica*. The class society was undermined but did not disappear, as we saw in 1936 when the awarding of the royal distinctions was strictly along the lines of

63 Letter from Ringer to Board of Trustees, 18 October 1925, UUA 59/1446.

the social order of the time and the hierarchy within the university.[64] But nevertheless, the support staff seems to have gained prestige due to their growing significance.

Acknowledgements

The author is grateful to Robert Smith for his translations.

About the Author

Bas Nugteren studied economic and social history at Utrecht University. He has been working in various staff (policy, strategic planning) and management positions in- and outside the university. From 2011 to 2020 and from 2021 to 2022, he was director of 'Sonnenborgh – public observatory and museum for astronomy and meteorology' in the city of Utrecht. For some years he has been conducting PhD research into the history of the support staff at Utrecht University between 1876 and 2011, and the professionalization of that support.

64 This chapter focused specifically on the support personnel at the Rijksuniversiteit Utrecht, but the uniform culture among Dutch universities at the time and the general pattern of development in science and technology leads one to suspect that similar developments occurred at the other universities in the Netherlands during this period.

11 A Revolution in Genetics at Gendered Experimental Venues: Cambridge and Berlin, 1890–1930

Ida H. Stamhuis

Abstract

If attention is only focused on 'iconic' university laboratories, the tentative beginning of genetics, including women's important role, would escape attention. At first, research was carried out in a domestic setting and the group of researchers formed an extended family. Furthermore, the organizational setting was an institution not belonging to a university. Genetics developed successfully thanks to the contributions of women. Women were accepted because of their competence and their low costs. They remained, however, in the shadow of the male head and could not become authorities. In that period the situation in genetics apparently did not allow for an exception.

Keywords: women in science, emergence of genetics, institutional setting, Berlin, Cambridge

Introduction

In 1914 the first Institute for Genetics (*Institut für Vererbungswissenschaft*) in Germany was founded in Berlin-Dahlem at the semi-academic Agricultural College (*Landwirtschaftliche Hochschule*). Professor of genetics Erwin Baur (1875–1933) was its first director. The scientific staff consisted almost exclusively of women: there were five of them, to be joined from 1921 by only a single male. In 1930, as the historian Werner Pfarre wrote in 1987, 'Thanks to the efforts of Baur and his colleagues the Dahlem Institute had

Berkel, Klaas van, and Ernst Homburg (eds), *The Laboratory Revolution and the Creation of the Modern University, 1830-1940*. Taylor & Francis Group, 2023
DOI: 10.5117/9789463720434_CH11

reached a position of world renown by the time Baur left.'[1] Nevertheless, a potential successor of Baur, Theodor Roemer, who headed the institute of plant breeding at the University of Halle, was not very keen to take over the institute. When asked whether he was interested in succeeding Baur, he wrote: 'The fact that I ... will in a certain sense take over a convent of noble women ... is most discouraging. A possible solution to this problem will have to be discussed.'[2] Although the research results of the almost exclusively female staff had brought about the prestige of the institute, its gender composition apparently made the directorship of the institute unattractive to this professor. He would not become Baur's successor.

During the second half of the nineteenth century, universities all over Europe and the United States expanded. Science was set on a new experimental footing. Data collection, mostly by means of experiments, was considered essential. Research and teaching became inextricably connected. Within the universities revolutionary new organizational entities emerged in which this new research and teaching could take place. Such a new entity was a laboratory, a fairly independent unit, in which experiments under controlled circumstances were performed. Petri dishes, test tubes, Erlenmeyer flasks, and microscopes became iconic instruments. It was tacitly assumed that the members of the laboratory staff were men in white coats. Knowledge produced in a laboratory came to be considered as trustworthy and genuinely scientific. This is how the developments in academia in this period are often interpreted.[3] The Englishman William Bateson (1871–1924), a founding father of genetics, could not escape its glamour. In a lecture in 1904, he stated: 'The breeding-pen is to us what the test-tube is to the chemist—an instrument whereby we examine the nature of our organisms and determine empirically what ... I may call their genetic properties.'[4]

1 Werner Pfarre, 'A Contribution to the History of Science of Heredity in Berlin', *Englera*, 7 (1987), 147–217 (p. 172).

2 Letter from Roemer, 5 December 1929, Geheimes Staatsarchiv Preußischer Kulturbesitz, Berlin. I. HA. Rep. 87. B. Nr. 20095. Min. LDF. Akten betreffend Landwirtschaftliche Hochschule in Berlin. Professoren (Dozente) Band 7. 1929–30. All translations are by the author, unless otherwise indicated.
See also: Ida H. Stamhuis and Annette B. Vogt, 'Discipline Building in Germany: Women and Genetics at the Berlin Institute for Heredity Research', *British Journal for the History of Science*, 50 (2017), 267–95 (p. 270).

3 Robert E. Kohler, 'Lab History: Reflections', *Isis*, 99 (2008), 761–68; Graeme Gooday, 'Placing or Replacing the Laboratory in the History of Science?', *Isis*, 99 (2008), 783–95.

4 Beatrice Bateson, *William Bateson, F.R.S., Naturalist: His Essays and Addresses, Together with a Short Account of His Life* (Cambridge: Cambridge University Press, 1928), p. 243.

When the history of the laboratory is studied, it becomes clear that in the past, on the one hand, a laboratory was not such a narrowly defined phenomenon and that, on the other hand, the characteristics ascribed to the laboratory were not restricted to laboratories.[5] What made a laboratory into a laboratory was also determined by time and space. According to Greame Gooday, we should take a much more inclusive look at the kind of spaces that served as venues of experimental epistemology, whether they were called laboratories or something else.[6] That is what will be done in this chapter, which focuses on genetics. If attention is only focused on 'iconic' university laboratories, the tentative important beginning of genetics, including the important role of women, would escape attention. In this chapter women's presence will be connected with the peculiar organizational setting in which genetics grew. We will look at the kind of research that was done, the organizational setting, the composition of the staff, and the social dynamics within those research groups, especially between the male director and the mainly female staff. To put this in context, we will start with how genetics initially developed and sketch the slow entrance of women into academia.

The Emergence of Genetics

To understand how research in genetics developed, the roots of genetics must be taken into account: evolutionary theory, on the one hand, and plant and animal breeding in agriculture and horticulture, on the other hand. After the publication of the *Origin of Species* by Charles Darwin in 1859, a discussion emerged about the role of heredity in evolution.[7] In the final decades of the nineteenth century, several theories of heredity were formulated. Experimental evidence was needed to support these theories. The Dutchman Hugo de Vries tried to convince his colleagues of the applicability of his theory of *intracellular pangenesis*, and Bateson wanted to demonstrate that, contrary to Darwin's view, the values of characters do, in principle, not blend in the offspring; in the offspring they have the values of one of the parents. De Vries, Bateson, and others performed hybridization

5 About the variety of spaces where science was done, see: Aileen Fyfe and Bernard Lightman, eds., *Science in the Marketplace. Nineteenth-Century Sites and Experiences* (Chicago: University of Chicago Press, 2007).

6 Gooday, 'Placing or Replacing the Laboratory', p. 785.

7 Peter J. Bowler, *The Mendelian Revolution: The Emergence of Hereditarian Concepts in Modern Science and Society* (London: Athlone Press, 1989); Staffan Müller-Wille and Hans-Jörg Rheinberger, *A Cultural History of Heredity* (Chicago: University of Chicago Press, 2012).

experiments. Mendel's laws were 'rediscovered' in 1900.[8] Thereafter, the applicability of Mendel's laws was investigated. After the publication of De Vries's *Mutation Theory* in 1901 and 1903, genetics also came to include the study of mutations, sudden permanent changes in the hereditary material. The American Thomas Hunt Morgan (1866–1945) and his co-workers put mutations on the research agenda with their work on fruit flies.[9] An international research programme of Mendelian–Morganian genetics developed.

Other reasons for the increasing interest in genetics were the efforts in plant and animal breeding to improve the yields in agriculture and horticulture. To do this successfully, knowledge of the genetics of relevant characters of certain plants and animals was needed. Hybridization and selection, the methods that were common in plant and animal breeding, were the ways to acquire that knowledge. The first geneticists cooperated closely with plant and animal breeders. Subjects that were studied were the external characters of organisms, like flower colour or the number of eggs from chickens. Extensive plant hybridization or animal breeding experiments had to be done. These investigations required sizeable plots of land for plants and barns and coops for animals. Although the necessary facilities were not expensive, they could not be set up easily in the vicinity of the university.[10] Researchers like Hugo de Vries and Tine Tammes, both Dutch, were able to do their genetics experiments in a relatively small university botanical garden, but in general, universities could not offer the necessary facilities needed by larger research groups.

The Entrance of Women into Academia

Before the early decades of the nineteenth century the—mostly aristocratic—home had been a location for scientific research.[11] Women

8 Ida H. Stamhuis, Onno G. Meijer, and Erik J.A. Zevenhuizen, 'Hugo de Vries on Heredity, 1889–1903: Statistics, Mendelian Laws, Pangenes, Mutations', *Isis*, 90 (1999), 238–67.

9 Garland E. Allen, *Thomas Hunt Morgan: The Man and His Science* (Princeton, NJ: Princeton University Press, 1978); Robert E. Kohler, *Lords of the Fly: Drosophila Genetics and the Experimental Life* (Chicago: University of Chicago Press, 1994).

10 Universities were in general not generous towards new fields because they feared requests for new research facilities. This was especially visible in Germany. See: Jonathan Harwood, *Styles of Scientific Thought: The German Genetics Community 1900–1933* (Chicago: University of Chicago Press, 1993), pp. 145–8, 196.

11 Donald L. Opitz, '"This House is a Temple of Research": Country-House Centres for Late-Victorian Science', in David Clifford, Elisabeth Wadge, Alex Warwick, and Martin Willis, eds.,

sometimes participated in this activity. As a result of the development of the research university, however, male researchers worked in the universities, but women could not work in these institutions, which had been male bastions from the start. During the first feminist wave, a growing group of women wanted the opportunity to receive a full education and to find a respected job. These women often belonged to families in which study was an accepted occupation. In the late nineteenth century, universities began to open their doors to women, first as students and later also as staff members.

When the first academically educated women were seeking opportunities to contribute to science, it seems that relatively many entered disciplines that were in an early stage. Radioactivity, biochemistry, ecology, and genetics were such fields in the beginning of the twentieth century.[12] It usually took time and required a struggle before new areas of knowledge gained an accepted and respected place within the university structure. Therefore, research in these new fields was originally often not carried out at the traditional universities. And these fields did not immediately appeal to ambitious scientists because the road was not yet paved, and it was uncertain whether an academic career would run smoothly, as the division of power within such fields had not yet been settled. As a result, there were more opportunities for women in these fields than in established ones.

This did not mean, however, that all positions and functions in these new fields were accessible to women. It seems that their options were limited. The historian of science Timothy Lenoir has stated that to gain authority in science, one must first do what he characterizes as 'technoscience', by which he understood research that is considered properly scientific and that can be published. But this is only a first stage in a scientific career. To become an acknowledged authority in a field, 'institution-oriented activities' are also required, such as directing an institution, founding a journal, creating a scientific society, or organizing a conference.[13] It seems that in the early twentieth century, the position of women was not yet such that

Repositioning Victorian Sciences: Shifting Centres in Nineteenth-Century Thinking (Cambridge: Anthem Press, 2006) pp. 235–59.

12 For radioactivity see: Marlene F. Rayner-Canham and Geoffrey W. Rayner-Canham, eds., *A Devotion to Their Science: Pioneer Women of Radioactivity* (Quebec: McGill-Queen's University Press, 1997); Maria Rentetzi, 'Women Pioneers in Radioactivity Research. Introduction', in Soňa Štrbáňová, Ida H. Stamhuis, and Kateřina Mojsejová, eds., *Women Scholars and Institutions. Proceedings of the International Conference, Prague, June 8–11, 2003*, 2 vols. (Prague: Research Centre for the History of Science and Humanities, 2004), vol. 2, pp. 579–89.

13 Timothy Lenoir, 'The Discipline of Nature and the Nature of Disciplines', in Ellen Messer-Davidow, David R. Shumway, and David J. Sylvan, eds., *Knowledges: Historical and Critical Studies in Disciplinarity* (Charlottesville/London: University Press of Virginia, 1993), pp. 70–102 (p. 79).

they could become engaged in institution-oriented activities and then become acknowledged authorities, not even in scientific fields that were not yet well-established. By and large, the only activity that was attainable for women was technoscience.

Margaret Rossiter, however, has indicated a discipline that was an exception. In the first decades of the twentieth century in the US, institution-oriented activities became attainable for women in the new scientific field of home economics. However, this field was viewed as 'female'. 'It was the only field where a woman scientist could hope to be a full professor, department head, or even a dean in the 1920s and 1930s.'[14] There were also exceptions on the individual level. I will mention two. In 1906 Johanna Westerdijk, the first female professor in the Netherlands, became the director of a small (private) laboratory for phytopathology and was successful in making it into a sizeable, flourishing institute.[15] She established a journal and had many PhD students. We shall return later to another well-known exception, Marie Curie, who became director of the Curie institute in Paris in 1914.[16] Here the question is what was possible in roughly the same period in early genetics. We will see whether women were only engaged in technoscience or if they were also engaged in institution-oriented activities.

The Beginning of Genetics in Cambridge

The first genetics research group I will discuss is the one in Cambridge (UK) between 1890 and 1910, headed by William Bateson.[17] Bateson's father was Master of St. John's College, Cambridge. His mother served as a member of the governing council of Newnham College, one of the two women's colleges in Cambridge (the other was Girton College). Like his mother, William was a keen supporter of the movement for higher education for women.[18] In

14 Margaret Rossiter, *Women Scientists in America. Struggles and Strategies to 1940* (Baltimore/London: Johns Hopkins University Press, 1982), pp. 65–70 (p. 70).

15 Patricia Faasse, *Een beetje opstandigheid. Johanna Westerdijk. De eerste vrouwelijke hoogleraar van Nederland* (Amsterdam: Atlas Contact, 2012).

16 Susan Quinn, *Marie Curie. A Life* (London: Mandarin, 1996).

17 Marsha L. Richmond, 'Women in the Early History of Genetics: William Bateson and the Newnham College Mendelians, 1900–1910', *Isis*, 92 (2001), 55–90; Marsha L. Richmond, 'The "Domestication" of Heredity: The Familial Organization of Geneticists at Cambridge University, 1895–1910', *Journal of the History of Biology*, 39 (2006), 565–605.

18 Marsha L. Richmond, '"A Lab of One's Own": The Balfour Biological Laboratory for Women at Cambridge University, 1884–1914', *Isis*, 88 (1997), 422–55 (p. 450, n. 70). See also Bateson, *William Bateson*, pp. 59–60.

the 1880s the young Bateson joined the new programme in experimental biology at the School of Morphology of the University of Cambridge. Unhappy with the unsatisfactory results of embryologically based speculation about evolution, he began to explore the origin of natural variation. His extensive descriptive study of 1894, which argued that variation of characters is usually discontinuous, was not well received by biometricians, because Bateson did not agree with Darwin that evolution takes place gradually.[19] Experimental evidence seemed necessary to convince his opponents. Bateson hoped that by crossing varieties with their nearest relatives, he would be able to demonstrate the discontinuous nature of variation.[20]

Such a programme would best be accomplished in a university institute with adequate facilities. Not only researchers but also plots of land, barns, and pens were needed to be able to do the extensive hybridization and breeding experiments. Bateson made tireless efforts to obtain these, but the members of the School of Morphology were unwilling to facilitate this kind of work, which was so different from their own embryology-based research. The facilities needed were not available anyway. And because of Bateson's contrary ideas, they were also rather unwilling to accept Bateson fully into their circle.[21]

Bateson sought other ways. He managed to find researchers who were willing to cooperate with him. Because women did not have official access to the university but were well-educated through their membership at the women's colleges, they were available. The first female staff members were his sister Anna (1863–1928) and her friend Dorothea (Dora) Pertz (1859–1939), both members of Newnham College. Dora Pertz came from a distinguished scientific family; she was the granddaughter of Leonard Horner and a niece of Charles Lyell.[22]

From the early 1890s, the Newnham botanist Edith Rebecca (Becky) Saunders (1865–1945) developed into Bateson's long-time colleague. She came from a lower-middle-class background, and after she was admitted to Newnham College, she was awarded several scholarships. She was appointed Demonstrator in Botany in 1888 and in 1890 became Director of the Balfour Laboratory for Women. Saunders was Director of Studies at

19 William Bateson, *Materials for the Study of Variation Treated with Especial Regard to Discontinuity in the Origin of Species* (London: Macmillan and Co., 1894).

20 About the so-called biometricians versus Bateson: William Provine, *The Origins of Theoretical Population Genetics* (Chicago: University of Chicago Press, 1971), pp. 25–89.

21 Richmond, 'Women in the Early History of Genetics', p. 59.

22 Richmond, 'Women in the Early History of Genetics', pp. 59–60; Richmond, 'The "Domestication" of Heredity', p. 569.

Figure 11.1 Edith Saunders in her garden allotment. Courtesy: Newnham College, Cambridge.

Girton College from 1904 to 1919 and at Newnham College from 1918 until her retirement in 1925.[23] (Fig. 11.1) Saunders was a very capable contributor to Bateson's research programme, but she was also independent, with her own career and income. Bateson rented a small allotment from the Cambridge Botanic Garden for her to use, and by conducting extensive hybridization experiments, Saunders demonstrated the discontinuity of the heredity of certain characters for various plant species.

Bateson wanted more evidence and needed more workers. He managed to get some grants from the Evolution Committee of the Royal Society. But he had to find other ways to fully implement his research programme. In 1899 he and his wife bought Merton House in the village of Grantchester with five acres of land, about three miles outside Cambridge. They set up poultry pens and experimental gardens.[24] In 1899 Bateson was asked

23 Richmond, 'Women in the Early History of Genetics', p. 60; Richmond, 'The "Domestication" of Heredity', pp. 569–70.

24 Bateson, *William Bateson*. See also: Reginald Crundall Punnett, 'William Bateson', *Edinburgh Review*, 244 (1926), 77; David Lipset, *Gregory Bateson: The Legacy of a Scientist* (Englewood Cliffs, NJ: Prentice-Hall, 1980).

to lecture as a substitute for the ailing zoology professor Alfred Newton (1829–1907). He eagerly took the opportunity to gain access to students by offering an extracurricular course on variation, heredity, and evolution, after 1900 based on Mendelian principles. Through his lectures, Bateson attracted a number of students, particularly women. Among them were Muriel Wheldale (1880–1932), Hilda Sollas (1877–1965), Dorothea Marryat (1880–1928), Bateson's sister-in-law Florence Durham (1869–1948), Hilda Killby (1877–1962), and after some time Emma Darwin (later Barlow) (1885–1989), a granddaughter of Charles Darwin.[25] Wheldale was the daughter of a Birmingham barrister. Sollas (1877–1965) was the daughter of the geology professor at Oxford. Durham and her sister Beatrice Bateson were the daughters of a surgeon.

The facilities Bateson could offer to these female college students were modest. The grounds at the Botanical Garden were only available on an ad hoc basis. Saunders and Marryat used the Newnham College garden for some of their plant breeding work. Sollas and Killby reared guinea pigs in an attic and goats in a field behind the college grounds. All in all, work was carried out using whatever means were available, including Merton House, Bateson's family home. Family life and research activities there were mutually adapted. Bateson's wife Beatrice later recalled:

> During these years, from early spring to late autumn we never left home together, except on 'Flower Show' day. Then we rose early, 'did' the incubators, and bicycled to the station for an 8 o'clock morning train. ... We tried once or twice to finish with a theatre but the midnight ride back to Grantchester from Cambridge station – the eggs still to be turned, the lamps adjusted – taught us to be content without this extra pleasure.

Her participation in his research becomes clear from a request by Bateson to Reginald Punnett (1875–1967), who was of lower-middle-class origin. In 1904 Punnett became the first man to join the group. He had a master's degree in zoology from Cambridge and was elected a Fellow of Gonville and Caius College in 1901. Bateson asked him to join the work because his wife was pregnant with their third child: 'My wife has hitherto done a large part both of the recording and of the many menial operations that such work involves, but I am sorry to say she will be more or less incapacitated this next season, so that help in some form or other I must get.'[26]

25 Richmond, 'Women in the Early History of Genetics', pp. 65–6.
26 Bateson, *William Bateson*, pp. 85–6; Richmond, 'The "Domestication" of Heredity', pp. 584–5.

For the Batesons, scientific and family life were intermingled.[27] Rooms used for incubating chicken eggs could quickly be transformed into a bedroom for the children's governess. Indoors the children lived in what amounted to a museum, and outdoors they played among rows of peas, greenhouses, and poultry pens.[28] 'I grew up in the middle of natural history and beetle collecting,' his son Gregory would later say. Bateson taught his children as he did his students.[29] Co-workers were left to themselves during the day. 'One of the things that used to be characteristic', Gregory also recalled, 'was that somewhere around three o'clock, the chickens had to be fed, and W.B. would personally conduct [a tour], showing the peas, etc., with considerable pride.' Dinner was a particularly important occasion, during which Bateson would hold forth at the head of the table while Beatrice 'sat opposite him', serving as a foil to his dominating personality.[30] The research group worked in a domestic environment and formed a kind of extended family with Bateson as the father figure.[31]

In the first decade of the twentieth century, this group was at the forefront of the emerging discipline of genetics, as can be seen from key figures' references to their work and the role the group played at the first international genetics conferences. Bateson relied on the work of his co-researchers, and the findings of Saunders, Durham, and Wheldale drew a great deal of attention.[32] In 1906 he referred to Saunders's work as providing 'the most complex illustration yet known of the effects of interaction between factors'. Such solutions offered hope that more complicated cases would soon join 'the simpler properties or structures' in discovering their Mendelian basis.[33] Saunders tried to understand the occasional appearance of double flowers in stocks and petunias. Although the genetic basis of this phenomenon took several years to work out, it drew international attention and was of great

27 This situation fitted in to a certain extent with a nineteenth-century tradition. Van Berkel has shown that Dutch scientists in the nineteenth century thought that, for a professor, the division between university and home should not be sharp; the professor should have his home at or near the university site. Klaas van Berkel, 'Het huis van de professor. De Koninklijke Akademie van Wetenschappen en de scheiding tussen werk en privé in de negentiende eeuw', in Esther van Gelder, Eric Jorink, Ilja Nieuwland, Marlise Rijks, and Alice Spruit, eds., *'Dingen die ergens toe dienen.' Verhalen over materiële cultuur van wetenschap* (Hilversum: Verloren, 2017), pp. 166–9.

28 Lipset, *Gregory Bateson*, p. 44.

29 Lipset, *Gregory Bateson*., p. 45.

30 Lipset, *Gregory Bateson*, pp. 56–7.

31 Richmond, 'The "Domestication" of Heredity', pp. 587–8.

32 Richmond, 'Women in the Early History of Genetics', p. 77.

33 William Bateson, 'The Progress of Genetic Research', in W. Wilks, ed., *Report of the Third International Conference, London 1906* (London: 1907), pp. 90–7 (p. 94).

interest to commercial plant breeders as well as geneticists.[34] The German Baur later praised Saunders's efforts.[35] Her fundamental contributions to the new field of genetics were recognized by the Royal Horticultural Society, which in 1906 awarded her the Banksian Medal at the Third International Conference on Genetics in London.

Geneticists were also widely interested in the work on the genetics of colour inheritance produced by Durham and Wheldale. Morgan, for example, wrote to Bateson in May 1909: 'Miss Durham's paper on mice interested me very much. It is a most convincing piece of evidence.'[36] Morgan was also struck, as were Baur and the German geneticist Richard Goldschmidt (1878–1958), by Durham and Marryat's investigation of the apparent coupling of different genes responsible for colour.[37] It was Wheldale's study of the genetics of flower coloration that gained the widest recognition. She described the early results of her study on *Antirrhinum* at the third International Conference on Genetics in London in 1906.[38] This line of research greatly interested geneticists, including De Vries. Baur actually worried about her encroachment onto his own turf. He published his results on *Antirrhinum* along with a German translation of one of Wheldale's papers.[39]

The Cambridge group was prominent at the early international conferences. The first, in 1899, was initiated and organized by the then still small Cambridge group and held in London. Due to the distance, Bateson was the only group member present at the conference in New York in 1902, but he referred extensively to Saunders's work. At the 1906 meeting in London, members of the Cambridge school were dominant as organizers and delegates, and for the first time they were visible as speakers. Saunders was then the first and—for the time being—the only female speaker.[40] In his presidential address Bateson referred as usual to the work of his co-workers, and he took the opportunity to acknowledge their contributions

34 Richmond, 'Women in Early History of Genetics', pp. 76–8.

35 Erwin Baur, *Einführung in die Experimentelle Vererbungslehre* (Berlin: Verlag von den Gebrüder Borntraeger, 1911), p. 107.

36 Letter from Morgan, 31 May 1909, Bateson Correspondence, #828. Quoted in: Richmond, 'Women in Early Genetics', p. 79.

37 Morgan referred to this work in: Thomas Hunt Morgan, *Heredity and Sex* (New York: Columbia University Press, 1913), as did Baur in: Baur, *Einführung*, p. 144. See: Richmond, 'Women in the Early History of Genetics', p. 79, n. 54.

38 Marsha L. Richmond, 'Muriel Wheldale Onslow and Early Biochemical Genetics', *Journal of the History of Biology*, 40 (2007), 389–426 (pp. 398, 400–11).

39 Richmond, 'Muriel Wheldale Onslow', pp. 402–4.

40 In addition to Bateson, who was president of the conference, E.R. Saunders was also a speaker. See the list of delegates in: Wilks, ed., *Report of the Third International Conference*.

in helping to establish the validity of Mendelism, particularly noting the work of Saunders:

> Had it not been for the work that has been done by my friends and pupils--first of all by my colleague, Miss Saunders ... there would have been nothing at all to justify me in speaking of the significance of the work of inheritance; but for that vast reservoir of work they have piled together, I could never have dared, without that force behind me, to have asserted that Mendelian research has been and is of the importance that we now know it must possess.[41]

This 'force' was visually represented by the ten members of the Cambridge school present at the conference, including five women: Saunders, Durham, Marryat, Sollas, and Wheldale. Punnett was also present. Members of this group would also be present at the conference in Paris in 1911, where five of the six women delegates came from England: Saunders, Wheldale, Durham, D.M. Cayley, and Caroline Pellew.[42]

Notwithstanding all the results attained and his ultimate appointment to the professorship of biology at the University of Cambridge in 1908, Bateson by 1909 had a sense of failure because he had not been able to institutionalize genetics at the University of Cambridge. He accepted the directorship of a new private institute, the John Innes Horticultural Institute in Merton, which provided him with the desired experimental research facilities. With a team of female and male assistants, he continued his research programme there.[43] Punnett took over his professorship in biology and in 1912 became the first professor of genetics at Cambridge.

The result of the work of the group in Cambridge, amongst others, was the emergence of a discipline of genetics that was still young and fragile but that was considered so promising that several researchers were motivated and indeed successful in establishing institutes elsewhere. At that time there were others working in what was later to be called genetics, but the

41 See Bateson's address given at the banquet of the 3rd International Conference on Genetics, London, 2 August 1906, in: Wilks, ed., *Report of the Third International Conference*, p. 75.

42 Ph. De Vilmorin, ed., *IVe Conférence Internationale de Génétique, Paris, 1911. Comptes Rendus et Rapports* (Paris: Masson, 1913), pp. 7–14.

43 Donald L. Opitz, 'Cultivating Genetics in the Country: Whittinghame Lodge, Cambridge', in David N. Livingstone and Charles W.J. Withers, eds., *Geographies of Nineteenth-Century Science* (Chicago: University of Chicago Press, 2011), pp. 73–98; Marsha L. Richmond, 'Women in Early Genetics. Comparisons of Private Research Institutions: Cold Spring Harbor (US) and John Innes Institute (UK)', PowerPoint presentation (2009).

Cambridge group was considered pivotal. This result had been made possible by the loose organization of the research group: an extended family in a domestic environment, with Bateson as the father figure. Women had ample opportunities to participate. This would not have been possible in the context of a university institute, department, or laboratory because, at the time, women had no access to any of the institutional research positions at the University of Cambridge. Although at other universities in Western Europe and the United States it was sometimes possible for women to become staff members, we shall see that women geneticists more often had the opportunity to contribute to science in venues outside traditional universities.

Genetics on a Firm Footing in Berlin

Stronger organizational structures would be needed to turn the fragile new discipline into a recognized one. These organizations could be part of universities, but more often they were not. In addition, other researchers scattered here and there were working on genetics, but they usually belonged to research groups not exclusively dedicated to genetics. It was in this context that genetics developed, and women played an essential role in this process. In the US the zoologist Charles Benedict Davenport (1866–1944) established the Station for Experimental Evolution in 1904. It was an independent institution, but its funding was provided by the Carnegie Institution. Women made a considerable contribution to the research.[44] In 1904 Thomas Hunt Morgan became professor of experimental zoology at Columbia University in New York. Through work with fruit flies, Morgan's research group would become central in the development of Mendelian–Morganian genetics. Women contributed significantly to the research programme but mostly after the 1930s, and they had on average lower positions than the male staff members.[45] In England women played an important role in the previously mentioned independent John Innes Horticultural Institute, founded in 1910 with Bateson as the director.[46] At the first private university in Moscow, the Shaniavskii City People's University, a laboratory of experimental biology

44 Richmond, 'Women in Early Genetics. Comparisons'.

45 Michael R. Dietrich and Brandi H. Tambasco, 'Beyond the Boss and the Boys: Women and the Division of Labor in *Drosophila* Genetics in the United States, 1934–1970', *Journal of the History of Biology*, 40 (2007), 509–28.

46 Richmond, 'Women in Early Genetics. Comparisons'.

in which genetics was an important subject was set up in 1913 by Nikolai Konstantinovich Koltsov (1872–1940). He also established a laboratory focusing on genetics at the Bestuzhev Advanced Courses for Women.[47] Female researchers were employed in large numbers at the Russian genetics research institutes.[48]

We will take a closer look at the Institute for Genetics established at the Agricultural College in 1914 in Berlin under the directorship of Erwin Baur.[49] Baur, who was of middle-class origin, studied at several universities, gaining a medical degree in 1900 in Kiel and one in botany in 1903 in Freiburg. He received his *Habilitation* in botany at the Friedrich Wilhelm Universität in Berlin.[50] In 1904 he worked as first assistant at the Botanical Institute there. He soon moved into the niche of Mendelian genetics, which in Germany was still rather empty, and began to cross different races of *Antirrhinum*. Soon he also hybridized grains, potatoes, and cabbages, apparently chosen because of the agricultural relevance of these plants.[51]

Baur thought he would never manage to become a professor because he 'lacked the family connections'.[52] And it is indeed true that he did not become a professor at the Friedrich Wilhelm Universität. In 1911 he was, however, appointed to a chair for genetics at the Agricultural College in Berlin. In Germany at the time, agricultural colleges did not have the prestige of a university.[53] The one in Berlin had been established in 1881, and it would be incorporated in 1934 into the Friedrich-Wilhelm Universität. The Agricultural College assumed that genetics would have promising agricultural applications, and Baur was expected to study the heredity of

47 Mark B. Adams, 'Science, Ideology, and Structure: The Kol'tsov Institute, 1900–1970', in Linda Lubrano and Susan Gross Solomon, eds., *The Social Context of Soviet Science* (Boulder, CO: Westview Press, 1980), pp. 173–204; Nikolai Krementsov, *International Science Between the World Wars. The Case of Genetics* (London: Routledge, 2005), esp. p. 20.

48 Olga Valkova, 'The Conquest of Science: Women and Science in Russia, 1860–1940', *Osiris*, 23 (2008), 136–65; Roman A. Fando, 'Women at the Beginnings of Russian Human Genetics', unpublished paper (2015); Elena Aronova, 'Women in the Early History of Genetics in Russia', ISHPSSB meeting, Vienna (July 2003); Ida H. Stamhuis, Elena Aronova, Roman Fando, and Marsha L. Richmond, 'Women in Early Genetics in Russia', unpublished draft chapter for book on women in early genetics (2020).

49 Stamhuis and Vogt, 'Discipline Building in Germany'.

50 Elisabeth Schiemann, 'Erwin Baur', *Berichte der Deutschen Botanischen Gesellschaft*, 52 (1935), 51–114 (pp. 51–64); Stamhuis and Vogt, 'Discipline Building in Germany', pp. 274–5.

51 Stamhuis and Vogt, 'Discipline Building in Germany', p. 275.

52 E. Baur to C. Correns, 21 November 1910, cited in: Harwood, *Styles of Scientific Thought*, p. 292, n. 32.

53 Jonathan Harwood, *Technology's Dilemma. Agricultural Colleges Between Science and Practice in Germany, 1860–1934* (New York: Peter Lang, 2005), pp. 112–13.

agricultural plants.[54] In fact, none of Germany's universities were carrying out Mendelian genetics research at the time. The botanical and zoological institutes of the universities were in general unable and unwilling to provide facilities for the intensive hybridization and breeding experiments.[55]

These kinds of facilities were not immediately available at the Agricultural College either. The means that Baur received were not nearly in line with his ambition. He therefore sought and found additional finances in the agricultural sector. In 1912 he made Gerta von Ubisch (1882–1965) his first assistant. Ubisch, from an intellectual Jewish background, had studied physics but switched to botany.[56] She found Baur to be 'full of real enthusiasm for his field.'[57] She did a Mendelian factorial analysis on varieties of barley. In the spring of 1914, she had to leave because, according to Baur, he was unable to appoint her as an assistant paid by the Agricultural College, as she lacked the required agricultural qualifications. Instead, Baur was able to hire Elisabeth Schiemann (1881–1972) as an official assistant.[58] She was a daughter of Theodor Schiemann, professor of East European history in Berlin. In 1912 Elisabeth Schiemann prepared a doctoral thesis on plant genetics under Baur.[59] She would develop into the assistant on whom Baur could rely and who would run the institute during his frequent absences, such as when Baur, as a medical doctor, had to assist in the Imperial Navy during the First World War.

In 1915 the Agricultural College took a step towards meeting Baur's wishes. The college acquired a plot of land for Baur and his staff to use. Schiemann took care that provisional facilities—including a building, greenhouses, grain storerooms, and sheds for rabbits and mice—were made suitable.[60] She continued the work on barley that had been started by Ubisch and investigated other important cereals, as well as tomatoes and strawberries.[61] In 1915, financed by Baur's agricultural contacts, Luise von Graevenitz (1877–1921) came to the institute and began to work on

54 Schiemann, 'Erwin Baur', pp. 72–3

55 Harwood, *Styles of Scientific Thought*, p. 196

56 Gerta von Ubisch, 'Aus dem Leben einer Hochschuldozentin', *Mädchenbildung und Frauenschaffen* (1956), pp. 413–22 (p. 420), pp. 498–507, and (1957), pp. 35–45.

57 Ubisch, 'Aus dem Leben', pp. 420–1.

58 Schiemann, 'Erwin Baur', pp. 72–3, 77, 79–81.

59 Stamhuis and Vogt, 'Discipline Building in Germany', pp. 277–8.

60 Schiemann, 'Erwin Baur', p. 79.

61 See especially: Ekkehard Höxtermann, 'Die genetischen Arbeiten Schiemanns', in Reiner Nürnberg, Ekkehard Höxtermann, and Martina Voigt, eds., *Elisabeth Schiemann (1881-1972): Vom AufBruch der Genetik und der Frauen in den UmBrüchen des 20. Jahrhunderts* (Rangsdorf: Basilisken-Presse, 2014), pp. 204–35.

potatoes, and in 1917 her friend Emmy Stein (1879–1954) also joined. The two friends had earned doctorates in botany at the University of Jena.[62] When Stein began working at the institute, a Röntgen apparatus became available, and she took this opportunity to irradiate *Antirrhinum* to produce mutations. Stein compared her results with the results on radiated fruit flies from the Morgan group.[63]

In 1918 Baur was given funds to study fibre plants, and he asked Ubisch to return to the institute to carry out this new project. She did but left again in 1921, although she said that 'the cooperation with Baur ... was so dear and valuable to me.'[64] Baur travelled widely to collect seeds of interesting varieties of cultivated plants. These collections aroused Schiemann's interest in the origin and history of cultivated plants. She would develop into an expert on this subject and in 1932 would publish a well-received monograph.[65]

What positions did the female staff members around 1920 have? Stein worked for five years unsalaried. She could afford to do this because she came from a family of wealthy industrialists. Schiemann was Baur's official assistant, employed by the Agricultural College.[66] She was the only female assistant in the Agricultural College, and her salary was one of the lowest. Graevenitz and Ubisch were so-called 'fund assistants', of whom the college employed eight in 1920. They were paid from external sources, and their positions were more vulnerable and lower paid than the official assistants. Of the eight fund assistants of the college, three were women. Graevenitz and Ubisch were the only ones among them holding a doctorate.[67] All in all, Baur had highly qualified assistants at very low cost.

In 1921 not only Ubisch but also Graevenitz (due to her untimely death) no longer contributed to the institute's activities. Baur instead appointed the first male, Hans Nachtsheim (1890–1979), who came from an

62 Elisabeth Schiemann, 'Emmy Stein, 21.VI.1879 – 21.IX.1954', *Der Züchter*, 25 (1955), 65–7; Stamhuis and Vogt, 'Discipline Building in Germany', p. 278.

63 Stein, Emmy, 'Über den Einfluß von Radium Bestrahlung auf Antirrhinum', *Zeitschrift für Induktive Abstammungs- und Vererbungslehre*, 29 (1922), 1–15.

64 Ubisch, 'Aus dem Leben', p. 505.

65 Elisabeth Schiemann, *Entstehung der Kulturpflanzen. Handbuch der Vererbungswissenschaft, Band III.* (Berlin: Verlag von Gebrüder Borntraeger, 1932).

66 Geheimes Staatsarchiv Preußischer Kulturbesitz, Berlin: I. HA Rep 87. Min. Landw. Dom. Forsten. Abt. I A. LWH. Nr. 20103. Assistenten Band 1, pp. 262–3, which lists all the assistants of the Agricultural College in 1920.

67 Letter from the rector, 10 November 1920, Geheimes Staatsarchiv Preußischer Kulturbesitz, Berlin: I. HA Rep 87. Min. Landw. Dom. Forsten. Abt. I A. LWH. Nr. 20104. Assistenten Band 2, pp. 1–5.

upper-middle-class background.[68] Nachtsheim was a zoologist and had translated Morgan's book *The Physical Basis of Heredity* into German.[69] He became Baur's second official assistant. Nachtsheim began to study the genetics of rabbits. His research culminated in a monograph in 1935.[70]

We know about what happened behind the scenes from a six-page document by Nachtsheim, which he circulated widely among his colleagues in 1930 when he was no longer dependent on Baur. He ended that document with the words: 'Nothing has been fulfilled of the promises made to me more than eight years ago.'[71] What had happened? In January 1920 Baur had sent him a letter saying that he was looking for a head of an envisaged zoological department in his institute.[72] The position needed final approval by Parliament, but everything would be settled before the winter semester.[73] Nachtsheim resigned from his job in Munich. However, in early September Baur told him to stay in Munich until spring. Nachtsheim was desperate, as his job in Munich had been given to someone else.[74] When Nachtsheim finally started work at the institute in Berlin after several complications in the spring of 1921, he discovered that his appointment and the salary were that of an assistant and not of a department head and that this would not change despite promises from Baur.[75] This demonstrates how eager Baur was to intensify the institute's research and how much he was willing to do in order to achieve this.

In 1921 the zoologist Paula Hertwig (1889–1983), daughter of the Berlin professor of anatomy Oscar Hertwig, joined the institute. After her PhD in 1916, she worked as a voluntary assistant in her father's institute. In 1919 she habilitated as the first woman in Berlin with the qualification *magna cum laude*. Despite her accomplishments, her appointment at the genetics institute was that of 'supernumerary assistant', which was a temporary

68 Günther Klaus, Walter Hirsch und Dorothea Opitz, *Bibliographie und Biographie Hans Nachtsheim zum 80. Geburtstag, 13 Juni 1970* (Berlin/Munich: Freie Universität/Max-Planck-Gesellschaft, 1970); Stamhuis and Vogt, 'Discipline Building in Germany', pp. 280–1.

69 Thomas Hunt Morgan, *Die Stoffliche Grundlage der Vererbung*. Translation by Hans Nachtsheim of *The Physical Basis of Heredity* (Berlin: Borntraeger, 1921).

70 Hans Nachtsheim, *Von Wildtier zum Haustier* (Berlin: Alfred Metzner, 1935).

71 Hans Nachtsheim, *Auszug aus meinen Verhandlungen mit Professor Baur*. Geheimes Staatsarchiv Preußischer Kulturbesitz, Berlin: Ministerium Landwirtschaft and Archiv Max Planck Gesellschaft, Berlin: Nachlass Max Hartmann (Dir. K-W-I Biologie Berlin) 1929, p. 6.

72 Nachtsheim, *Auszug aus meinen Verhandlungen*, p. 1.

73 Nachtsheim, *Auszug aus meinen Verhandlungen*, p. 2.

74 Nachtsheim, *Auszug aus meinen Verhandlungen*, p. 3.

75 Nachtsheim, *Auszug aus meinen Verhandlungen*, p. 4.

and poorly paid position.[76] Hertwig investigated irradiated mice and rats, chosen because the interest in human genetics had increased, and mice and rats were model organisms for humans.[77]

In 1923 Baur succeeded in getting an impressive new building erected for his institute in Berlin-Dahlem. The institute had greenhouses and an experimental plot of 3.5 hectares, facilities for carrying out microbiological as well as chemical experiments, and pens for breeding animals like chickens and rabbits. The non-academic staff consisted of around fifteen male and female workers.[78] With respect to facilities, Baur now seemed to have all he needed for his genetic experiments.[79] The year 1927 was very important. In that year the Fifth International Congress of Genetics was held in Berlin with no fewer than 660 attendants. All the members of the institute contributed towards making the conference a success (fig. 11.2). Baur's prestige was significantly enhanced by the congress. 'He reaped great successes at that time. His name was known all over the world and his voice was heard everywhere.'[80]

Baur felt that, notwithstanding these successes, he could not sufficiently spread his wings. He was, after all, still connected to an institution that was less prestigious than a university. After 1910 institutes had been established by the Kaiser Wilhelm Society. The Society created research institutes with a particular focus on funding areas not well accommodated by the universities.[81] Baur became the head of the new Institute for Plant Breeding, which opened in 1928.[82] He promised Schiemann that she would become

76 Sybille Gerstengarbe, 'The Geneticist Paula Hertwig (1889–1983) – A Female Scientist under Various Regimes.' in Štrbáňová, Stamhuis, and Mojsejová, eds., *Women Scholars and Institutions*, pp. 295–317; Sybille Gerstengarbe, 'Paula Hertwig – Genetikerin im 20. Jahrhundert. Eine Spurensuche', in *Acta Historica Leopoldina* 58 (Stuttgart: Wissenschaftliche Verlagsgesellschaft Stuttgart, 2012); Stamhuis and Vogt, 'Discipline Building in Germany', pp. 282–3.

77 She detailed the major findings of those geneticists who had taken the lead in chicken genetics, including Bateson and Punnett: Paula Hertwig, 'Der bisherige stand der erbanalytischen Untersuchungen an Hühnern', *Zeitschrift für Induktive Abstammungs- und Vererbungslehre*, 31 (1922–3), 183–254.

78 This was the composition of the non-scientific staff in 1929: Letter to Roemer, 02 May 1929, Geheimes Staatsarchiv Preußischer Kulturbesitz, Berlin:. I HA. Rep 87 Ministerium für Landwirtschaft, Domänen und Forsten. B Nr. 20095. Akten betreffend: Landw. Hochschule in Berlin. Professoren (Dozenten) Band 7, nr. 4 1919–30.

79 Schiemann, 'Erwin Baur', p. 82; Harwood, *Styles of Scientific Thought*, p. 200.

80 Hans Stubbe, 'Gedächtnisrede auf Erwin Baur gehalten am 25. Todestag (2. Dezember 1958) in Müncheberg/Mark', *Der Züchter*, 29 (1959), 1–6 (p. 4).

81 Bernard von Brocke and Hubert Laitko, eds., *Die Kaiser-Wilhelm-/Max-Planck-Gesellschaft und ihre Institute. Studien zu ihrer Geschichte: das Harnack Prinzip* (Berlin/New York: De Gruyter, 1996).

82 Stubbe, 'Gedächtnisrede', p. 5.

Figure 11.2 Elisabeth Schiemann (1881–1972) in white coat, demonstrating *Antirrhinum* mutants to the attendants of the International Conference for Genetics in 1927 in Berlin. Courtesy: Archives of the Max Planck Society, Berlin.

the head of a department for the history of cultivated plants of that new institute. However, ultimately none of the staff members of the Institute for Genetics were appointed there.[83]

A long procedure to find a new director followed.[84] Schiemann was not taken into consideration. The reason cannot have been that she lacked recognized leadership qualities. After all, in 1928, in a letter to the Rector in which Baur announced that he intended to travel to Russia, he said that 'in the daily work in my institute as well as in the Colloquium the private teacher Miss Dr. Schiemann can very well replace me.'[85] Ultimately the relatively unknown Hans Kappert (1890–1976) became the new director. Soon thereafter Elisabeth Schieman left the institute, which would develop into just one among several institutes, in Germany and abroad.

83 Stamhuis and Vogt, 'Discipline Building in Germany', pp. 287–8.

84 Letter from Roemer, 05 December 1929, Geheimes Staatsarchiv Preußischer Kulturbesitz, Berlin: I. HA. Rep. 87. B. Nr. 20095. Min. LDF. Akten betreffend Landw. Hochschule in Berlin. Professoren (Dozente) Band 7. 1929–30. See Stamhuis and Vogt, 'Discipline Building in Germany', pp 288–9.

85 Letter of Baur to the Rektor 12 December 1928, Geheimes Staatsarchiv Preußischer Kulturbesitz, Berlin: I HA. Rep 87 Ministerium für Landwirtschaft, Domänen und Fürsten. B Nr. 20283. Band 3. Akten betreffend: Landw. Hochschule in Berlin. Institut für Vererbungsforschung, 1927 bis 1934, 183; Stamhuis and Vogt, 'Discipline Building in Germany', p. 289.

Women Researchers: Extended Family Members in Cambridge and Chess Pieces in Berlin

What were the positions and opportunities of the aforementioned women in Cambridge and Berlin? What was, in particular, the relationship with their leader and what kind of leader was he? To start with the last point: when we compare Baur and Bateson we see that both were dominant personalities, both were ambitious, and both were creatively successful in forming a dedicated group of researchers around them using minimal means.[86] Both eagerly accepted women into their group, who were in turn eager to join. In both cases these women were talented, well-educated, often from upper-middle-class or intellectual backgrounds, and willing and able to work for low renumeration.

When we look more closely at the personalities of both leaders and their attitudes towards their staff members, the differences are striking. Bateson in Cambridge supported the emancipation of women in academia. He praised his female co-workers openly and stated that through their work, he had been able to argue convincingly that Mendelian genetics was important. He called them 'his friends and pupils'. Several of the members of his group wrote in positive tones about their experiences with him. He shared insights he gained from his visits to other geneticists abroad and sometimes gave advice about their publications. His relationship with Edith Saunders was on an equal basis. They travelled in each other's company to conferences. According to Cyril Dean Darlington (1903–1981), a cytologist at the John Innes Institute from 1923, Bateson liked to surround himself with slightly subservient assistants once he was in his fifties. Bateson's leadership style was, in short, informal but attentive: he behaved like the dominant but friendly father of an extended family.[87]

We have no direct information as to what extent and in what way Baur guided his staff members in their research in Berlin, but there are some clues. They all took up a line of work that fitted in with the overall research programme of the institute. Schiemann later characterized the research

86 Schiemann wrote that Baur had the character of a leader, a 'Führersnatur'. Schiemann, 'Erwin Baur', p. 100.

87 Darlington remarked that Bateson also created an 'extended family' atmosphere at the John Innes Institute. C.D. Darlington interview by Brian Harrison, 25 July 1979. Cited in Rosemary Harvey, 'Bateson's Ladies', unpublished manuscript, Colney, Norwich, UK, John Innes Centre Archives, p. 31. Thanks to Marsha Richmond for this information.

that was done as 'joint work'.[88] We also know, however, that Baur was usually not present at the institute, which suggests that his staff worked very independently. An important source on Baur's personality and the way he behaved towards people is the obituary that Schiemann wrote after Baur's sudden death in 1933. She wrote that Baur 'quickly gained people's sympathies, not only in public life, but also in more intimate personal contact.' And that 'he often talked to the first person he came across as if he were a close friend'.[89] It is clear that this text must be read against the background of Schiemann's own experiences. When she met him, she was impressed; it is said that she was 'entirely under his spell'.[90] This charismatic aspect of his personality would also explain Ubisch's positive attitude towards Baur. It is plausible that Nachtsheim was initially also charmed by Baur's charisma. However, Baur also had a different side. Schiemann stated that he was so deeply immersed in his own pursuits that he wasn't aware of the effect they might have on others. She concluded that Baur's behaviour could only result in 'conflicts, in which human being opposes human being' and in 'unnecessary and unhappy fights'.[91] Schiemann came to the conclusion that Baur was like a chess player who treated his staff members like 'chess pieces', saving them as long as they could improve his position but not hesitating to sacrifice them if that would further his game. The relationship between Baur and his staff was clearly different from that of Bateson and his assistants. But in both cases the result was a dedicated, successful, and productive research group.

Conclusions

In genetics revolutionary developments already occurred before the discipline found a place in genetics laboratories that were part of traditional universities. It was not a university laboratory where genetics as a discipline came into being. During this first stage, the organization of the group was informal: research was carried out in a domestic setting, and the group of researchers formed an extended family with the head as a father figure. Although this was a venue of experimental epistemology, it was neither an

88 Schiemann, 'Erwin Baur', p. 78. This is discussed in: Stamhuis and Vogt, 'Discipline Building in Germany', p. 288.
89 Schiemann, 'Erwin Baur', p. 101.
90 Hermann Kuckuck, ,'Elisabeth Schiemann 1881 bis 1972', *Berichte der Deutsche Botanische Gesellschaft*, 93 (1980), 517–37 (p. 520).
91 Schiemann, 'Erwin Baur', pp. 77 8.

institution nor a laboratory. In the second stage, the institutional contexts of the several groups dedicated to genetics, of which one was discussed, were more structured. The organizational setting was an institution, which could also be called a laboratory. But it usually did not belong to a university; it was independent or part of a newer organization. Due to typically less strict rules and the lower status of these organizations, the positions were less attractive to men, and women were therefore welcomed. Women were accepted because of their competence and their low costs. The result was that genetics went through the first two stages, to a large extent successfully, thanks to the contributions of women. Both groups discussed here were successful, although the male heads had very different attitudes towards their staff. Women were available because the first generations of female scientists had university educations during the period of the emergence and establishment of genetics, and these women wanted to use their capabilities, even under modest conditions. The women could often do without a full salary because they were from wealthy or intellectual backgrounds and had no families to maintain.

What positions were attainable for women? Did they only produce techno-science in the sense of Lenoir, or could they also develop institution-oriented activities and gain authority? Although in Cambridge Bateson gave Saunders and other co-workers public recognition on several occasions, and Saunders came to be considered as an authority in botanical genetics, the situation for women scientists in Cambridge did prevent Saunders's prestige from materializing into a career at the University of Cambridge. When Bateson left in 1910 it was his male assistant Punnett who succeeded him. And in Berlin Schiemann attained a certain scientific independence through starting her own research programme. She developed institution-oriented activities by undertaking the daily management of the institute during Baur's frequent absences. But she nonetheless remained under Baur's wings and was not considered as a possible successor. Saunders and Schiemann gained recognition, but they could not develop into authorities. They remained in the shadow of the male head who received the recognition.[92] This was not because of their lack of capabilities but because this was the situation for women in that period, and genetics apparently did not allow for an exception to this rule.

92 The phenomenon of 'over-recognition of those at the top of the profession' was first recognized by Robert Merton and called the Matthew effect in science. Margaret Rossiter has pointed out that this phenomenon is often sex-linked and called it the Matilda effect in science: Margaret Rossiter, 'The Matilda Effect in Science', *Social Studies of Science*, 23 (1993), 325–41.

Can this be generalized? As we have suggested already, it seems plausible that similar processes occurred in other novel fields. If they emerged in the same period as genetics and found opportunities outside the university, it is possible that they also benefitted from the availability of competent new scientists, women, who were at this time only rarely appointed at the traditional universities. The field of radioactivity emerged in the same period and originally outside the traditional university, and women—think of Marie Curie, but there were many more—played an enormously important role.[93] Curie was successful in developing institution-oriented activities and won two Nobel Prizes. She became an acknowledged authority. Historical investigations can reveal to what extent the processes were similar and different in the fields of radioactivity and genetics and will reveal the processes of exclusion and inclusion of women in other fields of knowledge that emerged at the same time.

After 1930 genetics became more and more based in laboratories at traditional universities. However, the role of women then became less prominent for the time being.[94] When we look more broadly at university laboratories, it seems likely that the threshold for women academics to become staff members was lower in university laboratories than in other parts of the university organization. In the university system the laboratory was new, and, as argued before, a new organizational setting is usually less rigid and more open to new participants. In addition, the everyday life that became common in the laboratory was perhaps more congenial to women, although there were traps as well.[95] More historical investigations must explore the tenability of these conjectures more deeply.

93 Rayner-Canham and Rayner-Canham, eds., *A Devotion to Their Science*; Rentetzi, 'Women Pioneers in Radioactivity Research'.

94 They usually had no women in staff positions. See: Helga Satzinger, 'Women's Places in the New Laboratories of Genetic Research in Early 20th Century: Gender, Work, and the Dynamics of Science', in Štrbáňová, Stamhuis, and Mojsejová, eds., *Women Scholars and Institutions*, pp. 265–94; Dietrich and Tambasco, 'Beyond the Boss and the Boys'.

95 At least these were the provisional results of the 2003 conference 'Women Scholars and Institutions', organized by the Commission on Women of DHST/IUHPS, the international organization of history of science. In the introduction to the conference volume, I wrote a section dedicated to laboratories: Ida H. Stamhuis, 'Historical Considerations', in Štrbáňová, Stamhuis, and Mojsejová, eds., *Women Scholars and Institutions*, pp. 36–40. That the story is multifaceted is clear from Margaret Rossiter's statement: 'The rise of the laboratory, generally acclaimed in the history of the physical sciences, can be seen as a new level of exclusion, creating new male retreats or preserves to which women gained entry only by special permission.' Margaret Rossiter, 'A Twisted Tale: Women in the Physical Sciences in the Nineteenth and Twentieth Centuries', in Mary Jo Nye, ed., *The Physical and Mathematical Sciences. The Cambridge History of Science, vol. 5*, 8 vols. (Cambridge: Cambridge University Press, 2003), pp. 54–71 (p. 55).

Acknowledgements

I thank Marsha Richmond (Wayne State University, Detroit, MI), with whom I developed my ideas on women in early genetics. For the part of this chapter on Cambridge, I rely on her publications. I also thank her for comments on a previous version of this chapter. I also took advantage of the comments from the editors. Thanks to Bas and Tina Jongeling-Utley for their improvement of my English.

About the Author

Ida Stamhuis is a retired associate professor at the Vrije Universiteit (VU) Amsterdam and honorary professor at the University of Aarhus. Her expertise is history of statistics, of genetics, and of women in science. See: Ida H. Stamhuis and Annette B. Vogt, 'Discipline Building in Germany: Women and Genetics at the Berlin Institute for Heredity Research', *British Journal for the History of Science*, 50 (2017), 267–95.

12 Serialized Laboratories: Laboratory Journals and the Making of Modern Science and Scientific Publishing, 1840s–1950s

Dorien Daling

Abstract

From the 1860s to the mid-twentieth century, many academic and private research laboratories issued their own serial publications. These 'house journals' seem to have proliferated especially in the European periphery. On the basis of a few (mainly Dutch) examples, this chapter explores the emergence and contents of house journals, their purposes and functions, and their place in the evolution of the scientific periodical press. It aims to shed light on the historical importance and meaning of the genre of laboratory journals, particularly in relation to transformations in the nature of the university, scientific authority and scientific authorship, knowledge communities, and the media landscape of science.

Keywords: laboratory journals, scientific publishing, serialized print, scientific communities, scientific modernity

Introduction

In 2018 the 200th birth anniversary of the illustrious Dutch physiologist and ophthalmologist Frans Donders (1818–1889) was celebrated at several institutes in the Netherlands.[1] In a commemorative essay published for the occasion, Donders is presented as 'the epitome of the values of modern

1 Mostly at Utrecht University, where Donders had been a professor for 40 years, and at the Donders Institute for Brain, Cognition and Behaviour at Radboud University, Nijmegen.

Berkel, Klaas van, and Ernst Homburg (eds), *The Laboratory Revolution and the Creation of the Modern University, 1830-1940*. Taylor & Francis Group, 2023
DOI: 10.5117/9789463720434_CH12

physiology'.[2] Highlighted are his achievements as laboratory scientist, laboratory leader, author of hundreds of research reports, and co-founder of various journals. In 1891, writing at the height of science's transition to modernity, Barend Stokvis made a similar assessment. In his obituary of Donders, issued by the Royal Academy of Arts and Sciences, he went to great lengths to praise the publishing strategies of his former tutor. Although the physiologist had read close to a hundred communications at meetings of the Royal Academy, only six of his papers appeared in the Academy's proceedings. His preference for non-society journals (including foreign ones) had, according to Stokvis, not only enhanced his own prestige but also the scientific reputation of his laboratory and his country. One such journal was the *Onderzoekingen, gedaan in het Physiologisch Laboratorium der Utrechtsche Hoogeschool*,[3] launched in 1848 by Donders himself.[4] Stokvis emphasized the value of this periodical, while at the same time wondering if series of collected papers of laboratories could be considered 'real journals'.[5]

By the end of the nineteenth century, institutional laboratories had become essential components of all the major scientific disciplines. Many academic and private research laboratories issued their own serial publications, which were often nearly as old as the institutes themselves. This chapter probes into their format and role. It does not set out to answer the question posed by Stokvis. In hindsight this is an interesting problem, however, as laboratory journals have been rendered analytically invisible since the early twentieth century. Ironically, this invisibility has been the result of both the construction of a heroic narrative around the scientific journal and the subsequent attempts to deconstruct that narrative. The myth implied that the scientific journal had remained largely unchanged throughout its existence and that its invention in the 1660s had set the stage for the birth of modern experimental science. Beginning in the late 1970s,

2 Douwe Draaisma, *The Age of Precision: F.C. Donders and the Measurement of Mind* (Nijmegen: Donders Institute for Brain, Cognition and Behaviour, 2018), p. 27. This essay was originally published in 2002 on the occasion of the official opening of the Donders Institute (then F.C. Donders Centre for Cognitive Neuroimaging). In 2018 the institute issued an updated version.

3 The Appendix at the end of this article contains the titles of all journals that have been consulted or referred to. There, the Dutch journal titles are followed by an English translation (some official, some by the author).

4 In Draaisma's commemorative essay of 2018, this journal only appears in a reference. It was here that Donders presented the results of his pioneering research on the speed of thought: 'Over de snelheid van psychische processen', *Onderzoekingen*, ser. 2, 2 (1868–9), pp. 92–120.

5 B.J. Stokvis, 'Levensbericht F.C. Donders', in *Jaarboek Koninklijke Akademie van Wetenschappen 1891* (Amsterdam: Koninklijke Akademie van Wetenschappen, 1891), pp. 17–8, 33–4. Here and throughout the article, all quotations from Dutch sources have been translated by the author.

this interpretation became increasingly contested, which led scholars to shift their attention from the serialized print genres[6] and publishing practices of elite science to the making of natural knowledge, the legitimacy of science, and modes of communication (print and oral) that had been marginalized and were associated with so-called popular science.[7] Laboratory journals have escaped analysis because they were linked with neither public-oriented genres nor elite ones (mostly journals of learned societies and academies).

Admittedly, laboratory journals, or 'house journals', have by and large ceased to exist (in academia). In the second half of the twentieth century, international disciplinary communities, which were trans-institutional in nature, came to prevail over institutional orientations. This article argues, however, that the disappearance of laboratory journals is not an indication of marginalization. Rather, they made themselves superfluous when their values and practices became the norm instead of the niche. From the 1840s and especially the 1860s onwards, house journals helped to produce the formats and uses of learned journals that were beginning to correspond to present-day conceptions of a 'scientific journal'. This perspective is in line with recent historiography, which has stressed that the scientific paper—as a somewhat cohesive genre—did not achieve privileged status until the later nineteenth century. Throughout the century, the media landscape remained highly diverse, also within the boundaries of expert science.[8]

6 On seriality, see: Nick Hopwood, Simon Schaffer, and Jim Secord, eds., *Seriality and Scientific Objects in the Nineteenth Century*, special issue of *History of Science*, 48(3/4) (2010).

7 On ideas that created the myth of the scientific journal and opposing studies that focused on other print genres, see: Alex Csiszar, *The Scientific Journal: Authorship and the Politics of Knowledge in the Nineteenth Century* (Chicago: Chicago University Press, 2018), pp. 9–15. Some scholars have also pointed to 'a Foucauldian fear and loathing of all dominant institutions'. See: Robert E. Kohler, 'Lab History: Reflections', *Isis*, 99 (2008), pp. 761–8 (p. 764). In any case, interest in publications associated with elite science has only picked up in the last ten or fifteen years (see note 8). An important exception is: David A. Kronick, *A History of Scientific & Technical Periodicals. The Origin and Development of the Scientific and Technical Press, 1665–1790* (Metuchen, NJ: Scarecrow Press, 1976; originally publ. 1962), although the perspective of this book is primarily taxonomic.

8 Influential studies include: Jim Secord, 'Publishing Science, Technology and Mathematics', in David McKitterick, ed., *The History of the Book in Britain, vol. 6. 1830–1914* (Cambridge: Cambridge University Press, 2009), pp. 443–74; Alex Csiszar, 'Seriality and the Search for Order: Scientific Print and its Problems During the Late Nineteenth Century', *History of Science*, 48 (2010), pp. 399–434; Melinda Baldwin, *Making* Nature: *The History of a Scientific Journal* (Chicago: University of Chicago Press, 2015); Aileen Fyfe, Julie McDougall-Waters, and Noah Moxham, eds., *350 Years of Scientific Periodicals*, special issue of *Notes and Records: The Royal Society Journal of the History of Science*, 69(3) (2015); Sally Shuttleworth and Berris Charnley, eds., *Science Periodicals in the Nineteenth and Twenty-First Centuries*, special issue of *Notes and Records: The Royal Society Journal of the History of Science*, 70(4) (2016).

Alex Csiszar in particular has shown that expert and elite practices of legitimizing knowledge claims were gradually transformed by the adoption of once peripheral values, genres, and formats.[9]

This article makes a case for studying laboratory journals to further our understanding of the 'laboratory revolution' and its effects on the nature of science between the 1840s and 1950s, specifically in relation to modernization and community-building processes. What follows this introduction is a modest contribution to what such a study of house journals might entail. It is not based on a comprehensive or systematic evaluation of the various types of serial publications issued by laboratories. Problems of historicity, cultural context, and disciplinary variation, for instance, are addressed only briefly. Instead, on the basis of a few (mainly Dutch) examples, this article explores the emergence and contents of house journals, their purposes and functions, and their place in the evolution of the scientific periodical press.

Although the sample of analysed house journals is relatively small, it does shed light on the historical importance and meaning of this serialized print genre. In laboratory journals two defining elements of the 'modernist package'[10], the research laboratory and the research journal, were co-produced. This process affected epistemic virtues and values; the scientific discipline, career, and publication system; and public meanings of knowledge. By committing to new conditions for making and legitimizing knowledge and for training science students—in laboratory practices and research as well as scientific writing and publishing—the genre was implicated in the gradual transformation of scientific expertise, authority, and authorship, and also in the gradual transition from the traditional teaching university to the modern research university. This study furthermore reveals the crucial role of house journals in the performance of laboratories as social institutions. They bolstered cohesion and group identity within the laboratory, underscored the developing reciprocal relationships between the laboratory and the nation-state (also between science and society in general), and were instrumental in the construction of specialized transnational knowledge communities. Lastly, this article proposes to perceive laboratory journals as a 'parallel genre'. These periodicals were intended to complement the existing system of serialized print, but they also paved the way for the dominance of a new type of research journal from the 1950s onwards.

9 Csiszar, *The Scientific Journal.*
10 Kohler, 'Lab History', p. 765.

'Gems of Our Literature': The Emergence and Contents of Laboratory Journals

Probably the earliest example of a Dutch house journal is found in chemistry. In the early 1840s, Gerrit Jan Mulder (1802–1880) started the *Scheikundige onderzoekingen, gedaan in het Laboratorium der Utrechtsche Hoogeschool* (1842–51). Mulder was the founding father of the discipline of chemistry and of laboratory-oriented teaching in the Netherlands. Around the middle of the century, his institute was the only one in the country that conducted chemical research at an international level. It was the first laboratory that could be considered 'modern'.[11]

The first volume of the *Scheikundige onderzoekingen* comprised almost 600 pages—of which just one page included illustrations—and 23 research reports, all in Dutch. Many of the reports were concerned with the composition of different substances or the protein theory that Mulder had developed in the second half of the 1830s. Thirteen of them were written by the director himself, signed at the end with the initial of his last name. The remaining ten reports were produced by assistants and advanced students, seven in total, whose names appeared on the first page of their papers. In a short preface to the volume, Mulder asked the reader to judge 'the work of young lads' benevolently. The purpose of their presence in his laboratory was 'to be trained in practical chemistry', but 'every now and then something is discovered that is worthwhile to communicate to the public.'[12] Additionally, the first volume contained a notice presenting Mulder's stance on the medical regulations of the state regarding the education of pharmacists, and the speeches that he gave at the opening of the chemical lessons in the years 1841 and 1842.

Presumably, Mulder had found at least some of his inspiration for launching a periodical in the German journal *Annalen der Chemie und Pharmacie* (1832),[13] not a house journal but closely affiliated with the famous laboratory of its founder and editor Justus Liebig (1803–1873). The two men exchanged

11 H.A.M. Snelders, *De geschiedenis van de scheikunde in Nederland, vol. 1. Van alchemie tot chemie en chemische industrie* (Delft: Delft Universitaire Pers, 1993), p. 84. On Mulder, see: Klaas van Berkel, Albert Van Helden, and Lodewijk Palm, eds., *A History of Science in the Netherlands: Survey, Themes, and Reference* (Leiden: Brill, 1999), pp. 535–7.

12 G.J. Mulder, 'Voorwoord', in *Scheikundige onderzoekingen, gedaan in het Laboratorium der Utrechtsche Hoogeschool*, 1 (1842), pp. iii–iv.

13 When the journal was founded in 1832, it was named *Annalen der Pharmacie*. In 1840 the name was changed to *Annalen der Chemie und Pharmacie*. Later the journal came to be known as *Liebigs Annalen*, which was absorbed by the *European Journal of Organic Chemistry* in 1998.

letters, and in 1840 Liebig had recommended Mulder for appointment as professor of chemistry at the University of Utrecht (six years before their well-known and fierce dispute over the concept of protein).[14] Liebig encouraged his students to publish in the *Annalen* and to do so under their own names. In the mid-1830s he allegedly said that 'nothing enthuses young people more than to see their names in print.' But he also added a less generous reason: 'If it is something good, a part of the credit is ascribed to me and I do not need to defend the mistakes.'[15]

Rationales behind launching house journals might therefore have ranged from more obvious ones, such as providing students a platform to showcase their work and advertising the laboratory's productivity, to implicit motivations: to push a conceptual agenda, affirm hierarchy (as testified, perhaps, by Mulder's practice to sign his papers with 'M'), construct an institutional identity, convey the legitimacy of the institute (to both scientific and political authorities), produce responsible professionals and citizens, negotiate relations with 'the public', and introduce writing as a technique to learn science, learn about science, and learn to do science (not only communicate it).[16] Financial reasons may have been involved too. A digitized copy of the first volume of the *Scheikundige onderzoekingen* reveals that in 1862 Harvard University 'bought with the gift of William Gray, of Boston, Mass.' the complete set of six volumes for 19.88 dollars (roughly equivalent to 590 dollars in 2022).[17]

The first full generation of laboratory journals in the Netherlands evolved in physiology, starting in 1848 with the above-mentioned *Onderzoekingen, gedaan in het Physiologisch Laboratorium der Utrechtsche Hoogeschool*. Donders most likely modelled the first series of his house journal after the *Scheikundige onderzoekingen*.[18] In terms of research and reputation, he

14 See: H.A.M. Snelders, ed., *The Letters from Gerrit Jan Mulder to Justus Liebig (1838-1846)* (Amsterdam: Rodopi, 1986); E. Glas, 'The Liebig-Mulder Controversy: On the Methodology of Physiological Chemistry', *Janus*, 63 (1976), pp. 27–46.

15 As cited in: Joseph S. Fruton, 'The Liebig Research Group: A Reappraisal', *Proceedings of the American Philosophical Society*, 132(1) (1988), pp. 1–66 (p. 19).

16 On the last aspect, see: Antonio García Belmar and José Ramon Bertomeu-Sánchez, 'Learning by Writing: Chemistry Students Notebooks and Lecture Demonstrations in Early 19th Century France', *Archives Internationales d'Histoire des Sciences*, 65 (2015), pp. 599–615.

17 Available via 'Mulder, G.J.', Chemie Historische Groep (CHG), KNCV, <https://chg.kncv.nl/geschiedenis/biografieen/m/mulder,-g.j.> (accessed 12 November 2022).

18 On Donders, see: Draaisma, *The Age of Precision*; Bert Theunissen, 'Turning Refracting into a Science: F.C. Donders' "Scientific Reform" of Lens Prescription', *Studies in History and Philosophy of Science, Part C: Studies in History and Philosophy of Biological and Biomedical Sciences*, 31 (2000), pp. 557–8.

would ultimately even surpass Mulder's accomplishments, but between 1842 and 1847 he had served as an assistant to the chemist (together they established physiological chemistry). Donders thus witnessed the birth of the *Scheikundige onderzoekingen*, in which he also published. Shortly thereafter, prior to getting appointed as a professor and head of a laboratory,[19] he began founding and editing journals himself: the second series of the *Nederlandsch Lancet* (1845) and the *Holländische Beiträge zu den anatomischen und physiologischen Wissenschaften* (1846).[20] The latter periodical was initiated by Donders and his friends Izaac van Deen and Jacob Moleschott in the context of their joint mission to advance science-based medicine and the emancipation of physiology as an autonomous discipline.[21]

The first series of the periodical of Donders's laboratory presented findings of research conducted in the institute between 1848 and 1856. During these years the laboratory was small and modest. It could accommodate ten persons and the only available instruments were microscopes and tools for dissection and vivisection of animals.[22] The publication series was essentially a collection of reprints taken from the *Nederlandsch Lancet*. This 'journal for the medical sciences' was, however, discontinued after 1856. It could not sustain itself after the merger of five other medical journals and the subsequent establishment of the *Nederlandsch Tijdschrift voor Geneeskunde* (1857). With a distribution of some 50 copies free of charge, the *Onderzoekingen* had been financially dependent on the *Nederlandsch Lancet*. Hence, it ceased to exist as well. In the ensuing period, moreover, Donders founded the Netherlands Hospital for Eye Patients (1858) and devoted most of his energy to ophthalmology.[23]

In 1866, though, a newly built physiological laboratory was opened, which led Donders to revive the *Onderzoekingen*. This laboratory was equipped with modern apparatus for measuring physiological processes, both physical

19 His talent for scientific work earned him an extraordinary professorship at the University of Utrecht in 1848. When in 1862 the chair of physiology fell vacant, Donders accepted it.

20 From 1855 Donders was also co-editor of the German journal *Archiv für Ophthalmologie*, which had been founded a year earlier by Albrecht von Graefe. It still exists today under the title *Graefe's Archive for Clinical and Experimental Ophthalmology*.

21 Stefan van der Poel, *Tussen zieken, boeken en kikkers: De fysiologie van een leven: Izaac van Deen (1804-1869)* (Groningen: Barkhuis, 2012), pp. 23–5.

22 Harm Beukers, 'Een nieuwe werkplaats in de geneeskunde: De opkomst van laboratoria in de geneeskundige faculteiten', *Tijdschrift voor de Geschiedenis der Geneeskunde, Natuurwetenschappen, Wiskunde en Techniek*, 9 (1986), pp. 266–77 (pp. 267–8).

23 F.C. Donders, 'Voorwoord', *Onderzoekingen, gedaan in het Physiologisch Laboratorium der Utrechtsche Hoogeschool*, ser. 2, 1 (1867–8), pp. i–ii.

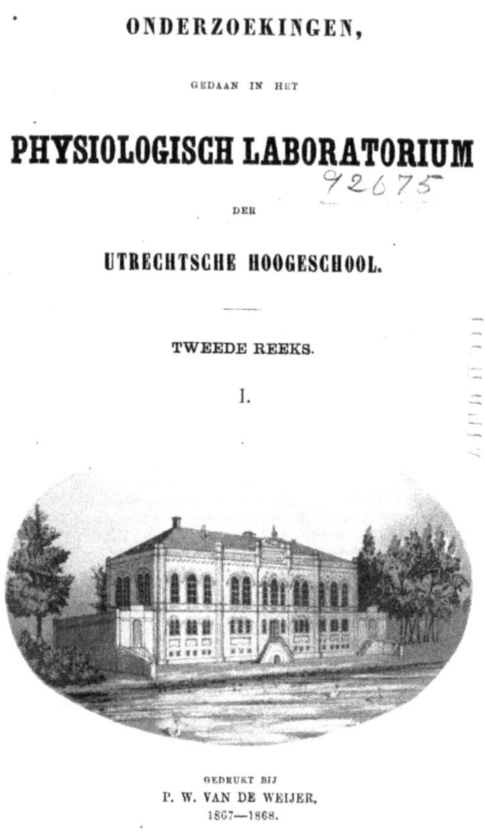

ONDERZOEKINGEN,

GEDAAN IN HET

PHYSIOLOGISCH LABORATORIUM

92675

DER

UTRECHTSCHE HOOGESCHOOL.

TWEEDE REEKS.

I.

GEDRUKT BIJ
P. W. VAN DE WEIJER,
1867—1868.

Figure 12.1 Title page of the first volume (1867–8) of the second series of the *Onderzoekingen*. Donders thanked all those who had contributed to the establishment of a laboratory that 'is certainly not surpassed in efficiency and elegance by any of the existing ones' (p. iv). The first volume (1872) of the third series contains a detailed description and ground plan of the building and its interior spaces. Although still very pleased with the laboratory, Donders lamented the lack of a residence for the director, which abroad was 'considered the conditio sine qua non' (p. xi).

and chemical.[24] These changes were reflected in the journal. A picture of the institute's building was proudly printed on the title page, and the second (and later) series contained more material of a technical nature, such as detailed reports of experimental data or methodological issues—including a lot of tables, figures, and illustrations. The volumes usually comprised around 400 pages, presenting research results acquired by Donders, his assistants, and his students. The second series (1867–70) published both original contributions and reprints derived from yet another periodical established by Donders, the *Nederlandsch Archief voor Genees- en Natuurkunde* (1865–70). The print circulation was larger than that of the first series, and some of the

24 Beukers, 'Een nieuwe werkplaats in de geneeskunde', pp. 269–70; Draaisma, *The Age of Precision*.

copies were sold instead of gifted.[25] Beginning with the third series (1872–89), the *Onderzoekingen* became 'generally available' and entirely independent, notwithstanding occasional reprints taken from foreign periodicals.[26] The journal was continued long after Donders's death, until 1965.

The practice of the Utrecht laboratory was copied by the other modern physiological laboratories that were established in the second half of the 1860s. Adriaan Heynsius in Leiden and Dirk Huizinga in Groningen started publishing the collected papers of their institutes—also called 'Onderzoekingen'—in 1869 and 1874, respectively. (Huizinga was the successor of Van Deen, the first director, who had died in 1869.) Heynsius had actually issued a house journal already between 1856 and 1861 as the leader of the new (but crude) Pathological-Physiological Laboratory in Amsterdam, a position he had obtained two years after finishing his (second) doctorate under Donders.[27] He had even published a volume in German with extracts from the original series, under the title *Studien des Physiologischen Instituts zu Amsterdam* (1861).

Some physiologists in Germany and Switzerland also began to publish the 'Untersuchungen' of their laboratories, for instance Eduard Pflüger in Bonn (1865), Albert von Bezold in Würzburg (1867), and Adolf Fick in Zürich (1869). The case of Würzburg is a fine example of how ideas and practices can travel across borders and wide networks: Bezold was married to the sister of Theodor Engelmann, who was assistant to Donders and co-editor of the Utrecht *Onderzoekingen*. (In 1869 he even married Donders's daughter.) The Würzburg series was furthermore published by his father—Bezold's father-in-law—Wilhelm Engelmann, who owned a prominent scientific publishing company in Leipzig.[28]

Apart from the types of contents described previously, house journals of physiological laboratories also included doctoral dissertations and short

25 Donders, 'Voorwoord', pp. iii–iv.

26 F.C. Donders, 'Bericht', *Onderzoekingen*, ser. 3, 1 (1872), pp. 1–3. In this notice Donders also explained why the *Nederlandsch Archief* had been discontinued after 1870. The journal had issued original papers ('researches in extenso') as well as surveys of other primary literature in the fields of medical and natural science that had been published in the Netherlands. However, shortly after its establishment, the Holland Society of Sciences and Humanities in Haarlem had founded the *Archives Néerlandaises des Sciences Exactes et Naturelles* (1866–1933), which was able to cover a broader range of subjects and was more successful in reaching foreign publics.

27 Beukers, 'Een nieuwe werkplaats in de geneeskunde', pp. 272–4.

28 Martin D. Kamen, 'On Creativity of Eye and Ear: A Commentary on the Career of T.W. Engelmann', *Proceedings of the American Philosophical Society*, 130(2) (1986), pp. 232–46 (pp. 234–5); Heinz Sarkowski, *Springer-Verlag: History of a Scientific Publishing House, part 1, 1842–1945: Foundation, Maturation, Adversity*, transl. by Gerald Graham (Berlin: Springer-Verlag, 1996), pp. 231–3.

accounts thereof, as well as, to a very limited extent, short announcements of findings to be published later (perhaps in order to claim priority), news about the institute or the journal, institutional memories, and obituaries of laboratory heads. To name but two examples, Heynsius opened the series of the Leiden institute with a description of the genesis, design, and facilities of the laboratory, and the fourth series of the Utrecht journal (1891–7) began with an in memoriam for Donders. Variations in the contents did occur, of course. The first volume of the Groningen *Onderzoekingen* consisted entirely of reprints sourced from the German journal *Archiv für die gesamte Physiologie des Menschen und der Tiere*, founded in 1868 by Pflüger. Consequently, the actual papers were in German, whereas the title of the series and the table of contents were in Dutch.

From the 1880s onwards, more laboratories for physiology—sometimes combined with related fields such as pathology and bacteriology—took the initiative to issue their own serial publications (not all of them were long-lived). Examples include the laboratories of the Military Hospital in Weltevreden, Batavia (now Jakarta) and the National Veterinary School in Utrecht, which were led respectively by Christiaan Eijkman and Hartog Jacob Hamburger. Additionally, increasing specialization and the resulting 'twigging' of disciplines and institutes continued to stimulate the emergence of new laboratory journals. When in 1918 in Utrecht Wilhelm Ringer became the first professor in the Netherlands of physiological chemistry only (still in the medical faculty, though), he soon commenced the publication of the *Onderzoekingen* of his laboratory (1921). Certainly by the 1890s, but probably earlier, house journals had earned a respected place among the scientific press. In 1895 the *Nederlandsch Tijdschrift voor Geneeskunde* wrote:

> The Medical circles of the Netherlands are vibrant with life. Testimony to that fact are numerous publications, of which the Reports of our Physiological Laboratories are not the least valuable. Detail-researches, but with a perspective on the big general questions of the day; examples of clear reporting, which are gems of our medical literature.[29]

Laboratories in other scientific disciplines followed as well. Three years after being appointed as professor of experimental physics, Heike Kamerlingh Onnes (1853–1926) launched the English-language series

29 H. Zwaardemaker, 'Boekaankondiging', *Nederlandsch Tijdschrift voor Geneeskunde*, 31 (1895), pp. 1072–83 (pp. 1072–3).

Communications from the Laboratory of Physics at the University of Leiden (1885–1994).[30] During the first decade of its existence, the periodical consisted to a high degree of shortened and translated versions of already published material, mostly PhD theses and papers in various Dutch, German, and French journals. The proceedings of the Section of Sciences of the Dutch Royal Academy were a particularly important source. In the years around 1900, *Nature* reported several times on the research done in the laboratory of Kamerlingh Onnes (with pictures of the innovative apparatus of the cryogenics department). The *Communications* was described, among other things, as 'a permanent testimony to his skill and industry'.[31]

The second house journal in the Netherlands in the English language was the *Publications of the Astronomical Laboratory at Groningen* (1900–60). This series was established by Jacobus Kapteyn (1851–1922), who achieved international fame despite (or, in retrospect, thanks to) receiving very little funding and remaining without an observatory throughout his entire career.[32] Kapteyn had been a professor since 1878, but only in 1896 did he get the opportunity to open an institute of his own. He named it 'laboratory' (not 'observatory') because he worked with instruments for measuring and processing photographic recordings instead of telescopes. Kapteyn thus used material from astronomical institutes that had adopted the new technology of photography, especially the Royal Observatory at the Cape of Good Hope, and from 1908 onwards he periodically visited the Mount Wilson Solar Observatory in California.

In astronomy, beginning in the first half of the nineteenth century, all major institutes issued their own serial publications. This practice arose from the inherent necessity in this discipline to share results worldwide—due to the different parts of the sky—but at some point publishing 'annals' also came

30 On Kamerlingh Onnes and his laboratory, see: Van Berkel, Van Helden, and Palm, eds., *A History of Science in the Netherlands*, pp. 491–4; Dirk van Delft, *Freezing Physics: Heike Kamerlingh Onnes and the Quest for Cold*, transl. by Beverley Jackson (Amsterdam: KNAW Press, 2007); Dirk van Delft, 'The Physics Laboratory of Leiden University', in Ana Simões, Maria Paula Diogo, and Kostas Gavruglu, eds., *Sciences in the Universities of Europe, Nineteenth and Twentieth Centuries: Academic Landscapes* (Dordrecht: Springer, 2015), pp. 311–26.

31 L.C. Jackson, 'Contributions to Physics and Chemistry', *Nature*, 147 (1941), 163–4 (p. 164). See also: *Nature*, 54 (1896), 345–7; 64 (1901), 208–11; and 69 (1904), 259–61.

32 On Kapteyn, see: Van Berkel, Van Helden, and Palm, eds., *A History of Science in the Netherlands*, pp. 495–7; Pieter C. van der Kruit and Klaas van Berkel, eds., *The Legacy of J.C. Kapteyn: Studies on Kapteyn and the Development of Modern Astronomy* (Dordrecht: Kluwer Academic Publisher, 2000); Pieter C. van der Kruit, *Jacobus Cornelius Kapteyn: Born Investigator of the Heavens* (Cham: Springer International Publishing, 2014).

to mean 'proof of existence and worth'.[33] As the astronomical annals mainly recorded the observations and the general activities of the institutes, they were of a different category than the house journals in the chemical, medical, and physical sciences. Yet the house journal of Kapteyn resembled those in physiology more than the annals in astronomy. It was used for presenting extensive material, including PhD theses and observational data (with many tables), as well as discussions of such data, results derived from them, and methodological issues.[34] The reason for that resemblance is explained above: Kapteyn conceived of his institute as a laboratory. (Remarkably enough, prior to having an institute, he was invited to use two rooms of Huizinga's physiological laboratory.) Historians have agreed moreover that the observatory, as a general category, became more and more like a laboratory. Mechanization of observation went hand in hand with the introduction of factory-like procedures, a strict hierarchy of staff, and a rigid division of labour (particularly in astrophysics).[35] Although these changes did not occur in Kapteyn's own institute, where he worked with one assistant and a few human computers, they were definitely happening at the California observatory, and through those channels they were also embodied in the *Publications*.

Both the research and the house journal of the Groningen laboratory provided a model of modernization for the observatories of the other Dutch universities: Leiden, Utrecht, and Amsterdam. In 1921, the year of Kapteyn's retirement, the four institutes collaborated in founding the *Bulletin of the Astronomical Institutes of the Netherlands*, initially alongside their own series. David Baneke has disclosed that the *Bulletin* was intended as an alternative to the 'graceful tomb' of the *Proceedings* of the Royal Academy (even though these were published in English as well as Dutch since 1898) and that it made a substantial contribution to 'the renaissance of Dutch astronomy' after World War I.[36]

Lastly, a non-Dutch and non-academic example of an influential house journal can be found in Denmark: *Comptes Rendus des Travaux du Laboratoire*

33 David H. DeVorkin, 'Internationalism, Kapteyn and the Dutch Pipeline', in Van der Kruit and Van Berkel, eds., *The Legacy of J.C. Kapteyn*, pp. 129–31; Aileen Fyfe, 'Journals and Periodicals', in Bernard V. Lightman, ed., *A Companion to the History of Science* (Chichester: Wiley, 2016), pp. 387–99 (p. 389).
34 See: Van der Kruit, *Jacobus Cornelius Kapteyn*, pp. 251–3 and Appendices A.1 and A.3; 'The Astronomical Laboratory at Groningen', *Publications of the Astronomical Society of the Pacific*, 33(192) (1921), pp. 94–7 (p. 96).
35 Robert W. Smith, 'The Observatory', in Lightman, ed., *A Companion to the History of Science*, pp. 196–209 (pp. 202–6).
36 David Baneke, 'Teach and Travel: Leiden Observatory and the Renaissance of Dutch Astronomy in the Interwar Years', *Journal for the History of Astronomy*, 41 (2010), pp. 167–98 (pp. 188–91).

Carlsberg (1878–1989), although the original title was *Meddelelser fra Carlsberg Laboratoriet.* The Carlsberg Laboratory in Copenhagen was established in 1876 by J.C. Jacobsen, a rich philanthropist and founder of the Carlsberg Breweries. It was, however, an institute for basic research, independently administered by the Royal Danish Academy of Sciences and Letters. The laboratory was split into a department for chemistry and a department for physiology, directed respectively by Johan Kjeldahl (1849–1900) and Emil Chr. Hansen (1842–1909). The house journal only included articles reporting on research done in the laboratory, either by staff or visiting scientists. At first it featured papers in Danish, but each issue contained a résumé in French. After only a few years, the résumés were also published as separate issues, under the French title, which increased the paid circulation from 550 to 650 copies. In 1901 the *Comptes Rendus* became a journal with full reports (instead of summaries) in French, German, or English. The Danish edition was terminated in 1933. Thereafter, the *Comptes Rendus* appeared almost exclusively in English. Partly through its house journal, the Carlsberg Laboratory became a major player in both Danish and international science.[37]

'Matrix of the Modern': Purposes and Functions of Laboratory Journals

It is not surprising that the first generation of laboratory journals developed in physiology, as this field was at the forefront of making modern science. In its attempt to carve out a scientific discipline of its own outside the framework of medicine, physiology came to be a model science for many other fields by emulating the experimental approach of physical scientists and pursuing specialization and institutionalization.[38] The laboratory became the primary locus of both experimental research and the epistemic, political, and social shifts associated with scientific modernity—or even modernity at large.

Robert E. Kohler has asserted that the laboratory as a social institution was involved in every element of the 'matrix of the modern'.[39] It provided

37 H. Holter and K. Max Møller, eds., *The Carlsberg Laboratory, 1876–1976* (Copenhagen: The Carlsberg Foundation, 1976), pp. 283–7; Helge Kragh, 'From Ørsted to Bohr: The Sciences and the Danish University System, 1800–1920', in Simões, Diogo, and Gavroglu, eds., *Sciences in the Universities of Europe*, pp. 31–47 (pp. 40, 45).

38 See for example: Peter J. Bowler and Iwan Rhys Morus, *Making Modern Science: A Historical Survey* (Chicago: Chicago University Press, 2005), pp. 165–88; Merriley Borell, 'Instrumentation and the Rise of Modern Physiology', *Science & Technology Studies*, 5 (1987), pp. 53–62.

39 Kohler, 'Lab History', p. 765.

public education and entry to careers on a mass scale, constituted the institutional source of disciplinary communities, and embodied a bundle of cultural concepts that were congruent with those of the bureaucratic state, industrial capitalism, and the middle class. These moral and epistemic virtues and values included self-denial, a strong work ethic, the social role of the scientist, progress, mechanical objectivity, the universal scientific method, and a 'logic of trust' that relied on 'public transparency—through publishing—to establish truth and merit.[40] While Kohler has hinted at co-production[41] of the laboratory, laboratory science, and the nation-state, Bruno Latour has even gone so far as to argue that laboratories are 'such an irreplaceable source of political strength' owing to their ability of 'transforming society into a vast laboratory'.[42] According to Latour, the only extraordinary feature of laboratories is their proficiency in writing specific types of texts, which make their statements easily readable and limit the number of counter-arguments via the use of 'inscriptions' such as diagrams, tables, and pictures.[43]

This article suggests that publishing a house journal was one of the (possible) practices that laboratories relied on in their performance as social institutions. The *raison d'être* of laboratory journals should not be sought in the circulation of scientific information or the strife for individual power positions (unlike the case of Liebig, 'the chemical gatekeeper',[44] and his *Annalen*). As has been shown, much of the contents of house journals had already appeared elsewhere, and many of their founders already held influential roles, either as editors of other, more prestigious journals (most notably Donders and Pflüger) or as prominent members of the national

40 Kohler, 'Lab History', pp. 765, 768. See also: Andrew Cunningham and Perry Williams, 'De-centring the "Big Picture": *The Origins of Modern Science* and the Modern Origins of Science', *The British Journal for the History of Science*, 26 (1993), pp. 407–32; Lorraine Daston and Peter Galison, *Objectivity* (New York: Zone Books, 2007); the chapter by Frans van Lunteren in this volume.

41 On co-production, see: Sheila Jasanoff, ed., *States of Knowledge: The Co-Production of Science and Social Order* (London: Routledge, 2004).

42 Bruno Latour, 'Give Me a Laboratory and I Will Raise the World', in Karin D. Knorr-Cetina and Michael Mulkay, eds., *Science Observed: Perspectives on the Social Study of Science* (London: Sage Publications, 1983), pp. 141–70 (pp. 161, 166).

43 Latour, 'Give Me a Laboratory and I Will Raise the World', p. 161. Simplified inscriptions are produced through 'inscription devices': laboratory instruments that can translate material substances (into diagrams, tables, pictures, etc.). See also: Bruno Latour and Steve Woolgar, *Laboratory Life: The Construction of Scientific Facts* (Princeton, NJ: Princeton University Press, 1987; originally publ. 1979).

44 William H. Brock, *Justus von Liebig: The Chemical Gatekeeper* (Cambridge: Cambridge University Press, 1997).

academies of their countries. Rather, the point of launching a house journal was to put the laboratory in a position of power (too).

Bringing together its research output in a serial publication was one way of forging social cohesion in the laboratory and connecting the institute to the world at large. The former concern grew more urgent as new laboratories were built on sites scattered across the city, which caused the university to become spatially and socially dispersed.[45] The nature of research and teaching in the science and medical faculties as well as the distance from their colleagues in other faculties and disciplines—and from the university as a whole—led the science professors and students to focus their lives around their laboratories. This process was further strengthened when the laboratories started having their own social customs, events, festivities, student associations, libraries, and publication series. The latter concern of connecting the laboratory to the world should not only be understood in the conventional sense of establishing relationships with other scientific institutes, at home and abroad, but also as translation processes in the Latourian sense of the term. As a genre, moreover, laboratory journals contributed to the transformation of scientific expertise after the 1850s and the renegotiation of the public status of scientists within a wider political landscape.[46]

At first glance, there is not a whole lot of evidence to be found in the house journals themselves. Given the diversity and the vast scope of the subject (extensive serial publications of many laboratories in different disciplines, periods, and cultural contexts), it would take a large and systematic study before any general claims can be made. Nevertheless, the examples discussed here have in common that they consisted almost exclusively of research reports. Mission statements or other types of editorials were not included. The first volumes of the second and third series of the Utrecht *Onderzoekingen* did contain a preface, for instance, but Donders merely used them to give the reasons for starting a new series (which have been described previously) and to explain that the volumes were 'mainly gifted to societies and laboratories'.[47]

However, taking the observations of Kohler and Latour into consideration, it can be assumed that it was exactly the lack of general contents—both editorials and varieties of reviews and news—that makes house journals historically important. The logic of trust was conveyed

45 See the chapter by Klaas van Berkel in this volume.
46 Csiszar, *The Scientific Journal*, pp. 3–4.
47 Donders, 'Bericht', p. 3.

by their complete focus on research, and their persuasive power came from the incorporation of laboratory practices and inscriptions, which 'fostered interest groups' and 'attracted the attention of groups that are the mouthpiece of larger social movements'.[48] In 1895 the *Nederlandsch Tijdschrift voor Geneeskunde* made a somewhat related observation when it complimented the house journals of the physiological laboratories on their 'strictly logical' and 'precise' reporting style and their role in increasing awareness of the usefulness of various apparatus.[49] The author of the review repeatedly expressed the wish that provisions would be made for upgrading the equipment when needed.

In the light of these reflections, Mulder's house journal constitutes a special case. On the one hand, the *Scheikundige onderzoekingen* offered an alternative to the scientific practices and periodicals of the learned societies, 'where a lot happens, but little that contributes to the advancement of science'.[50] It displays deep commitment to laboratory research and training. In his notice on the education of pharmacists, Mulder stipulated that 'the undergraduate students spend their hours in between classes in the laboratory, the more advanced students many hours per day; the doctoral students practically live there.'[51] The notice also testifies, even though the chemist was very critical of the state regulations,[52] to the involvement of laboratories and laboratory scientists in projects of the nation-state: moulding effective professionals and reforming professions to serve the needs of the state (in this case the medical profession and the need for healthy citizens, workers, and soldiers).[53]

On the other hand, both the format of the *Scheikundige onderzoekingen* and the moral values that it embodied were (much) less modern than those of the other laboratory journals analysed here. It contained more general contents, did not include pictures or other inscriptions, and was used to promote conservative ideas about knowledge, higher education, and the

48 Latour, 'Give Me a Laboratory and I Will Raise the World', p. 150.

49 Zwaardemaker, 'Boekaankondiging', pp. 1073, 1079.

50 G.J. Mulder, 'Toespraak bij het openen der scheikundige lessen, 1841', *Scheikundige onderzoekingen*, 1 (1842), pp. 1–16 (p. 10).

51 G.J. Mulder, 'Pharmaceutisch onderwijs', in *Scheikundige onderzoekingen*, 1 (1842), pp. 589–91.

52 See: Bert Theunissen, '*Bildung* of beroep: Gerrit Jan Mulders "mannen van karakter"', in Bert Theunissen, *'Nut en nog eens nut': Wetenschapsbeelden van Nederlandse natuuronderzoekers, 1800-1900* (Hilversum: Verloren, 2000), pp. 80–97.

53 Kohler, 'Lab History', pp. 767–8. See also: Steve Sturdy and Roger Cooter, 'Science, Scientific Management, and the Transformation of Medicine in Britain, c. 1870–1950', *History of Science*, 36 (1998), pp. 421–66.

relationship between science and society. In his speech to his students in 1841, printed in the first volume, Mulder insisted that the higher purpose of practising science was not 'usefulness' for society or 'comfort and wealth' but 'the development of the noble faculties of the mind'.[54] He also stressed the importance of the unity of science and the unity of the university ('*Universitas*').[55] Mulder's ideal community consisted of 'academic citizens' ('Academieburgers'),[56] and his logic of trust was paternalistic and personal. As such, his conceptions and values are hardly reconcilable with the disciplinary communities and the impersonal language of scientific expertise that were to become characteristic of modernity.

Therefore, the *Scheikundige onderzoekingen* represented both innovation and tradition. Discrepancies with other house journals are largely explained by the fact that experimental laboratory science had only just emerged in the 1840s, as well as the fact that the chemical laboratory journal had ceased to exist long before the genre caught on in the 1860s and 1870s. In this period Mulder criticized developments in higher education and the social structure of science even more fervently. He remained strongly opposed to (the fragmenting effects of) specialization and institutionalization.[57] But this process was only intensifying, partly due to the emergence of more laboratory journals. Consequently, these periodicals helped to introduce new conditions (epistemic, moral, and social) for making and legitimizing knowledge, which were not only competing with the conditions favoured by conventional elite science but also with the practices and values of the founding father of the genre of house journals. This does not mean, however, that the evolution of laboratory journals—or any other serialized print genre—was linear nor that they replaced older genres. Their formats, purposes, and functions were not uniform.

One such area where it is hard to distinguish clear chronological or other patterns is the language policies of house journals and underlying ideologies regarding 'the national' and 'the international'. There was a wide variety between periodicals, and some of them altered their policies in the course of their existence. Strategies ranged from publishing exclusively in the national language to publishing exclusively in (a) foreign language(s), and everything in between. Translating and reprinting could be involved across the whole spectrum. As these were pertinent issues for the entire

54 Mulder, 'Toespraak', p. 2.
55 Mulder, 'Toespraak', p. 3.
56 Mulder, 'Toespraak', pp. 3, 7, 9.
57 Theunissen, '*Bildung* of beroep', pp. 89–97.

scientific periodical press,[58] this article only explores them insofar as they pertain to the construction of disciplinary communities beyond national boundaries. The circulation of laboratory journals between institutes in different countries was, again, not initiated for the reason of disseminating research findings, at least not initially. Their use in other countries was often limited because of language barriers. But this had not stopped societies and universities either. They had been exchanging periodicals for a long time already as common currency, as proof of their existence and legitimacy, and to advertise the national language, culture, or scientific style.[59]

Laboratories thus copied an established practice, with the difference that their networks were inherently specialized and transnational in nature (as opposed to international). At first this merely meant that material objects moved from one institute to another. (The journal volumes did probably have agency, though, especially if they carried a lot of visual representations, tables, and formulae.) Later, laboratory journals were involved in the creation of social networks across borders: people moving between institutes and actually working together. The capacity to attract attention in the English-speaking world and to inspire travel is most evident in the house journals of Kamerlingh Onnes, Kapteyn, and the Carlsberg Laboratory. This capacity was enhanced by favourable reviews published in other periodicals. Reviews in *Nature* explicitly stated that Kamerlingh Onnes 'will be very glad' if physicists come to visit his laboratory.[60] In the case of Kapteyn, it was the other way around: he sought collaboration with observatories where *he* could work, particularly in the United States. Both men were pioneering in this respect, as Dutch scientists and science remained mostly Germany-oriented well into the twentieth century. Personal networks were usually already in place there.

58 The literature on these issues is quite extensive. On language and translation, see for example: Michael D. Gordin, *Scientific Babel: How Science Was Done Before and After Global English* (Chicago: Chicago University Press, 2015); Jenny Beckman, 'The Publication Strategies of Jöns Jacob Berzelius: Negotiating National and Linguistic Boundaries in Chemistry', *Annals of Science*, 73 (2016), pp. 195–207. On reprinting, see: Iain Watts, '"We Want No Authors": William Nicholson and the Contested Role of the Scientific Journal in Britain, 1797–1813', *The British Journal for the History of Science*, 47 (2014), pp. 397–419.

59 On this topic in relation to the Dutch Royal Academy, see: Klaas van Berkel, *De stem van de wetenschap: Geschiedenis van de Koninklijke Nederlandse Academie van Wetenschappen*, 2 vols. (Amsterdam: Bert Bakker, 2008–11), *vol. 1. 1808-1914*, pp. 532–41. From 1898 the *Proceedings* of the Academy were published in English as well as Dutch.

60 'The Physical Laboratory at Leiden (Holland)', *Nature*, 54 (1896), pp. 345–7 (p. 347).

During and after World War I, issues related to (inter)nationalism took on new meaning, which also touched on the *raison d'être* of laboratory journals (as well as reprints). In 1922 Hamburger, professor of physiology in Groningen since 1901, re-established the Groningen *Onderzoekingen*. The first volume not only contained the papers of the period 1919–22, but also 'a list of all publications that issued from the laboratory' since the very beginning (1851–1919).[61] Hamburger explained his initiative in the preface to the volume, printed in Dutch, German, English, and French:

> Not hampered by traditions of my two predecessors, I had until not long ago, considered such a publication superfluous, believing the experimental work, carried out in my laboratory, would by means of generally accessible journals come sufficiently within reach of an international circle of readers. The situation created by the world war has made an end to this illusion. ... It has moreover the advantage of making men of different nationalities acquainted with the literature quoted in these papers. For is it not true that great peoples read but insufficiently what is produced by other equally great nations? Workers of small, neutral countries are differently situated in this respect. ... Those who might feel interested in a particular research will on application, to the best of my abilities, be provided with a reprint of that investigation.[62]

Had it not been for the war, Hamburger would not have seen the point of collecting the papers of his laboratory. Nevertheless, after 1900 house journals were still being launched. An example that has not been mentioned yet is the *Mededeelingen uit het Phytopathologisch Laboratorium Willie Commelin Scholten* (1910–38, 1957–81). This periodical is interesting for several reasons. First of all, it has cultural significance because it was founded by a woman. Appointed in 1906, Johanna Westerdijk (1883–1961) was the second director of the laboratory.[63] In 1917 she became the first female professor in the Netherlands: extraordinary professor of phytopathology at the University of Utrecht. In 1930 she obtained the same position at the University of Amsterdam. Under Westerdijk's leadership, the laboratory was transformed from a small private organization

61 H.J. Hamburger, 'Preface', *Onderzoekingen, gedaan in het Physiologisch Laboratorium der Rijks-Universiteit te Groningen*, 1 (1922), pp. 5–6.
62 Hamburger, 'Preface'.
63 On Westerdijk, see: Van Berkel, Van Helden, and Palm, eds., *A History of Science in the Netherlands*, pp. 603–5.

Figure 12.2 The 1923 International Conference of Phytopathology at the laboratory of Johanna Westerdijk (seated in the second row, fourth from the right) in Baarn. The conference was held from 24 to 30 June at various institutes in the Netherlands. It hosted over 65 phytopathologists from 26 countries, in addition to some 30 Dutch participants. The last two days were spent at Westerdijk's institute, where the final papers were read and the visitors were treated to a festive reception. Like the *Mededeelingen*, the conference was important in enhancing the scientific status of the Willie Commelin Scholten Phytopathological Laboratory. The papers, including several by Westerdijk and her staff, were published in the *Report of the International Conference of Phytopathology and Economic Entomology* (1923). Courtesy: Collectie IAV-Atria.

(established in 1894 in Amsterdam) into an internationally renowned centre for the study of plant disease (located in Baarn since 1920). Out of the 55 students who received their PhD under her supervision, 26 were women.[64]

Second, the *Mededeelingen* is an example of a house journal that served the purpose of moving away from an industrial context and a wider public. It helped to convert phytopathology into a laboratory science and an autonomous discipline, to build a new community of professional phytopathologists, and to earn the laboratory a reputation of doing 'pure' science. As the institute also housed the Central Bureau of Fungal Cultures, it needed to discard its reputation as solely an information service. Patricia Faasse has revealed Westerdijk's successful strategies and the role of the house journal in them. Through the use of both foreign and technical

64 Patricia E. Faasse, *In Splendid Isolation: A History of the Willie Commelin Scholten Phytopathology Laboratory, 1894–1992* (Amsterdam: KNAW Press, 2008), p. 3.

language(s), the periodical aimed at a public of phytopathologists, not farmers and growers, and 'represented a break with the past'.[65] Furthermore, Faasse writes:

> By placing every *Communication* in the context of ongoing research, Westerdijk helped to build up continuity in phytopathology. Invoking earlier research, and presenting the findings of the present as contributions to solving problems stated in the past, created the impression that phytopathology as a science was steadily advancing. ... By selecting methodically when citing the earlier research, she created a certain tradition and placed herself in it; thus intentionally excluding certain other possibilities in the process. Imposing order on the past, she provided a justification for the present.[66]

Third, the story of this house journal sheds light on the functioning of laboratories as microsocial institutions. In her communications Westerdijk never referred to the work of her predecessor Jan Ritzema Bos. His *Tijdschrift over Plantenziekten* (1895–1962) was scorned at the institute for being an instrument of self-justification and self-glorification.[67] Given the fact that Ritzema Bos had left the institute on bad terms, erasing and mocking his legacy must have been a means of forging meaning and a new group identity around a shared negative experience and antipathy.

Other evidence, too, suggests that a microhistorical approach to the laboratory revolution can yield new insights into the nature of modern scientific life. Former (visiting) scientists of the Carlsberg Laboratory, for instance, have identified themselves as 'an unofficial but highly chauvinistic "club"'[68] and have compared laboratory life to 'a lodge membership', with unwritten rules, rites, conferences, colloquia, dinners, parties, specific humour, and playing music and games or drawing caricatures in between experiments.[69] The laboratory 'library'—arguably one of the pioneers of the modern scientific library—had a similar function. Especially after 1900, when networks of knowledge, communication, and journal exchange grew

65 Faasse, *In Splendid Isolation*, pp. 79, 262–3.

66 Faasse, *In Splendid Isolation*, p. 79.

67 Faasse, *In Splendid Isolation*, pp. 67–8, 73, 79.

68 Christian B. Anfinsen, 'The International Influence of the Carlsberg Laboratory on Protein Chemistry', *Perspectives in Biology and Medicine*, 29(3) (1986), pp. 87–9 (p. 88).

69 John Walsh, 'Carlsberg Laboratory: Ferment over Future of Copenhagen Lab', *Science*, 175 (1972), pp. 973–4 (p. 974). See also: Holter and Max Møller, eds., *The Carlsberg Laboratory*, pp. 290–330.

wider and denser, and laboratories grew in size, reading rooms often came to be at the heart of them, spatially as well as socially.[70]

Closer scrutiny of laboratory journals in their microsocial context might also deepen our understanding of science writing practices, their role in the production of knowledge, and their incorporation into the training of future scientists (both learning to write and writing to learn). In particular, the genre may provide a window into laboratories' strategies for offering assistants and students instruction and experience in writing and publishing papers. The availability of a house journal allowed publication without prior approval of senior scientists other than the supervisor and enabled a wider dissemination of doctoral dissertations. Conceivably, writing in the laboratory developed into a pedagogical tool and a collective or collaborative practice, which differed from traditional elite practices of writing and also from the relatively new custom of reading communications on behalf of junior scientists at society meetings.[71] By embracing the emerging academic ideal of integrated research and teaching, and providing a means for acting on it, house journals facilitated the transition from the teaching university to the research university.

Thus, whereas Kohler has highlighted the laboratory as a macrosocial institution that reflects and also produces the conventions of other major social institutions, microhistories can, in the words of Catherine M. Jackson, provide 'fully situated accounts of the production of scientific knowledge' that integrate the 'institutional essence [of laboratories] with what goes on within their walls'.[72] A promising approach to laboratory history that offers the possibility to consider the two levels (macro and micro) in combination is the study of commemorations (textual, oral, visual, and material) in the context of modern laboratory life. Obituaries, and to a lesser degree institutional memories, were typically the only type of general content that house journals permitted themselves. Truus van Bosstraeten has analysed the role of celebrations in the social life of a pharmacological laboratory (1902–70) at Ghent University in Belgium. She has disclosed how private commemorative practices bolstered group cohesion and acknowledged the laboratory's accomplishments as being teamwork, while public commemorations—published in the house journal or as separate booklets—conveyed

70 Beukers, 'Een nieuwe werkplaats in de geneeskunde', p. 270; Holter and Max Møller, eds., *The Carlsberg Laboratory*, pp. 48–9; Faasse, *In Splendid Isolation*, p. 114.

71 On this custom in the Dutch Royal Academy, see: Van Berkel, *De stem van de wetenschap*, vol. 1, pp. 353–4, 458–61.

72 Catherine M. Jackson, 'The Laboratory', in Lightman, ed., *A Companion to the History of Science*, pp. 296–309 (p. 304).

continuity and legitimacy, conferred the group's professional identity, and paid tribute to the 'brilliant' leader of the laboratory.[73] In general, moreover, both private and public celebrations convey epistemic, moral, and gendered values, instilling them in the multi-layered community that constitutes a laboratory as well as in society at large. As shown above, revealing erased legacies can also be informative in analysing the matrix of the modern and the periodical press's implication in it.

'Real Journals'? Laboratory Journals and the Evolution of the Scientific Periodical Press

Over the course of the nineteenth century, scientific authority and scientific authorship went through profound transformations. Around 1800 the validation of scientific knowledge rested predominantly on learned societies and academies. These institutions derived their authority from royal protection and published voluminous memoirs and transactions. Serialized print did not carry much epistemic weight. At the end of the century, it constituted the main source of the legitimacy of the scientific enterprise. The metamorphosis began with joint initiatives of entrepreneurs and individuals who, inspired by the French Revolution, committed themselves to making natural knowledge accessible to broader publics. From the 1830s onwards, mounting pressure and competition from commercial journals forced elite institutions to get into the business of short serialized genres too. But in order to counterbalance the broadening of the audience for science, those expert genres became increasingly specialized, technical, and exclusive.[74] Simultaneously, societies and academies lost their monopoly on publishing expert scientific information. Particularly in Germany, in the last quarter of the century, commercial publishers seized opportunities provided by the rise and differentiation of research in reorganized universities and newly founded industries.[75]

Amidst these transformations, laboratory journals entered the scene. Their place in the evolution of the scientific periodical press is best understood by approaching that history not through the lens of change and

73 Truus van Bosstraeten, 'Dogs and Coca-Cola: Commemorative Practices as Part of Laboratory Culture at the Heymans Institute Ghent, 1902–1970', *Centaurus*, 53 (2011), pp. 1–30. On commemoration in science, see: Pnina G. Abir-Am and Clark A. Elliott, eds., *Commemorative Practices in Science*, special volume of *Osiris*, 14 (1999).

74 Csiszar, *The Scientific Journal*, pp. 1–9; Fyfe, 'Journals and Periodicals', pp. 389–91.

75 See for example: Sarkowski, *Springer-Verlag*.

continuity, but as a 'hybrid' history.[76] Analysis of the rise, contents, and functions of house journals underlines the permanent state of transition of the serialized media landscape of science, in which hybridity seems to be the only constant factor. Laboratory journals were, on the one hand, 'more modern' and 'more scientific' than many older and contemporary formats, especially regarding their relatively technical reporting style, their embodiment of experimental laboratory science, and the large share of research papers in the total journal content. On this ground, looking back in 1976, the editor of the *Comptes Rendus* of the Carlsberg Laboratory proudly stated that it 'has during most of its long history ... been a purely scientific journal'.[77] On the other hand, house journals were 'less advanced' with respect to their habits of publishing reprints and attributing responsibility for the quality of the contributions to the directors of the institutes, without consultation of external referees. Before 1900, however, these habits were not exceptional. Awareness of the potential problems of duplicate publication began to grow around that time, while established peer-review procedures developed only in the second half of the twentieth century.[78]

More relevant within the context of hybrid history is the fact that house journals were never intended as substitutes for society journals or periodicals such as Pflüger's *Archiv*. It might be clarifying to typify them as a 'parallel genre', invented to showcase the output of individual research institutes and to promote the kind of science that laboratories represented, without denying the need for journals with broader scopes and other purposes and functions. This is most likely the reason why Stokvis, in 1891, questioned if serial publications of laboratories were 'real journals'. In his obituary of Donders, with the intention of inspiring emulation of the physiologist's publishing attitudes, Stokvis identified differences between journals of scientific societies and non-society journals. Periodicals of the first type appeared 'with solemn coupon frequency' ('met stemmige coupon-regelmatigheid'),[79] served as depositories of knowledge, were authoritative but ponderous, and published in the national language. By contrast, other types of periodicals appeared irregularly and relatively fast, functioned as vehicles for the dissemination and accumulation of knowledge, were easily acquirable, and

76 On hybridity, see for example: Andrew Chadwick, *The Hybrid Media System: Politics and Power* (Oxford: Oxford University Press, 2013).

77 Holter and Max Møller, eds., *The Carlsberg Laboratory*, p. 286.

78 Fyfe, 'Journals and Periodicals', pp. 394–5.

79 Stokvis, 'Levensbericht F.C. Donders', p. 18.

aimed at strengthening the country's position in international science by publishing in (an) internationally common language(s).[80] Yet laboratory journals shared features with both society and non-society periodicals.

As a special and specialized category of non-society periodicals, laboratory journals tapped into the rise of community building along institutional, disciplinary, and transnational lines (while publication was still largely a national practice). In this respect, house journals were important precursors to the trans-institutional, specialized, and international research journals that came to prevail from the 1950s onwards. From a synchronic perspective, the evolution of the scientific periodical press was more like a variational process with continual branching and extinction. The genre of house journals seems to have caught on mostly in countries that were in the periphery of continental Europe but near to its centre: Germany. As such, they emerged and developed in reference to and in interaction with German journals. These became particularly influential from the 1870s onwards. Commercial publishers introduced what has been called the 'German model of science publishing', which was based on a careful layering of different types of publications (encyclopaedic handbooks, monographs, and four types of journals[81]) as well as an entrepreneurial approach to incipient research specialisms.[82]

In the years before World War II, and with the help of Jewish émigrés, the German model was adopted by newly established publishing companies in the Netherlands and the United States.[83] Their reinvention of the scientific journal proved to be crucial in achieving leading positions in the rapidly growing international market for scientific literature after 1945. It cannot be a coincidence that the two founding editors of one of the earliest examples of the new, post-war type of research journal, Elsevier's *Biochimica et Biophysica Acta* (1946), were directors of physiological-chemical laboratories: Hendrik Westenbrink at the University of Utrecht and Kaj Linderstrøm-Lang at the Carlsberg Laboratory.[84]

80 Stokvis, 'Levensbericht F.C. Donders', pp. 17–8.

81 Research journals, abstract and review journals (*Zentralblätter*), progress journals (*Ergebnisse*), and professional journals.

82 On the German model, see: Hendrik Edelman, 'Maurits Dekker and Eric Proskauer: A Synergy of Talent in Exile', *Logos*, 15 (2004), pp. 188–93; Sarkowski, *Springer-Verlag*.

83 See: Dorien Daling, 'The Encyclopaedia as Pioneer of the Journal: The Early Years of Elsevier's Scientific Publishing Company, 1936–1956', in Marieke van Delft, Frank de Glas, and Jeroen Salman, eds., *New Perspectives in Book History: Contributions from the Low Countries* (Zutphen: Walburg Pers, 2006), pp. 31–48.

84 On the post-war 'reinvention' of the scientific journal by Dutch commercial publishers, see: Dorien Daling, *Stofwisselingen: Nederlandse uitgevers en de heruitvinding van het natuurwetenschappelijke tijdschrift, 1945-1970* (Zutphen: Walburg Pers, 2011).

Conclusion

Adhering to Csiszar's call 'to bring the sensibilities of book history to the politics of scientific knowledge',[85] this article has shown that genres and formats for the dissemination of research findings have epistemic, political, and social consequences. Serial publications of laboratories helped to introduce, (re)define, and (re)produce the forms of knowledge, central values, scientific practices, and social conditions of modern science, particularly between the 1860s and 1930s and in the European periphery.

Regarding the performance of laboratories as social institutions, the study of house journals allows glimpses of the social life in laboratories and of how these institutes were entangled in national political communities and transnational knowledge communities. It also reveals a wide range of specific agendas that scientists tried to pursue through the periodical press, such as promoting 'pure' science, emancipating disciplines, attracting financial resources, and repairing international relations after wars.

The study of laboratory journals can furthermore make a valuable contribution to the historiography on publishing and publishers of expert serialized print in the nineteenth and twentieth centuries. Recent scholarship has convincingly connected the changing role of the scientific journal to transformations in the demarcation, meaning, and significance of scientific expertise, but it has mainly concentrated on Britain. The German model of scientific publishing deserves more attention too, including the commercial, specialized, international, fast, and peer-reviewed research journals that were established by Dutch and American exponents of that model after World War II.

Appendix. Journals Discussed or Mentioned

Laboratory Journals

Communications from the Laboratory of Physics at the University of Leiden (1885)
Comptes Rendus des Travaux du Laboratoire Carlsberg (1878)
Mededeelingen uit het Physiologisch-Pathologisch Laboratorium van 's Rijks Veeartsenij School te Utrecht [Communications from the Physiological-Pathological Laboratory of the National Veterinary School in Utrecht] (1889)

85 Csiszar, *The Scientific Journal*, p. 13.

Mededeelingen uit het Phytopathologisch Laboratorium Willie Commelin Scholten [Communications from the Willie Commelin Scholten Phytopathological Laboratory] (1910)

Onderzoekingen, gedaan in het Laboratorium voor Physiologische Scheikunde der Rijksuniversiteit te Utrecht [Researches, carried out in the Laboratory for Physiological Chemistry of the University of Utrecht] (1921)

Onderzoekingen, gedaan in het Physiologisch Laboratorium der Leidsche Hoogeschool [Researches, carried out in the Physiological Laboratory of the University of Leiden] (1869)

Onderzoekingen, gedaan in het Physiologisch Laboratorium der Utrechtsche Hoogeschool [Researches, carried out in the Physiological Laboratory of the University of Utrecht] (1848)

Onderzoekingen, gedaan in het Physiologisch Laboratorium te Amsterdam [Researches, carried out in the Physiological Laboratory in Amsterdam] (1856)

Onderzoekingen uit het Physiologisch Laboratorium te Groningen [Researches from the Physiological Laboratory in Groningen] (1874)

Publications of the Astronomical Laboratory at Groningen (1900)

Scheikundige onderzoekingen, gedaan in het Laboratorium der Utrechtsche Hoogeschool [Chemical researches, carried out in the Laboratory of the University of Utrecht] (1842)

Studien des Physiologischen Instituts zu Amsterdam (1861)

Untersuchungen aus dem Physiologischen Laboratorium der Züricher Hochschule (1869)

Untersuchungen aus dem Physiologischen Laboratorium in Würzburg (1867)

Untersuchungen aus dem Physiologischen Laboratorium zu Bonn (1865)

Verslag over de onderzoekingen verricht in het Laboratorium voor Pathologische Anatomie en Bacteriologie te Weltevreden [Report on the researches carried out in the Laboratory for Pathological Anatomy and Bacteriology in Weltevreden] (1888)

Other Journals

Annalen der Chemie und Pharmacie (1832)

Archiv für die gesamte Physiologie des Menschen und der Tiere (1868)

Archiv für Ophthalmologie (1854)

Archives Néerlandaises des Sciences Exactes et Naturelles (1866)

Biochimica et Biophysica Acta (1946)

Bulletin of the Astronomical Institutes of the Netherlands (1921)

Holländische Beiträge zu den anatomischen und physiologischen Wissenschaften (1846)

Nederlandsch Archief voor Genees- en Natuurkunde [Dutch Archives of Medical and Natural Science] (1865)
Nederlandsch Lancet [Dutch Lancet] (1845)
Nederlandsch Tijdschrift voor Geneeskunde [Dutch Journal of Medicine] (1857)
Proceedings of the Section of Sciences, Koninklijke Akademie van Wetenschappen te Amsterdam [Royal Academy of Arts and Sciences in Amsterdam] (1898)
Tijdschrift over Plantenziekten [Journal on Plant Diseases] (1895)

About the Author

Dorien Daling is Lecturer in Modern History at the University of Groningen. Her research focuses on scientific publishing and scientific print from the nineteenth century to the present, including forms and norms of knowledge communication in the era of open science. She published a book on the post-World War II 'reinvention' of the scientific journal by Dutch commercial publishers: *Stofwisselingen: Nederlandse uitgevers en de heruitvinding van het natuurwetenschappelijke tijdschrift, 1945-1970* (Zutphen: Walburg Pers, 2011).

13 Images of the Laboratory in the Popular Press

Geert Vanpaemel

Abstract

During the Laboratory Revolution of the late nineteenth century, the popular press was instrumental in disseminating an increasing number of visual representations of the laboratory, which contributed to the rising public authority of laboratory research. This chapter offers a typology of these images, based on a sample of pictures taken from the pages of the French journal *La Nature* between 1873 and 1914. It argues that there was a shift in epistemological meaning attributed to these representations, in which the laboratory images tended to emphasize the dehumanized, impersonal nature of strict scientific procedures, in opposition to but also complementary with the portrayal of the individual genius of scientists.

Keywords: laboratory, German Model, popular press, visual culture, Marie Curie

Introduction

Science popularizers have always appreciated the value of images in conveying messages about complicated and difficult scientific issues.[1] From the early modern technical treatises to the eighteenth-century *Encyclopédie* of Diderot and D'Alembert, illustrations have been used to explain theoretical arguments or practical procedures to non-scientific

1 Although images can be considered in many different forms, ranging from any type of visual display (e.g. museum dioramas) to a tabular presentation of research data, in this chapter I focus on printed images of people, objects, and places as they may appear in popular books and magazines.

Berkel, Klaas van. and Ernst Homburg (eds), *The Laboratory Revolution and the Creation of the Modern University, 1830-1940.* Taylor & Francis Group, 2023
DOI: 10.5117/9789463720434_CH13

audiences. But different from scientific images, which are designed and produced by and for scientists, popular images do not necessarily reflect the current status of science. On the contrary, according to Berd Hüppauf and Peter Weingart, they are 'reflections of popular views and prejudices' and activate 'memories of previous, similar images ... often compressed to a stereotype not dissimilar to an archetypical image'. These popular images form a 'collective imaginary space of science', 'the product of a complex combination of pre-modern mythology and recent science images', 'reflecting the audience's beliefs, expectations, and anxieties'.[2] Although popular images often support the narrative of an accompanying text, they also carry an independent metaphorical message, that 'connote[s] or symbolize[s] meanings and values over and above what they literally represent'.[3] To unearth these messages, an analysis of forms and formats is needed, looking beyond the pictorial content of the image. In this chapter we focus on the representation of the laboratory during the Laboratory Revolution. The aim is to formulate a typology of images, which may bear on the understanding of the Laboratory Revolution, and its impact on the perception of authority and the nature of science.

The Laboratory Revolution is hard to visualize as a single historical event or artefact. Many historians agree that a number of developments in the last quarter of the nineteenth century contributed to push forward the laboratory as a privileged space for research, but a clear conception of the revolution that could be disseminated to popular audiences is hard to come by.[4] Even if some features are common among the new laboratories founded in the late nineteenth century, there is still is a wide diversity in types, design, or function. The mere concept of the laboratory was not new, and images of laboratories have been available since the seventeenth century.[5] The most salient features of the new laboratory culture (scale, authority, function) are difficult to visualize. Moreover,

2 Berd Hüppauf and Peter Weingart, 'Images in and of Science', in Bernd Hüppauf and Peter Weingart, eds., *Science Images and Popular Images of the Sciences* (New York/London: Routledge, 2011); Nick Hopwood, *Haeckel's Embryos. Images, Evolution, and Fraud* (Chicago: University of Chicago Press, 2015), pp. 3–31, quotation on pp. 20–1.

3 Gunther Kress and Theo van Leeuwen, *Reading Images* (London: Routledge, 1996), p. 45; Harald Hornmoen, '"Making Us See Science": Visual Images in Popular Science Articles and Science Journalism', *Journalistica*, 2 (2010), 79–99.

4 As one example, see Catherine M. Jackson, 'The Laboratory', in Bernard Lightman, ed., A *Companion to the History of Science* (Chichester: Wiley Blackwell, 2016), pp. 296–309.

5 C.R. Hill, 'The Iconography of the Laboratory', *Ambix*, 22 (1975), 102–10; Davy Jacobs, ed., *De wijze en de dwaas. De alchemist in woord en beeld* (Bruges: Van de Wiele, 2007).

it is not self-evident that the Laboratory Revolution as such would be represented in the popular press. Bernard Lightman has convincingly argued that the British popular press preferred a more diverse presentation of the places of research, including zookeepers, fishermen, and all kinds of self-educated amateur scientists among the 'exemplary scientists'. According to Lightman, this attitude was connected to the desire among science popularizers to keep science and scientific research accessible to the general reader.[6]

This chapter consists of two parts. In the first part we will argue that since the end of the nineteenth century, the role of the laboratory became a central issue in the mindset of scientists. In the wake of the ongoing Laboratory Revolution, scientists became painfully aware of the material circumstances in which they had to work. Under pressure of national and international competition, laboratories were constantly changing, thereby developing new tools and standards and making older laboratories look obsolete. In the last quarter of the nineteenth century, pictures and floorplans of laboratories started to appear in scientific journals, with details on their architecture, technical installations, and scientific equipment.[7] The growing importance of the laboratory also affected the public image of science, as evidenced in the analysis of portraits of scientists.[8] To the extent that scientists became aware of and pointed to the laboratory as a crucial feature of their work, it is to be expected that the laboratory would also take a more prominent place in popular accounts.

In the second part of the chapter, we focus on the pictorial visualization of laboratories. Which features were made visible, how did the images influence the representation of science, and was there any coherent development that could be brought into line with our understanding of the Laboratory Revolution? For reasons of convenience, but also because of its pioneering status as an illustrated popular journal on science, our analysis will focus on the images in *La Nature* from its foundation in 1873 until the First World War. We will offer a tentative typology of laboratory pictures that covers

6 Bernard Lightman, 'Popularizers, Participation and the Transformations of Nineteenth-Century Publishing: From the 1860s to the 1880s', *Notes and Records: The Royal Society Journal of the History of Science*, 70 (2016), 343–59.

7 Henning Schmidgen, 'The Laboratory', *European History Online* (EGO), published by the Institute of European History (IEG) (2011), <http://www.ieg-ego.eu/schmidgenh-2011-en≥ (accessed 3 March 2022).

8 Gabriele Werner, 'Das Bild vom Wissenschaftler – Wissenschaft im Bild. Zur Repräsentation von Wissen und Autorität im Porträt am Ende des 19. Jahrhunderts', *kunsttexte.de*, Sektion BildWissenTechnik, 1 (2001), 1–11, <www.kunsttexte.de>.

most of the relevant images and that may help to connect them to the Laboratory Revolution.

The Myth of the Laboratory

In the spring of 1907, the German chemist Wilhelm Ostwald travelled to Paris to learn more about the new procedure for the liquefaction of air, developed by Georges Claude. While in the city, he asked to visit the laboratory in the rue Lhomond where Marie Curie, with her husband Pierre, had made her famous discovery of radium.[9] Although Marie Curie was absent at the time, Ostwald was allowed to have a look. He was flabbergasted: 'It was', he wrote, 'a cross between a stable and a potato-cellar, and, if I had not seen the worktable with the chemical instruments, I would have thought it a practical joke.'[10]

Over the years, the lamentable state of Marie Curie's laboratory has become part of the standard heroic representation of her path to discovery, not least because of her own emphasis on the difficult circumstances in which she had to work.[11] In her autobiography she described this 'miserable shed' in the following terms:

> Its glass roof did not afford complete shelter from the rain; the heat was suffocating in summer, and the bitter cold of winter was only a little lessened by the iron stove, except in its immediate vicinity. There was

9 This laboratory was part of the École municipale de physique et de chimie industrielles de la ville de Paris.

10 Wilhelm Ostwald, *Lebenslinien, eine Selbstbiographie*, 3 vols. (Berlin: Klasing und Co., 1927), vol. 3, p. 158. All translations from German and French in this chapter are by the author.

11 On the occasion of the 25th anniversary of the discovery of radium, Marie Curie acknowledged that the old 'hangar' had become 'dressed up with the forms of a legend'. See: *L'Oeuvre*, 27 December 1923, p. 2. Robert Reid also observed that she 'plainly felt a great bitterness at having to work in inadequate circumstances'. See Robert Reid, *Marie Curie* (Frogmore: Paladin. Granada Publishing, 1987), p. 81. Actually, before starting their work on radium, both Pierre and Marie Curie had other, albeit modest, working rooms at their disposal. However, the search for the sources of radioactivity necessitated a working procedure on an almost industrial scale, going through several tonnes of pitchblende ore. To handle this material and the obnoxious gases that were produced in the process, they had to relocate their research to an abandoned hangar of the École municipale de physique et de chimie industrielles. The working conditions were horrible, but the hangar could never really be considered as a scientific laboratory.

Figure 13.1 The laboratory of Madame Curie. Source: *La Nature*, 32 (1904), p. 217.

> no question of obtaining the needed proper apparatus in common use by chemists. We simply had some old pine-wood tables with furnaces and gas burners. We had to use the adjoining yard for those of our chemical operations that involved producing irritating gases; even then the gas often filled our shed.[12]

Complaints about the poor working conditions are commonplace among scientists. With every opening of a new laboratory, the previous years spent in more primitive surroundings were often painfully depicted or commemorated with pride and a good dose of nostalgia. One chemist recalled that before Ostwald's new laboratory was installed in 1897, the previous workplace had been 'in every way unfitted'.

> Research was carried on under countless difficulties; the light was bad, the room unventilated, the heating effected by means of stoves difficult to regulate and producing dust which caused much injury to the finer instruments; no precautions had been taken in laying the foundations to ensure the deadening of vibrations; thus many experiments were ruined;

12 Marie Curie, *Pierre Curie* (New York: MacMillan Co., 1923), p. 186. On Marie Curie's efforts at self-fashioning, see: Eva Hemmungs-Wirtén, *Making Marie Curie. Intellectual Property & Celebrity Culture in an Age of Information* (Chicago: The University of Chicago Press, 2015).

> the lack of space precluded the use of telescopes for reading scales, and altogether it would have been difficult to construct a laboratory worse adapted for physico-chemical investigations.[13]

And although the new laboratory was, according to the same source, 'equipped with all the accessories that modern ingenuity had devised', it was again considered to be insufficient already in 1906, and the laboratory had to be substantially expanded.[14]

As a result of the Laboratory Revolution in the second half of the nineteenth century, the state of the laboratory had become a matter of great importance to scientists. Ostwald's amazement at the dilapidated state of the Curie laboratory illustrates his tacit assumption that high-quality scientific research could only be performed in optimal laboratory surroundings. Inspection of a laboratory was the first and essential step towards an appreciation of the work that would take place there. But these anecdotes also show that, despite the so-called Laboratory Revolution sweeping the scientific world in the final decades of the nineteenth century, many laboratories were *not* the modern, well-equipped, and spacious facilities which we may associate with some glorious examples, in particular in the German universities. As often happened, the laboratories were situated in small and humble rooms, haphazardly assembled and awkwardly furnished.

It is good to keep in mind the problematic state of so many laboratories (at least in the eyes of later generations) when looking at laboratory pictures. Understandably, pictures of inadequate laboratories are hard to find. The unsatisfactory features of a laboratory were always situated in the past, and pictures were often made and disseminated when the laboratory was new and state-of-the-art. Already the very existence of these images necessarily reinforces the myth of the laboratory as a privileged space for research.

Even so, the study of historical images of laboratories can tell us much about the public perception of science. However, in doing so, some caveats are in place. Historians have long recognized that images cannot be taken as

13 F.H.N. (probably Francis Henry Neville), 'Some Scientific Centres. II. The Laboratory of Wilhelm Ostwald', *Nature*, 64 (29 August 1901), 428–30. It is not clear whether Neville ever visited Ostwald's laboratory.

14 Max Le Blanc, 'Das Physikalisch-Chemische Institut', in *Festschrift zur Feier des 500 Jährigen Bestehens der Universität Leipzig*, 5 vols. (Leipzig: S. Hirzel, 1909), vol. 4, pt. 2, pp. 85–106, quotation on p. 91.

Figure 13.2 Pierre and Marie Curie in their laboratory. Source: *Le Petit Parisien, Supplément littéraire illustré*, 16 (10 January 1904), front page.

straightforward representations of some objective reality.[15] This particularly applies when scientific objects but also science itself—its scientists, its activities, its sites—are portrayed.[16] Therefore, representations of the modern

15 Peter Burke, *Eyewitnessing: The Uses of Images as Historical Evidence* (London: Reaktion Books, 2001); Michael L. Wilson, 'Visual Culture. A Useful Category of Historical Analysis?', in Vanessa R. Schwartz and Jeannene M. Przyblyski, eds., *The Nineteenth-Century Visual Culture Reader* (New York/London: Routledge, 2004), pp. 26–33.

16 See e.g.: Ludmilla Jordanova, *Defining Features. Scientific and Medical Portraits 1660–2000* (London: Reaktion Books, 2000); Christopher Lawrence and Steven Shapin, eds., *Science Incarnate. Historical Embodiments of Natural Knowledge* (Chicago: The University of Chicago Press, 1998).

laboratory should not be regarded as realistic depictions of a particular space but need to be analysed as codified images that articulate the values and beliefs of the scientific community, and perhaps even more generally, of civil society at large.[17]

Using historical images must therefore start from the assumption that an image cannot be taken as a true representation of a factual state. Ostwald's impressions of his visit to Paris and Marie Curie's later descriptions of her 'miserable shed' are often cited as an accurate description of Curie's laboratory. Yet none of the contemporary representations of the radium laboratory seem to mention anything sub-standard.[18] Pictures of the Curies' laboratory appearing in the public media show a well-equipped and even agreeable workspace, which sometimes takes on a homely atmosphere with the presence of the couple Pierre and Marie. In the drawing, appearing in 1904 on the front page of *Le Petit Parisien*, some of the 'legendary objects', such as the large windows of the hangar, the central stove, and the bitumen floor, are visible, but the laboratory table is placed on a carpet and there are curtains at the windows. The laboratory table appears rather primitive, but the room is well equipped. Neither Pierre nor Marie Curie, pictured in their laboratory, seem to be hampered in their work by any inconvenience. The imagery of the dilapidated state of the laboratory only seems to have become a recurring theme later, after Marie Curie moved to a larger laboratory and when her efforts to gain financial support for her costly research had become part of her public image. Such divergences in the representation of the laboratory must be taken into account when interpreting the pictorial content of images.

Old Stereotypes, New Technology

When looking at images of laboratories in the popular press, one must also bear in mind that during the period of the Laboratory Revolution, the laboratories were not open to the public. Whereas lecture halls, natural history collections, or scientific museums could still be visited and inspected by members of the general public, laboratories were inaccessible, secluded, and private places, often even considered to be

17 A typology of chemical laboratories is given in Peter J.T. Morris, *The Matter Factory. A History of the Chemistry Laboratory* (London: Reaktion Books, 2015).

18 E.g.: Jacques Danne, 'Les Sels de Radium', *La Nature*, 32 (1904), 217. Danne was an assistant (*préparateur*) to Pierre Curie.

dangerous for the uninitiated. This created a sense of secrecy and mystery. Not surprisingly, the representation of laboratories must be considered as more a reflection of the imagination of the viewers than of the actual features of the laboratories. These images often retained something of the older iconography of the alchemical laboratories. Typically, the alchemist worked in a 'cave' with little outside light penetrating his working space. The room would be populated by objects with hidden, occult meanings, and the alchemist himself would be no less a strange creature. Some of these elements can be identified in the well-known watercolour images of Faraday in his laboratory at the Royal Institution, painted in the 1850s by Harriet Moore, as well as in the 1886 pictures of Michel-Eugène Chevreul posing in his laboratory when the press celebrated his 100th birthday.

The image of the alchemical cave continued to reappear in the public imagination. At the opening of the research laboratory at the Louvre in Paris (1930), the modern installation of the rooms was contrasted in the popular press with the reclusive alchemical laboratory, as if the readers would still have this image foremost in their mind when hearing the word laboratory.[19] In the 1931 Universal movie *Frankenstein*, featuring Boris Karloff as Victor Frankenstein, the central scene in which Frankenstein brings a dead body to life is an amalgam of modern elements (white coats, clean benches, electrical instruments) and alchemical symbols (the low 'underground' position of the laboratory, the intervention of cosmic agents in producing the effect, Frankenstein's frenzied expression).[20] This was partly inspired by Mary Shelley's original narrative (in which references to alchemy also played a central part), but at the same time it was a rhetorical mixing of alchemical imagery with the modern view of a scientific laboratory. The persistence of the alchemical laboratory in the public imagination needs to be further clarified. Its meaning may point to the inaccessibility of scientific research and the position of the scientist as knowledgeable and competent, but it may also suggest the futility and the unaccountability of scientific knowledge. The historical references contained in these images seem to suggest that this mode of representation was scarcely influenced by the Laboratory Revolution, but further study is needed to clarify this point.

19 Raymond Lécuyer, 'Comment sont examinés les tableaux dans le laboratoire du Louvre?', *Le Figaro*, (29 August 1934), 5.

20 Susan E. Lederer, *Frankenstein. Penetrating the Secrets of Nature* (New Brunswick, NJ: Rutgers University Press, 2002).

In contrast to the persistent conceptual frame of the secluded alchemical cave, the public image of the modern laboratory, as presented in nineteenth-century books, magazines, and newspapers, emphasized several new elements. These images roughly coincided with the global introduction of the 'German model' of university laboratories in the second half of the century, which is one element of the Laboratory Revolution at large.[21] Laboratories became the hallmark of modernity, the temples of science, the key to scientific progress. The laboratories concerned were not the traditional, small workplaces of individual scientists, but large-scale university laboratories where students worked together in spacious rooms under the supervision of a professor or one of his assistants. The laboratories were housed in new buildings, typically in the new urban quarters of old university towns. The design of the laboratory resembled a factory workspace, where everything happened according to plan. It was to these laboratories that the French chemist Adolphe Wurtz was sent to examine the state of German science.[22]

The emergence of a new type of laboratory pictures also coincided with a concurrent revolution in image printing. Throughout the nineteenth century, wood engraving was the most common technique for book illustration. It was easy to integrate in the main text of books, magazines, and newspapers; it could be used many times without noticeable degradation; and it could be mass-produced fairly rapidly, opening the way to pictorial journalism. Yet by the end of the century, photomechanical techniques started to take over. Whereas wood engraving still retained much of the artistic style of composition, the photomechanical image fostered a growing desire among the public for 'realistic' pictures. Until the First World War, the two techniques were often used side by side, with wood engravings being increasingly based on photographs. This blurred the distinctions between both modes of expression. But if the wood engraving was pushed by the growing popularity of photographs towards a more realistic, documentary representation, the modern photographic illustration also retained something of the narrative character of the wood engravings by allowing for a *mise-en-scène* of people and things.[23]

21 Geert Vanpaemel, 'The German Model of Laboratory Science and the European Periphery (1860–1914)', in Anna Simões, Maria Paula Diogo, and Kostas Gavroglu, eds., *Sciences in the Universities of Europe. Nineteenth and Twentieth Centuries.* Boston Studies in the Philosophy and History of Science, vol. 309 (Dordrecht: Springer, 2015), pp. 211–25.

22 Adolphe Wurtz, *Les hautes études pratiques dans les universités allemandes* (Paris: Imprimerie impériale, 1870).

23 Anthony Griffiths, *Prints and Printmaking. An Introduction to the History and Techniques* (London: The British Museum Press, 1980); A. Hyatt Mayor, *Prints & People. A Social History of*

A third development must be taken into account to understand the emergence of the new laboratory pictures. Apart from the new laboratories and the new printing techniques, the era also saw the emergence of the so-called New Journalism, in which the critical and intellectual aspects of journalism were being replaced by entertainment, sensationalism, and market-oriented material.[24] One characteristic of New Journalism was the increasing significance of illustrations. This does not, however, lead to the conclusion that illustrations were swiftly incorporated into the popular science literature. For one thing, the topic of science was deemed too serious to allow for superficial reading. Illustrations were often seen as superfluous or even harmful. One reader admonished the editors of the newly founded journal *L'Ami de la maison* in 1856 'not to lose whole pages with images', while another one pleaded not to make too many concessions to the current fashion, and a third one referred to the mixture of illustrations and text as a sort of 'decadence for lazy readers'.[25] The English journal *Nature* did advertise itself as a 'weekly illustrated journal of science', but its founder Norman Lockyer stipulated that the journal would publish full accounts of scientific discoveries, 'illustrated when necessary'.[26] A competing journal, *Knowledge. An Illustrated Magazine of Science*, added to its masthead the description 'plainly worded – exactly described' as if to make it clear to its readers not to expect a low-level popular publication.[27] Yet, by the end of the nineteenth century, illustrations had become the standard in popular scientific journals. For some journals, this equalled a shift towards some sort of New Journalism.

Printed Pictures (New York: The Metropolitan Museum of Art, 1971); Klaus Schlüpmann, 'Les avatars du réalisme à l'époque de la photo imprimée', in Bernadette Bensaude-Vincent and Anne Rasmussen, *La science populaire dans la presse et l'édition. XIXe et XXe siècles* (Paris: CNRS Editions, 1997), pp. 121–39; Rachel A. Mustalish, 'The Development of Photomechanical Printing Processes in the Late 19th Century', *Topics in Photographic Preservation*, 7 (1997), 73–87.

24 Peter Broks, *Media Science before the Great War* (Houndsmill: Macmillan, 1996), p. 15–16. The term New Journalism was coined in 1887 by Matthew Arnold and signalled the levelling down of serious journalism to the level of advertising, rather than up to literature or criticism. Laurel Brake and Marysa Demoor, eds., *Dictionary of Nineteenth Century Journalism in Great Britain and Ireland* (Gent: Academia Press, 2009), p. 443.

25 Daniel Raichvarg and Jean Jacques, *Savants et ignorants. Une histoire de la vulgarisation des sciences* (Paris: Seuil, 1991), p. 174.

26 Quoted in: Melinda C. Baldwin, *Making* Nature. *The History of a Journal* (Chicago: Chicago University Press, 2015), p. 35. The journal would quickly turn away from science popularization and focus on its role within the scientific community.

27 Baldwin, *Making* Nature, p. 72.

Picturing the Laboratory in *La Nature*

In the following we will focus on the images of the laboratory in the French popular science magazine *La Nature*. *La Nature* was founded in 1873 by the French chemist Gaston Tissandier (1843–1899). Although science popularization in France was in full swing at the time, Tissandier felt that there still was room for a popular magazine covering a broad range of scientific and technological topics and aimed at a middle-class, bourgeois, and non-scientific audience. To differentiate his project from the more 'serious' scientific journals, Tissandier announced that most of the articles would be written by professional popularizers, not by scientists. The magazine appeared in weekly issues until 1926, after which it became fortnightly. At the beginning it was published in an edition of 2,000 copies, but in 1885 this number had risen to 15,000. It was one of the few French popular science journals to make it into the next century. In 1972 *La Nature* merged with *La Recherche* and is still published under that name today.[28]

From the start, *La Nature* included images, which were on an equal footing with the text. This was rather unusual and quite new for a journal that aspired to an intellectual readership. The use of images still seemed to lower the intellectual status of the text. But Tissandier emphasized the didactical role of the images, at the same time stipulating that these images also served an aesthetic goal:

> It is difficult, indeed, to do without illustration in a journal of this kind. [However,] if one wanted to confine oneself to facilitating the intelligence of the text, simple figures would suffice, analogous to those published in technical books. We thought that the public would not complain about having more, that they would not criticize the engravings of a journal, however scientific it may be, for being executed with great care. ... What inconvenience would there be in embellishing a figure of science? Why would it not be a work of art if it still remains exact and serious? Why would one fear to animate the scenes, to represent a machine in action, without an *a priori* rejection of the picturesque? Why would the scientific journal be condemned to be arid, dry, and often boring? Would it not, on the contrary, benefit from taking on the appearance of an attractive,

28 Manuel Chemineau, *Fortunes de La Nature: 1873-1914* (Vienna/Berlin: LIT Verlag, 2012); Guy Vautrin, *Histoire de la vulgarisation scientifique avant 1900* (Paris: EDP Sciences, 2018), pp. 308–40.

> pleasant book, in order to attract readers and increase the number of those willing to study?[29]

In looking at the illustrations appearing in *La Nature*, we must keep in mind this double character of being didactical as well as pleasing. Most of the time, both aspects were present at the same time. In didactical images natural phenomena were shown or a machine was depicted with its inner mechanism visible. But these images could just as well be expanded to include a narrative aspect, e.g. by adding the presence of people or by showing details of a landscape. These pictures were obviously more inviting to the reader and increased the attractiveness of the journal.[30]

The first volume of *La Nature* contained 231 illustrations for a total of 424 pages, or 0.54 illustrations per page. The number of illustrations would increase over the years. In the tenth annual volume, the ratio had already reached 0.77 illustrations per page. Ten years later the ratio was 0.91. After 1905 the number of illustrations per volume (with still an equal number of pages) rose further to over 600, with an average ratio of 1.4 illustrations per page.[31]

Still, images of the laboratory were rather few in *La Nature*. From 1873 until 1914, 2,152 issues appeared, of which only 71 contained one or more pictures with a label mentioning 'laboratoire'. Of these images, some refer to laboratory instruments or experiments without showing the laboratory space; some refer to a portable instrument chest or even a photographer's dark room. Words can be misleading in digital searches. In some cases, laboratory pictures were found in instances where terms like 'cabinet de travail' or 'atelier' were used; in others a laboratory was shown as part of a larger picture without mentioning the word 'laboratoire' in the caption. As our sample was not meant to be representative, we have taken a broad view of laboratory images, including as many instances as possible. This yielded a sample of about 100 pictures, ranging from 1873 until the mid-1930s.

The depictions of actual laboratory settings can be grouped into four categories. The first one is the ground plan of the laboratory. The very first issue of *La Nature* contained an article, written by Gaston Tissandier himself,

29 Gaston Tissandier, 'Préface," *La Nature*, 1 (1873), v–viii (p. v–vi).

30 Although other journals did not include in the text as many illustrations as *La Nature*, they often showed a narrative picture on the cover of the journal issue.

31 Axel Hohnsbein, 'Les illustrations de la Nature: approche quantitative', *La science en mouvement* (2019), <https://lasciem.hypotheses.org/223>.

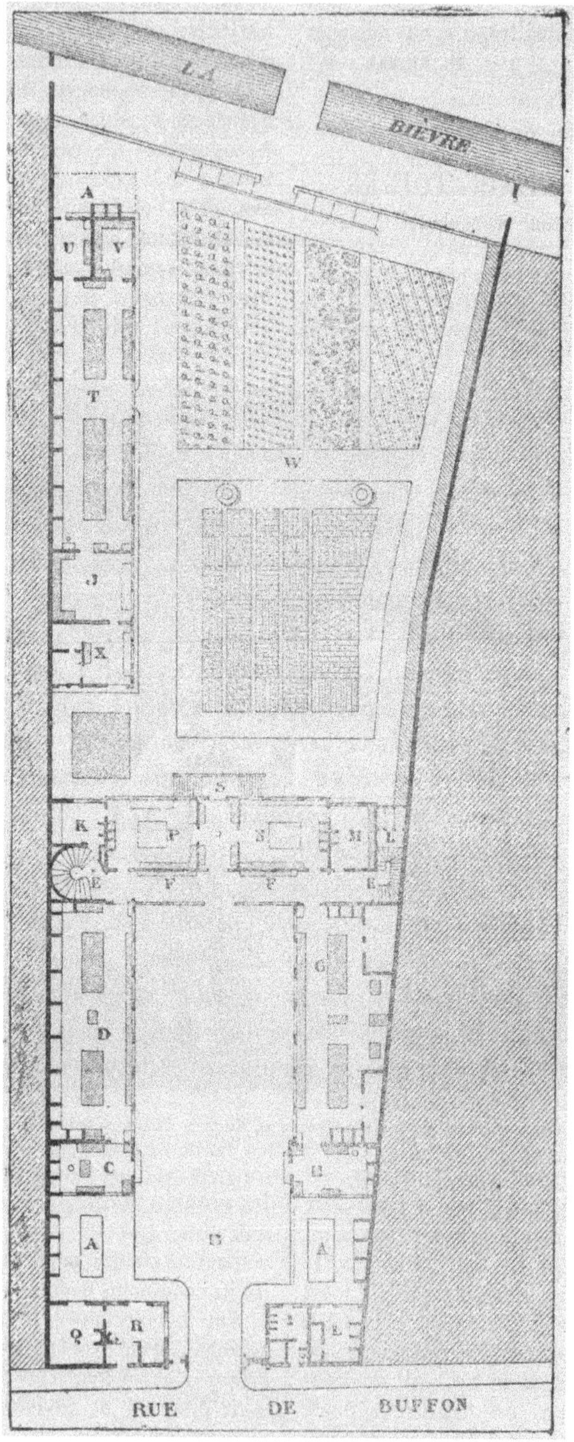

Figure 13.3 Floorplan of the ground floor of the new laboratories of the Muséum d'histoire naturelle in Paris. Source: *La Nature*, 1 (1873), p. 6.

on the new laboratories of the Muséum d'histoire naturelle. After lamenting the neglect of the Muséum in the French higher education system, Tissandier described the layout of the new buildings in the rue de Buffon. Tissandier did not mention the equipment of the laboratories nor the teaching programme for students but guided the reader on a walking tour through the buildings, praising each professor for his zeal and devotion. The article was illustrated by a plan of the site, with a legend indicating the precise location of every laboratory.

The ground plan was a regular feature of the presentation of laboratories. In the first decades of *La Nature*, thirteen floor plans of laboratories were published. Given the non-scientific audience of *La Nature*, one might wonder what the readers made of these plans. This type of laboratory picture all but disappeared after 1900. One possible explanation for the 'success' of the ground plan as a representation was the public debate on the state of French and German laboratories. From the ground plan, something of the grandeur and splendour of the laboratory could be deduced by inspecting its dimensions, the design pattern of a courtyard, or the presence of a majestic entrance hall. The ground plans also emphasized the conscious planning of a research institute, in which nothing was left to chance and everything was built to guarantee the best possible results. The ground plan was the expression of rationality governing the organization of scientific research. It therefore hinted most clearly at the organization behind the laboratory and the complexity of the laboratory as an institute, which included the preparation rooms, storage rooms, library, lecture halls, and even the private lodgings of the director of the laboratory, the housekeeper, and the research assistants.

A second type of representation was the interior view of the laboratory, as in the image of the *laboratoire de microbie générale* at the Institut Pasteur. Architecture still plays an important role in the design of the picture, as the images accentuate the spaciousness of the room, the large windows flooding the room with light and the evenly distributed work benches as repetitive elements creating depth. There is no central point of attraction in the room. Typically, several people are engaged in research, either absorbed in their own work or conversing with each other. The room is clean, well ordered, and functional. The interior view highlights the collective nature of research in the laboratory, which perhaps unintentionally diminishes the role of the individual scientist in the process of investigation. None of the researchers shows any sign of excitement or extreme mental concentration; all of them are engaged in what appears to be routine activities, which may need a lot of dexterity and skill but no particular talent. The researcher has

Figure 13.4 The laboratory of general microbiology at the Institut Pasteur. Source: *La Nature*, 17 (1889), p. 65.

become an integral part of the laboratory, obeying to the discipline of the scientific method.

The same elements return in a slightly different setting in an image of the small laboratory in the marine station at Roscoff (Brittany). One researcher is sitting at his desk but takes no notice of the magnificent sea view in front of him. The architecture of his little room and his own posture is repeated through the door, where a researcher sits in the next room. The picture stresses the repetitive, formatted work occurring in a research laboratory, with again no hint of individual genius or inspiration.

A third type of image is the interior view of an empty laboratory, such as the image of the Curie laboratory discussed earlier (fig 1). In this type of picture, all human elements have been removed. Only the architectural setting is preserved. In many cases the geometry of the room is particularly brought forward. With its strict design, its geometrical architecture of tables, chairs, flasks, and other apparatus, the emphasis falls on the objective nature of science. Human interference seems to play no part in the investigation process. The laboratory assumes the character of a privileged space for research, almost autonomous in its operations. The empty laboratory exudes power and authority. By not showing the laboratory as an actual workspace,

Fig. 2. — Vue intérieure d'une stalle d'étude au laboratoire de Roscoff. (D'après une photographie.)

Figure 13.5 Interior of a small study in the laboratory of the marine station at Roscoff (Brittany). Source: *La Nature*, 13 (1885), p. 345.

the image shows the laboratory as universal, not limited by local or human constraints.

A fourth category of images can be characterized as 'the scientist in his laboratory'. The paradigmatic example of this category is the famous picture of Louis Pasteur made in 1885 by the Finnish painter Albert Edelfelt. Also well-known are the early twentieth century pictures of Pierre and Marie Curie in their laboratory (see fig. 2). Surprisingly, this type of picture does not feature prominently in *La Nature.* Before 1914 only two pictures can be included in this category, both chemists: Michel Chevreul and Henri Moissan.[32] The laboratory now plays a subordinate role. The scientist is always and necessarily the focal point of the picture. He is shown performing a specific action, sitting in a position so that the viewer can read the expression on his face. The laboratory retreats to the background but is still very much present with its unruly disposition of flasks, glassware, and other instruments. However, in sharp contrast to the typology of the alchemical laboratory, where the chaos of the room mirrors the confusion in the alchemist's mind, the scientist commands order in his laboratory. In

32 A third image of Thomas Edison 'in his laboratory' actually only shows Edison talking to his co-workers. The laboratory is hardly visible.

Figure 13.6 A large student laboratory in the Chemical Laboratory of Leipzig University. Source: *La Nature*, 5 (1877), p. 169.

the middle of the imposing chaos, the scientist is shown to be in full control of the laboratory, which thereby enhances his special status as scientist. The laboratory appears in the background of the picture as an extension of the scientist's personality, an amplification of his creative powers. This view of the laboratory is not about routine skills and scientific method but about personal ingenuity and creativity. The laboratory background is not even essential in composing the image—in some cases, scientists are portrayed holding only one piece of equipment. But however little is visible of the laboratory, the positioning of the scientist in his laboratory does add to the epistemological status of the representation.

Images of the Laboratory Revolution?

Considering our small sample size of pictures, it is nearly impossible to find meaningful correlations between text and image. However, our brief overview still allows for some observations. First, the sample does not indicate a particular preference for laboratory images, nor could we indicate an increased attention to laboratories. *La Nature* also published pictures of other research environments, such as meteorological stations, geographical explorations, volcanic research, marine research ships, or hot air balloon ascents. These pictures, however, emphasized the unusual features of the

Figure 13.7 Henri Moissan busy preparing fluor substances in his laboratory at the École de pharmacie in Paris. Source: *La Nature*, 18 (1890), p. 177.

situation, such as danger and adventure. More stable research environments which could be viewed as an alternative to the laboratory, such as botanical gardens, libraries, agricultural experimental stations, or natural history museums, were much less represented, if at all. With the possible exception of the astronomical observatory, the laboratory seems to have been the most often portrayed facility for scientific research in the pages of *La Nature*. And like no other scientific environment, the laboratory became an asset in the representation of scientists. Whether this indicates a preference for laboratory science is hard to tell. Most of the scientific topics discussed in the journal concerned open-air phenomena or technical inventions. Laboratories were predominantly mentioned in the context of (new) buildings or in relation to the commemoration of scientists.

In the period under review, there seems to have been a shift between the four types of laboratory pictures. The ground floor type almost disappears near the end of the nineteenth century. This type seems to be connected to the early phase of the building of laboratories, when the mere existence of well-equipped and modern laboratories was a prerequisite for the revival of science in France.[33] This format closely resembled the scientist's view of

33 Harry W. Paul, *The Sorcerer's Apprentice. The French Scientist's Image of German Science 1840–1919* (Gainesville, FL: University of Florida Press, 1972).

the laboratory. Being familiar with the scientific aspect of laboratory work, a scientist would indeed pay more attention to the adequate arrangement of rooms than to the equipment or work procedures within the laboratory. The lay reader could just as well inspect the rationality of the floor plan, the arrangement of the different rooms, or the solidity of the building, irrespective of its adequacy for scientific work. The replacement of the ground plan image with interior views of the laboratory, which occurred towards the end of the century, may have actually increased the separation of the reader from the laboratory. In this setting, the laboratory was indeed reduced to one single room, hiding all other elements from view, such as the storage rooms, the office of the laboratory director, or the external environment. The reader was instead confronted with a specific narrative, showing people at work, doing things that he could not understand. This enhanced the status of the laboratory as a special place. Allowing the reader a glimpse of the work that went on in the laboratory, to which he had no access, fortified the authority of the laboratory.

The transition from wood engravings to photographs had a marked influence on the presentation of the laboratory. When pictured in wood engravings, the laboratory appeared through the engraver's imagination in an idealized form. Everything and everyone were evenly distributed in the picture, well in place. As photographs began being used to depict the interior, the former idealized narrative was gradually replaced by a documentary gaze, including the unintentional (?) presence of laboratory workers at their benches and even a certain measure of disorder. Consequently, the emphasis shifted from the rational ideal towards the routine aspects of a workshop organization. The importance of discipline and routine was further accentuated in the twentieth century by the presence of women laboratory workers, who to contemporary viewers would not be associated with the image of a creative or original scientist. As women were often perceived as occupying a lower level in the scientific hierarchy, their involvement in manual laboratory work could only lead to the conclusion that the real power of the laboratory was not situated *in* the actual laboratory room but rather in some hidden or invisible power behind the laboratory. These images may therefore indicate the low importance of laboratory workers in the research process and the independence of laboratory work from human intervention. In this way, the photographic, documentary style of representing the interior of a laboratory tended towards a dehumanized representation of the laboratory, as in the image of the empty laboratory.

A final observation concerns the use of the laboratory in representing the scientist. If this type was nearly absent from the pages of *La Nature* before the

Figure 13.8 Marcellin Berthelot in his study, after a photograph by 'Dornac' (Paul Cardon). Source: *La Nature*, 22 (1894), p. 281.

First World War, it did gain popularity in the 1930s. Between the wars, over fifteen illustrations show the scientist in his laboratory. An example is the picture of Marcellin Berthelot. Whereas in 1894 *La Nature* preferred to show Berthelot in a cosy armchair in his study, in 1927 he was depicted working

Figure 13.9 Berthelot in his laboratory at the Collège de France. Source: *La Nature*, 55 (1927), p. 338.

in a laboratory setting. The laboratory had clearly gained an authority which could be marshalled by the individual scientist to corroborate his own standing. Yet it must be pointed out that the two dominant images (the 'scientist in his laboratory' and 'the empty laboratory') represent two different and even antagonistic interpretations of the laboratory. Common to both images is the superhuman power of the laboratory, transcending the efforts of either individual and/or ordinary people. But whereas the empty laboratory situates this power in the routine procedures of work and the collective nature of research, the laboratory in the scientist's portrait enhances his individual and extraordinary genius.

The observations mentioned above, taken together, sketch a tentative outline of the Laboratory Revolution in *La Nature*. The revolution as such appears not to have been explicitly addressed, but an evolution towards the laboratory as the most important expression of a stable scientific research environment seems to have occurred. The laboratory was characterized by specific features, which set it apart from other types of workshops, in particular the small domestic laboratories which had been dominant in the early nineteenth century. Laboratories were almost always presented within the context of education and research, and—with a few exceptions—situated in France. What is most important is that the laboratory became more closely associated with the career of scientists and figured prominently in portraits of scientists. As such, the laboratory was presented as an institution of particular cultural authority, a special place characterized by a rational outlay not unlike a modern factory. The central place of the laboratory did not, however, restrict scientific research to the laboratory bench. Even if no other individual institution was put next to the laboratory, most of the scientific advances discussed in *La Nature* were made or concerned phenomena outside of the laboratory. In particular, *La Nature* showed an increasing interest for technological innovation and industrial machinery. This leads us to suggest that the laboratory was not perceived by the editors of *La Nature* as the major driving force of science; on the contrary, the modern laboratory was an adaptation of scientific research to the modernization of industry and society. What set the laboratory apart was that it was the main such adaptation to be made visible to the readers of *La Nature*.

About the Author

Geert Vanpaemel is emeritus professor for history of science at KU Leuven. His research concerned the history of universities, the history of science in Belgium since 1500, and the popularization of science. Until his retirement in 2021, he was also head of the Postgraduate Program on Science Communication and director of the Academic and Historical Office at KU Leuven.

Acknowledgements

Most of the contributions to this volume are part of a research collaboration titled 'The Laboratory Revolution. The Rise of the Laboratory and the Changing Nature of the University, 1850–1950' that began with a meeting in Groningen in October 2017. The project was prompted by the presentation of the second volume of the history of thc University of Groningen written by Klaas van Berkel, co-editor of this volume. In the years that followed the project was enriched by new contributions while others were dropped, resulting in a coherent volume that for the first time analyses the laboratory revolution in all its aspects. We are grateful to all those people and organizations that made this cooperation and the resulting project possible. First of all, we thank Dr. Ulco Kooystra and Dr. Sieger Vreeling, who were co-organizers in initiating this project. We further acknowledge the generosity of the University of Groningen for housing the first meeting and serving as the hub of our research. We are also thankful for the financial support we received from the University of Groningen and the Royal Society of Physics. Our gratitude extends to the colleagues from Belgium, France, Great Britain, Norway, the United States, and the Netherlands—authors, commentators and those who supplied illustrations—with whom we collaborated while working on the book. Special words of thanks go to Alan Rocke, the editors and publisher of *Ambix*, and the Society for the History of Alchemy and Chemistry for their perpmission to republish Rocke's seminal paper published in 2003. We are also grateful that several archives and holders of picture collections gave permission to publish illustrations supplied by them. Finally, a big thank you is due to Julie Benschop, our editor at Amsterdam University Press, who was always supportive of our endeavours.

Index

For Product Safety Concerns and Information please contact our EU representative GPSR@taylorandfrancis.com
Taylor & Francis Verlag GmbH, Kaufingerstraße 24, 80331 München, Germany

www.ingramcontent.com/pod-product-compliance
Ingram Content Group UK Ltd.
Pitfield, Milton Keynes, MK11 3LW, UK
UKHW020020200926
475702UK00014B/1982
* 9 7 8 1 0 4 1 1 8 8 3 3 9 *